MODERN PREDICTIVE CONTROL

MODERN PREDICTIVE CONTROL

DING BAO-CANG

CRC Press
Taylor & Francis Group
Boca Raton   London   New York

CRC Press is an imprint of the
Taylor & Francis Group, an **informa** business

CRC Press
Taylor & Francis Group
6000 Broken Sound Parkway NW, Suite 300
Boca Raton, FL 33487-2742

First issued in paperback 2017

ISBN 13: 978-1-138-11769-3 (pbk)
ISBN 13: 978-1-4200-8530-3 (hbk)

This book contains information obtained from authentic and highly regarded sources. Reasonable efforts have been made to publish reliable data and information, but the author and publisher cannot assume responsibility for the validity of all materials or the consequences of their use. The authors and publishers have attempted to trace the copyright holders of all material reproduced in this publication and apologize to copyright holders if permission to publish in this form has not been obtained. If any copyright material has not been acknowledged please write and let us know so we may rectify in any future reprint.

Trademark Notice: Product or corporate names may be trademarks or registered trademarks, and are used only for identification and explanation without intent to infringe.

Library of Congress Cataloging-in-Publication Data

Bao-Cang, Ding.
   Modern predictive control / Ding Bao-Cang.
      p. cm.
   Includes bibliographical references and index.
   ISBN 978-1-4200-8530-3 (hardcover : alk. paper)
   1. Predictive control. I. Title.

TJ217.6.B36 2010
629.8--dc22   2009034799

Visit the Taylor & Francis Web site at
http://www.taylorandfrancis.com

and the CRC Press Web site at
http://www.crcpress.com

Abstract

This book addresses industrial model predictive controls (MPC), adaptive MPC, synthesis approaches of MPC and two-step MPC, with emphasis on synthesis approaches and the relationship between heuristic MPC and synthesis approaches. Chapter 1 introduces the concepts of system, modeling and MPC, including the description of transition from classical MPC to synthesis approaches. Chapters 2, 3 and 4 are concerned with model predictive heuristic control, dynamic matrix control and generalized predictive control, respectively. Chapter 5 talks about two-step MPC for systems with input nonlinearities. Chapter 6 concludes the main ideas in synthesis approaches of MPC. Chapters 7, 8 and 9 are concerned with synthesis approaches when the state is measurable. The polytopic description is mainly considered. This is one of the first works to systematically address robust MPC. Chapter 10 looks at synthesis approaches of output feedback MPC. This book presents an unbiased account of the significance of various MPC.

Preface

Model predictive control (MPC) differs from other control methods mainly in its implementation of the control actions. Usually, MPC solves a finite-horizon optimal control problem at each sampling instant, so that the control moves for the current time and a period of future time are obtained. However, only the current control move is applied to the plant. At the next sampling instant, the same kind of optimization is repeated with the new measurements. One is used to compare implementing MPC to passing the street or playing chess, which has similar pattern with MPC: acting while optimizing. The pattern "acting while optimizing" is unavoidable in many engineering problems, i.e., in many situations one has to "act while optimizing." Thus, to a degree, for a lot of engineering problems the unique pattern of MPC is not artificial, but inevitable.

MPC is mostly applied in the constrained multivariable systems. For unconstrained nonlinear systems and unconstrained linear time-varying systems, applying MPC may also yield good control performance. For unconstrained linear nominal systems, there is no necessity to utilize the optimization pattern of MPC (i.e., finite-horizon receding horizon optimization) since solution of infinite-horizon optimal control is preferred. Moreover, if some satisfactory off-line control laws are obtained for a control problem, then utilizing MPC does not necessarily work much better. The applications of MPC should be on those control problems where off-line control laws are not easy, or are impossible, to achieve. The superiority of MPC is its numerical solution.

For a clear understanding of the above ideas, one should confirm that the plant to be controlled is usually very complex, where exact modeling is impossible. Thus, applying linear models, even applying off-line feedback laws, often presents as the key to simplifying the engineering problems and gaining profit. Linear models and off-line control laws avoid the complex on-line nonlinear optimization, such that MPC can be run in the current available computers. MPC based on a linear model is not equivalent to MPC for a linear system.

The development of MPC is somewhat unusual. Usually, one may suggest that MPC originated from a computer control algorithm in the late 1970s. Dynamic matrix control (DMC) and model predictive heuristic control (MPHC), which appeared at that time, have been gaining widespread acceptance. However, before that time (in the early 1970s), there was already investigation into receding horizon control (RHC). In the 1980s, there were attentive studies on the adaptive control, and D. W. Clarke *et al.* from Oxford was timely in proposing generalized predictive control (GPC). At that time, GPC was more flexible for theoretical studies than DMC and MPHC. There have been a large number of citations on the earliest paper of GPC.

From the 1990s, the main stream of theoretical MPC has been turned to the synthesis approach; the sketch of MPC with guaranteed stability based-on optimal control theory has been shaped. The early version of synthesis approach is RHC proposed in the early 1970s. In the early 1990s, some scholars commented on classical MPC (DMC, MPHC and GPC, etc.) as having the characteristic similar to "playing games" (not the game of playing chess, but the game of gambling); the main reason is that stability investigation of these algorithms is hard to develop, and one has to tune with "trial and error." All results based on theoretical deductions have their limitations.

Now, for synthesis approaches there are well established results; however, they are rarely applied in the real processes. The main reason is that synthesis approaches mostly apply state space models. Recall that in the late 1970s, industrial MPC was first proposed to overcome some deficiencies of the "modern control techniques" based on the state space model. One of the main challenges of synthesis approaches is the handling of unmeasurable state. Synthesis approaches based on the state estimator are still rather conservative. Another main challenge of the synthesis approach is its model adaptation, for which only a small progress has been achieved. In DMC, MPHC and GPC, the unmeasurable state won't be encountered; GPC was an adaptive control technique when it was first proposed by D. W. Clarke *et al.*

To understand the differences between industrial MPC, adaptive MPC and synthesis approaches, several aspects of the control theory, including system identification, model reduction, state estimation and model transformation, etc., are involved. This is rather complicated and the complication leads the research works to be undertaken from various angles in order to achieve any breakthrough. By utilizing a simple controller, such as DMC or MPHC, one can obtain a closed-loop system which is hard to analyze. By utilizing a complicated controller, such as a synthesis approach, one can obtain the closed-loop system which can be easily analyzable. GPC applies a controller not quite as simple (the identification inclusive), and obtains a closed-loop system which is more difficult for analysis; however, this is unavoidable for an adaptive control.

One who takes MPC as his/her research focus should know the differences between various methods and be well aware of the roots for these differences. One should first believe the usefulness of various methods and not let prejudice warp his/her judgment. When one writes a research paper, he/she should comment on the related MPC in a fair-minded fashion, and point out the creative idea in his/her paper. An engineer should know that there isn't an omnipotent MPC; any success or failure can have its deep reasons; he/she should know that the importance of model selection in MPC cannot be merely concluded as "the more accurate the better."

After I had thought of the above points, I wrote this book. I hope it can provide help for the readers. Chapter 1 introduces some basic concepts of the systems, modeling and predictive control, including the description for the development from classical MPC to synthesis approaches; it is suggested that the mismatch between a plant and its model is a general issue. Chapters 2, 3 and 4 study MPHC, DMC and GPC, respectively. Chapter 5 tells about "two-step MPC," which is a special class of MPC; a benefit is the introduction of the region of attraction and its computation. Chapter 6 is about the general ideas of synthesis approaches, which is an important chapter. Chapters 7, 8 and 9 study synthesis approaches when the state is measurable; it mainly considers the systems with polytopic uncertainty. Chapter 10 studies synthesis approaches of the output feedback MPC.

The constrained systems are mainly considered. Half of the content corresponds to linear uncertain systems. The results presented for linear uncertain systems have revealed some fundamental issues in MPC. The most important issue is given in Chapter 9 (open-loop optimization and closed-loop optimization); this issue can be inevitable in the real engineering problems.

I would like to take this opportunity to appreciate the support and guidance of Prof. *Pu Yuan* (China University of Petroleum), Prof. *Yu-geng Xi* (Shanghai Jiaotong University), Prof. *Shao-yuan Li* (Shanghai Jiaotong University), Prof. *Biao Huang* (University of Alberta, Canada), Prof. *Li-hua Xie* (Nanyang Technological University, Singapore) over these years. Prof. *Yue Sun* (Chongqing University) gave crucial help for the publication of this book. A Ph.D. candidate *Shan-bi Wei* helped me in editing the contents. Moreover, my research work has been supported by National Natural Science Foundation of China (NSFC grant no. 60934007, grant no. 60874046, grant no. 60504013) and by the Program for New Century Excellent Talents (NCET) in University of China. This book may have missed citing some important materials, I sincerely apologize for that.

Bao-cang Ding
P. R. China

Contents

Chapter 1

Systems, modeling and model predictive control

Model predictive control (MPC) was proposed in the 1970s first by industrial circles (not by control theorists). Its popularity steadily increased throughout the 1980s. At present, there is little doubt that it is the most widely used multivariable control algorithm in the chemical process industries and in other areas. While MPC is suitable for almost any kind of problem, it displays its main strength when applied to problems with:

- a large number of manipulated and controlled variables,

- constraints imposed on both the manipulated and controlled variables,

- changing control objectives and/or equipment (sensor/actuator) failure,

- time delays.

Some of the popular names associated with MPC are Dynamic Matrix Control (DMC), Model Algorithmic Control (MAC), Generalized Predictive Control (GPC), etc. While these algorithms differ in certain details, the main ideas behind them are very similar. Indeed, in its basic unconstrained form MPC is closely related to linear quadratic (LQ) optimal control. In the constrained case, however, MPC leads to an optimization problem which is solved on-line in real-time at each sampling interval. MPC takes full advantage of the power available in today's industrial computers.

In order to have a fundamental knowledge of MPC (especially for those new starters), Chapter 1 introduces the basic concepts of system, modeling and predictive control. Section 1.5 is referred to in [63]. Section 1.8 is referred to in [12].

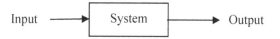

Figure 1.1.1: The interactions between the system and its environment.

1.1 Systems

In researching MPC, a system usually refers to the plant to be controlled or the closed-loop system including the controller.

A system exists independent of its "environment." Although a system is affected by its environment, it exists independently and affects its environment. The interaction between a system and its environment is shown in Figure 1.1.1. The effect of environment on the system is represented by the inputs of the system; the effect of a system on its environment is represented by the output of the system. The relationship between the input and output of a system exhibits the feature of this system. The input and output of a system, which change with time, are called input variable and output variable. If a system has one input and one output, it is called single input single output (SISO) system. If a system has more than one input and more than one output, it is called multi-input multi-output (MIMO) system.

The boundary of a system is determined by the function of this system and the target in studying this system. Therefore, a system has a relative relationship with its constituent parts (called subsystems). For example, for the management of a large oil corporation, each refinery is an independent system which includes all the production units in this refinery. For the management of a refinery, each production unit is an independent system, and each constituent part of this production unit, such as a chemical reactor, is a subsystem of this production unit. However, when one studies the reactor, this reactor is usually taken as an independent system.

In studying a system, in order to clarify the relationship between its different subsystems, one often utilizes the single directional information flow. Take the control system in Figure 1.1.2 as an example; it consists of two subsystems, i.e., the plant to be controlled and the controller. The output of the plant is the input of the controller, which is often called controlled variable. The desired value of the controlled variable (referred to as setpoint) is another input of the controller, which is the effect of the system's environment on the system. The output of the controller acts on the plant, is the input of the plant. The exterior disturbance is another input variable of the plant. Each input or output variable is marked with an arrow, which indicates the effect direction, so that the interaction between the system and its environment is easily recognized.

Note that, depending on the different boundaries, a system can have different inputs and outputs. Take the chemical reactor as an example. If one considers the energy (heat) conservation relationship, then the heat enter-

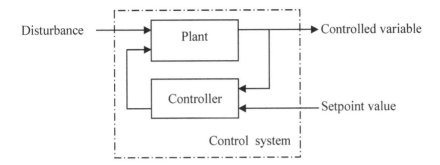

Figure 1.1.2: The interactions between the control system and its environment.

ing the reactor is the input of the system (reactor), and the heat taken out from the reactor is the output of the system. However, if the reactor is to be controlled, and the heat taken from the reactor is served as the mean for manipulation, and the control target is to sustain the reaction temperature at its desired value, then the reaction temperature is the output of the reactor, and the heat taken from the reactor is the input of the reactor.

For a specific system, if all the constituent parts and their input and output variables are determined, so that the block diagram as in Figure 1.1.2 is formed (note that, in Figure 1.1.2, the plant and the controller also consist of their subsystems; with these subsystems and their input and output variables determined, a more delicate block diagram can be obtained), then the relationships of the subsystems can be easily clarified.

Systems can be classified according to different rules:

(1) linear system and nonlinear system

(2) nominal system and uncertain system

(3) deterministic system and stochastic system

(4) time-invariant system and time-varying system

(5) constrained system and unconstrained system

(6) continuous state system and discrete state system

(7) continuous-time system and discrete-time system

(8) time driven system and event driven system

(9) lumped parameter system and distributed parameter system

(10) system with network included (networked system) and system without network included

etc. Moreover, if a system includes both the continuous and discrete states, or both the continuous-time and discrete-time, or both the time driven and event driven properties, then it is called a

(11) hybrid system

which is a very important kind of system. Note that this book mainly studies continuous-time, time driven, lumped parameter systems without network included.

For most of the system classes, there are corresponding subject branches. Nearly every plant in the process industry has nonlinearity, uncertainty, time-varying dynamics, constraint, and distributed parameters, and the continuous-time systems are prominent. Currently, since the digital computer is applied in the process control, usually users meet with sampled data systems. In the future, the computer network will be involved.

1.2   Modeling

In order to study a system, often one needs to set up a model of the system. Models can be classified into two categories. One is the physical model (e.g., experimental unit) or analog model (e.g., by utilizing the similarities, apply circuit and network to simulate the physical plant). The other is the mathematical model, i.e., the use of mathematical equations to describe the system. In the last decades, it has become more and more acceptable to use the mathematical model to study the systems. The mathematical model has developed, from the tool of theoretical studies, to the means of real applications. In this study, "model" usually refers to "mathematical model."

The real systems can be manifold and complex in different aspects. Moreover, the targets for analyzing the systems can be different, which make the styles of mathematical model different. The methods for constructing the mathematical model can be classified into two categories.

(I) Use of physical principles to construct a model (principle model). For example, for a production process, one can construct the mathematical model according to the material balance, energy balance and other relations. This not only gives the relationship between the input and output of the system, but also gives the relationship between the state and input/output of the system. By this kind of model, one has a clear understanding of the system. Hence, this kind of model is referred to as "white-box model."

(II) Suppose the system complies with a certain kind of mathematical equation. Measure the input and output of the system. Determine the parameters in the model via a certain mathematical method. The structure of the model can also be modified, such that the relationship between the input and output of the system is obtained (identification model). However, the dynamics of the state (i.e., the internal dynamics of the system) is not known. Hence, this kind of model is referred to as "black-box model."

The above two methods for model construction can be taken as two classes of subjects. The first method can obtain the detailed description of the system, but a deep investigation on the system has to be undergone, which constitutes the subject of "process dynamics." The second method has developed to the subject of "system identification." The detailed discussions of these two subjects are beyond the topic of this book. If the mathematical model is on-line identified, the corresponding controller belongs to "adaptive controller" and, when combined with predictive control, "adaptive predictive controller."

The models can be selected and distinguished according to the respective system features:

(1) linear model and nonlinear model

(2) nominal model and uncertain model

(3) deterministic model and stochastic model

(4) time-invariant model and time-varying model

(5) continuous state model and discrete state model

(6) continuous-time model and discrete-time model (e.g., differential equation and difference equation)

(7) time driven model and event driven model

(8) lumped parameter model and distributed parameter model (e.g., ordinary differential equation and partial differential equation)

(9) automaton, finite state machine

(10) intelligent model (e.g., fuzzy model, neural network model)

etc. For hybrid systems, there are more kinds of models, including

(11) hybrid Petri net, differential automaton, hybrid automaton, mixed logic dynamic model, piecewise linear model, etc.

Due to the limited knowledge for investigating a system, it never happens that one kind of model is fixed for one kind of system (e.g., a partial differential equation for a continuous-time distributed parameter system, etc.). The selection of model should be done according to both the system dynamics and the applicability and certain human factors have to be introduced. Thus, for continuous-time distributed parameter system, one can select discrete-time lumped parameter model; for nonlinear time-varying system, one can select linear uncertain model, etc.

In control theory, for different kinds of systems corresponding to different kinds of models, there are different branches. For example,

- robust control adopts uncertain model, but can be applied to different systems in cases where the system dynamics can be included in the dynamics of the uncertain model;

- stochastic control applies stochastic model, which takes advantage of some probability properties of the system;

- adaptive control with on-line model identification mainly adopts linear difference equation, where the effect of time-varying nonlinearity is overcome by on-line refreshment of the model;

- fuzzy control adopts fuzzy model for uncertain system and nonlinear system, etc.;

- neural network control applies neural network model for nonlinear system, etc.;

- predictive control widely adopts various kinds of models to study various kinds of systems (mainly, multivariable constrained systems).

1.3　State space model and input/output model

1.3.1　State space model

The output of a system is affected by its state and, sometimes, directly by the input. The output can be the state or function of the state. The variation of the state is affected by the variation of the input. In order to emphasize the variation in the state, in general the mathematical model of a system can be represented as

$$\dot{x} = f(x, u, t), \quad y = g(x, u, t) \tag{1.3.1}$$

where x is the state, y the output, u the input, t the time, $\dot{x} = dx/dt$.

Usually, x, u, y are vectors, $x = \{x_1, x_2, \cdots, x_n\}$, $y = \{y_1, y_2, \cdots, y_r\}$, $u = \{u_1, u_2, \cdots, u_m\}$. x_i is the i-th state. One can represent the state, output and input as

$$x \in \mathfrak{R}^n, \quad y \in \mathfrak{R}^r, \quad u \in \mathfrak{R}^m$$

where $\mathfrak{R}^n$ is the n-dimensional real-valued space.

If the solution of (1.3.1) exists, then this solution can be generally represented as

$$x(t) = \phi(t, t_0, x(t_0), u(t)), \quad y(t) = \varphi(t, t_0, x(t_0), u(t)), \quad t \geq t_0. \tag{1.3.2}$$

If the system is relaxed at time t_0, i.e., $x(t_0) = 0$, then the solution can be generally represented as

$$x(t) = \phi_0(t, t_0, u(t)), \quad y(t) = \varphi_0(t, t_0, u(t)), \quad t \geq t_0. \tag{1.3.3}$$

If the solution (1.3.3) satisfies the following superposition principle, then the system (1.3.1) is linear.

Superposition Principle Suppose $\phi_{0,a}(t, t_0, u_a(t))$ and $\phi_{0,b}(t, t_0, u_b(t))$ are the motions by applying the inputs $u_a(t)$ and $u_b(t)$, respectively. Then, the motion by applying $\alpha u_a(t) + \beta u_b(t)$ is

$$\phi_0(t, t_0, \alpha u_a(t) + \beta u_b(t)) = \alpha \phi_{0,a}(t, t_0, u_a(t)) + \beta \phi_{0,b}(t, t_0, u_b(t))$$

where α and β are arbitrary scalars.

For a linear system, if it is not relaxed, then the motion caused by the initial state $x(t_0)$ can be added to the motion with relaxed system, by applying the superposition principle. Satisfying the superposition principle is the fundamental approximation or assumption in some classical MPC algorithms.

If a system satisfies the superposition principle, then it can be further simplified as

$$\dot{x} = A(t)x + B(t)u, \ y = C(t)x + D(t)u \qquad (1.3.4)$$

where $A(t)$, $B(t)$, $C(t)$ and $D(t)$ are matrices with appropriate dimensions. If a linear system is time-invariant, then it can be represented by the following mathematical model:

$$\dot{x} = Ax + Bu, \ y = Cx + Du. \qquad (1.3.5)$$

A linear system satisfies the superposition principle, which makes the mathematical computations much more convenient and a general closed-form solution exists. In the system investigation and model application, complete results exist only for linear systems. However, a real system is usually nonlinear. How to linearize a nonlinear system so as to analyze and design it according to the linearized model becomes a very important problem.

The basic idea for linearization is as follows. Suppose a real system moves around its equilibrium point (steady-state) according to various disturbances, and the magnitude of the motion is relatively small. In such a small range, the relationships between variables can be linearly approximated.

Mathematically, if a system is in steady-state (x_e, y_e, u_e), then its state x is time-invariant, i.e., $\dot{x} = 0$. Hence, according to (1.3.1),

$$f(x_e, u_e, t) = 0, \ y_e = g(x_e, u_e, t). \qquad (1.3.6)$$

Let $x = x_e + \delta x$, $y = y_e + \delta y$, $u = u_e + \delta u$. Suppose the following matrices exist (called Jacobean matrices or Jacobian matrices):

$$A(t) = \left. \frac{\partial f}{\partial x} \right|_e = \begin{bmatrix} \partial f_1/\partial x_1 & \partial f_1/\partial x_2 & \cdots & \partial f_1/\partial x_n \\ \partial f_2/\partial x_1 & \partial f_2/\partial x_2 & \cdots & \partial f_2/\partial x_n \\ \vdots & \vdots & \ddots & \vdots \\ \partial f_n/\partial x_1 & \partial f_n/\partial x_2 & \cdots & \partial f_n/\partial x_n \end{bmatrix},$$

$$B(t) = \left. \frac{\partial f}{\partial u} \right|_e = \begin{bmatrix} \partial f_1/\partial u_1 & \partial f_1/\partial u_2 & \cdots & \partial f_1/\partial u_m \\ \partial f_2/\partial u_1 & \partial f_2/\partial u_2 & \cdots & \partial f_2/\partial u_m \\ \vdots & \vdots & \ddots & \vdots \\ \partial f_n/\partial u_1 & \partial f_n/\partial u_2 & \cdots & \partial f_n/\partial u_m \end{bmatrix},$$

$$C(t) = \frac{\partial g}{\partial x}\Big|_e = \begin{bmatrix} \partial g_1/\partial x_1 & \partial g_1/\partial x_2 & \cdots & \partial g_1/\partial x_n \\ \partial g_2/\partial x_1 & \partial g_2/\partial x_2 & \cdots & \partial g_2/\partial x_n \\ \vdots & \vdots & \ddots & \vdots \\ \partial g_r/\partial x_1 & \partial g_r/\partial x_2 & \cdots & \partial g_r/\partial x_n \end{bmatrix},$$

$$D(t) = \frac{\partial g}{\partial u}\Big|_e = \begin{bmatrix} \partial g_1/\partial u_1 & \partial g_1/\partial u_2 & \cdots & \partial g_1/\partial u_m \\ \partial g_2/\partial u_1 & \partial g_2/\partial u_2 & \cdots & \partial g_2/\partial u_m \\ \vdots & \vdots & \ddots & \vdots \\ \partial g_r/\partial u_1 & \partial g_r/\partial u_2 & \cdots & \partial g_r/\partial u_m \end{bmatrix},$$

where e indicates "at the equilibrium point." Then in the neighborhood of (x_e, y_e, u_e) the system (1.3.1) can be approximated by

$$\delta \dot{x} = A(t)\delta x + B(t)\delta u, \quad \delta y = C(t)\delta x + D(t)\delta u. \tag{1.3.7}$$

1.3.2 Transfer function model

Use of the input/output model to describe a system is, in general, carried out when the properties of the system are not clearly known. In general, it is taken as granted that the system is relaxed, or the initial sate is zero, when the output is uniquely determined by the input. If the initial state is not zero, for the linear system, the motion by the initial state should be added. If the input/output relationship is considered, it should be noted that it is an incomplete description.

In spite of this, by using the block diagram formed through the input/output relationships, the relations between various parts of the system can be shown clearly. Hence, the classical methodology based on the transfer function is still widely applied and continuously developed.

The basic idea of the transfer function is as follows. Apply Laplace transformation, to transform the differential equation into algebraic equation, such that the computation can be simplified. The computed results, when required, can be reversed to the time-domain by applying the inverse Laplace transformation.

The transfer function corresponding to (1.3.5) is

$$G(s) = C(sI - A)^{-1}B + D \tag{1.3.8}$$

where s is Laplace transformation operator and each element of $G(s)$ can be represented as the following fractional:

$$g_{ij}(s) = \frac{b_{ij,m}s^m + b_{ij,m-1}s^{m-1} + \cdots + b_{ij,1}s + b_{ij,0}}{s^n + a_{ij,n-1}s^{n-1} + \cdots + a_{ij,1}s + a_{ij,0}}.$$

$g_{ij}(s)$ represents the relationship between the j-th input and the i-th output. For a real system, $m \leq n$.

1.3.3 Impulse response and convolution model

The model of a system can be set up by applying the impulse response. Since the impulse response can be measured, this provides an alternative for modeling.
    Let

$$\delta_\Delta(t - t_1) = \begin{cases} 0, & t < t_1 \\ 1/\Delta, & t_1 \leq t < t_1 + \Delta \\ 0, & t \geq t_1 + \Delta \end{cases}.$$

For any Δ, $\delta_\Delta(t - t_1)$ always has a unitary area. When $\Delta \to 0$,

$$\delta(t - t_1) \triangleq \lim_{\Delta \to 0} \delta_\Delta(t - t_1) \tag{1.3.9}$$

which is called impulse function or δ function. The response to the impulse signal is called impulse response.
    Let $g_{ij}(t, \tau)$ be the i-th output response due to the j-th impulse input (where τ is the time point when the impulse is added). Denote

$$G(t, \tau) = \begin{bmatrix} g_{11}(t, \tau) & g_{12}(t, \tau) & \cdots & g_{1m}(t, \tau) \\ g_{21}(t, \tau) & g_{22}(t, \tau) & \cdots & g_{2m}(t, \tau) \\ \vdots & \vdots & \ddots & \vdots \\ g_{r1}(t, \tau) & g_{r2}(t, \tau) & \cdots & g_{rm}(t, \tau) \end{bmatrix}$$

as the impulse response of the MIMO system. If the system is relaxed, the input/output model is

$$y(t) = \int_{-\infty}^{+\infty} G(t, \tau) u(\tau) d\tau \tag{1.3.10}$$

which is called convolution model. As long as the impulse response is known, the response to any known input $u(t)$ can be calculated.
    For (1.3.5), when $x(0) = 0$, the Laplace transformation of the response to the impulse signal $\delta(t)$ is the transfer function. The Laplace transformation of the response to any input $U(s)$ is $Y(s) = G(s)U(s)$.

1.4 Discretization of continuous-time systems

The last section discusses the continuous-time system, where the input, output and state change continuously with the evolution of time. In the real applications, there is another kind of system whose every variable changes only after a certain time interval (e.g., bank interest counting system), rather than continuously with time, which is called discrete-time systems.
    Another kind of system is intrinsically continuous-time; however, when one observes and controls these systems, the action is only taken at some

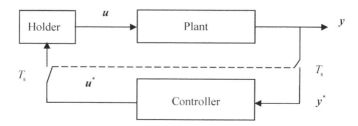

Figure 1.4.1: The structure of the sampling control system.

discrete time instants. The most commonly seen is the control system based on the digital computer, which is called sampling control system. MPC is usually based on the computers and, hence, is mainly for sampling control systems. The computers acquire the process variables with certain sampling intervals. The data sent from the computer to the controller are instant values with sampling intervals. The control moves are sent out also with certain time intervals. The procedure that transforms an originally continuous-time system to a discrete-time system is called the discretization of this continuous-time system.

The structure of the sampling control system is shown in Figure 1.4.1. The computer acquires the output y with certain time intervals, and sends out a discrete value y^*. It is as if there is a switch between y and y^*, which switches on instantly with sampling interval T_s. After each time interval T_s, the controller sends out a control move u^*, which is also an instant value. In order to control a continuous-time system, during the interval between the two sequential control moves, u^* is usually unaltered (which is called zero-order holder; there are other methods to calculate the u between two sequential control moves, which are not zero-order holders). Therefore, the output of the controller is usually taken as constituting a switch, switching on with certain time interval, and a holder.

For the controlled process in Figure 1.4.1, let us suppose

(i) the intervals for output sampling and the control sending-out are both equal to T_s, and the two switches work synchronously;

(ii) the switches switch on instantly, so that the time period when the switches are on can be overlooked;

(iii) the controller output u^* is restored by zero-order holder.

Based on the above assumptions, let us give the discretized forms of the continuous-time systems in section 1.3.

1.4.1   State space model

Simply denote kT_s as k, e.g., the state, output and input at time k are represented as $x(k)$, $y(k)$ and $u(k)$, respectively. The following is utilized to approximate the derivative of x:

$$\frac{dx}{dt} \approx \frac{x(k+1) - x(k)}{T_s}. \tag{1.4.1}$$

Thus, corresponding to (1.3.1) we obtain

$$x(k+1) = x(k) + T_s f(x(k), u(k), k), \quad y(k) = g(x(k), u(k), k) \tag{1.4.2}$$

and corresponding to (1.3.4),

$$x(k+1) = (I + A(k)T_s)x(k) + B(k)T_s u(k), \quad y(k) = C(k)x(k) + D(k)u(k). \tag{1.4.3}$$

Many linear systems have their general solutions, by which the values at different sampling instants can be calculated and the exact result of the discretization is obtained. For the linear time-invariant system (1.3.5) the exact result for discretization is

$$x(k+1) = e^{AT_s}x(k) + \int_0^{T_s} e^{At}dt Bu(k), \quad y(k) = Cx(k) + Du(k) \tag{1.4.4}$$

where the calculation of e^{AT_s} is a mature technique.

1.4.2 Impulse transfer function model

Since, in a discrete-time system, only the values in the sampling instant are taken, its Laplace transformation can be replaced with a special form, Z transformation. The effect of the Z transformation is to transform the difference equation into the algebraic equation, such that the computation can be greatly simplified. If the system is relaxed, i.e., the initial state is zero, then the input and output of the system can be linked by the impulse transfer function, so as to analyze the system and design the controller. If any variable value at the sampling instant is required, then the inverse Z transformation can be invoked, such that the solution to the system equation can be obtained.

Consider the linear time-invariant system

$$x(k+1) = Ax(k) + Bu(k), \quad y(k) = Cx(k) + Du(k). \tag{1.4.5}$$

Its corresponding discrete-time impulse transfer function is

$$G(z) = C(zI - A)^{-1}B + D \tag{1.4.6}$$

where z is the operator of Z transformation, and each element of $G(z)$ can be represented by the following fractional:

$$G_{ij}(z) = \frac{b_{ij,m}z^m + b_{ij,m-1}z^{m-1} + \cdots + b_{ij,1}z + b_{ij,0}}{z^n + a_{ij,n-1}z^{n-1} + \cdots + a_{ij,1}z + a_{ij,0}}.$$

$g_{ij}(z)$ represents the relationship between the j-th input and the i-th output. For a real system, $m \leq n$.

1.4.3   Impulse response and convolution model

Consider a linear time-invariant system

$$x(k+1) = Ax(k) + Bu(k), \ y(k) = Cx(k). \tag{1.4.7}$$

The recursive solution of the output is

$$y(k) = CA^k x(0) + \sum_{i=0}^{k-1} CA^{k-i-1} Bu(i). \tag{1.4.8}$$

Suppose $x(0) = 0$ and the input is $\delta(k)$, then $CA^{k-i-1}B = H(k-i)$ is the output at time $k - i$ and

$$y(k) = \sum_{i=0}^{k-1} H(k-i)u(i). \tag{1.4.9}$$

$H(k-i)$ is the impulse response matrix, and (1.4.9) or its equivalent form

$$y(k) = \sum_{i=1}^{k} H(i)u(k-i) \tag{1.4.10}$$

is called discrete convolution model. It is easy to obtain $H(i)$ via experiments. $H(i)$ is actually the output of the system, when at $k = 0$ an impulse input of magnitude 1 and width T_s is implemented. Notice that, under the zero order hold, a δ function becomes the square-wave impulse.

1.5   Model predictive control (MPC) and its basic properties

The background of MPC, at least for the earlier version of industrial MPC, is to substitute the "controller" in Figure 1.4.1 by "MPC." Therefore, when one investigates MPC, he/she should notice that usually there is a mismatch between the model (usually discrete-time) and the system. Actually, MPC was invented based on the fact that the traditional optimal control methods rely on the accurate mathematical models.

1.5.1   Streams and history

The studies on the MPC are manifold, which can be classified into several streams, including

(a) industrial MPC which is widely applied in the process industry, which adopts heuristic algorithms, with the relatively mature industrial software usually adopting linear nominal models, the representatives of which are the famous DMC and MAC;

(b) adaptive MPC which originated from the minimum variance control (MVC) and adaptive control, the representative of which is GPC;

(c) synthesis approach of MPC with guaranteed stability, which widely adopts the state space models and "three ingredients" are introduced, emerged because of the difficulties encountered in stability analysis of industrial MPC and adaptive MPC which make the existing stability results impossible to be generalized to nonlinear, constrained and uncertain systems;

(d) the method invented by borrowing techniques from other branches of control theory or applied mathematics, etc., which is largely different from the above (a)-(c).

For a long time, (a)-(c) were developed independently. MPC in (c) was the first to appear. In fact, many methods from the traditional optimal control can be regarded as MPC, which can be dated back to the 1960s. However, model predictive control, named as a process control algorithm, was formally proposed firstly in the form of (a), which dates back to the 1970s. In the 1980s, adaptive control was a hot topic; however, the famous MVC could not be successfully applied in the process control industry and, hence, the ideas from both the predictive control and adaptive control were combined.

The proponents of industrial MPC have not achieved intrinsic progress in the theoretical analysis of MPC; however, they are well aware of the importance of stability. Industrial MPC has no guarantee of stability. But, if the open-loop stable system is considered, and the optimization horizon is chosen to be sufficiently large, then usually the closed-loop system is stable. It is a reflection of the fact that infinite-horizon optimal control has guarantee of stability. For the MPCs in (a) and (b), it is difficult to apply Lyapunov method, up to now the most powerful stability tool. This theoretically most prominent deficiency propelled the development of MPC algorithms with guaranteed stability after the 1990s. Hence, from the beginning of the 1990s, people called the more extensive optimal control problems, including (a)-(c), predictive controls. Notice that, before the 1990s, MPC usually referred to (a), (b) and some special form of MPCs.

Remark 1.5.1. MPCs in (d) emerged from the 1980s. There were some algorithms, such as internal model control, nonlinear separation predictive control, predictive functional control, data driven predictive control, etc. Sometimes, it is difficult to distinguish (d) from (a)-(c). Sometimes, MPC in (d) does not have stability ingredients as in synthesis approaches; however, stability analysis is easier than industrial MPC and adaptive MPC.

1.5.2   Basic properties

Whatever stream it belongs to, MPC has the following important characteristics.

1. *MPC is based on model and the prediction model is utilized.*

MPC is a set of algorithms based on the models. MPC pays more attention to the function, than to the formulation, of the model. The function of a prediction model is based on the past information and the future inputs to predict the future output. Any collection of information, as long as it has the function of prediction, irrespective of the concrete form, can be the prediction model. Therefore, the traditional models, such as state space equation and transfer function, can be prediction models. For linear open-loop stable systems, even the non-parametric models, such as impulse response and step response, can be directly utilized as the prediction model. Further, nonlinear system and distributed parameter system, as long as it has the function of prediction, can be utilized as prediction model. Hence, MPC has no strict requirement on the model structure, which is different from the former control techniques. MPC pays more attention to the selection of the most convenient modeling methods, based on the information available.

For example, in DMC and MAC, the non-parametric models such as step response and impulse response, which are easily obtainable in the real industry, are adopted; in GPC, the parametric models such as Controlled Auto-Regressive Integrated Moving Average (CARIMA) model and state space model, are selected.

MPC discards the strict requirements on the model. A prediction model has the function to reveal the future behavior of the system. Thus, a prediction model can provide the a priori knowledge for the optimization, so that the control moves are decided such that the future output can comply with the desired output.

In the system simulation, by arbitrarily giving the future control strategies, the output of the system can be observed for any input (see Figure 1.5.1). This can be the basis for comparing different control strategies.

2. *The key point that MPC differs from other control techniques is that MPC adopts receding horizon optimization, and the control moves are implemented in a receding horizon manner.*

If one wants a unique difference between MPC and other control techniques, then it should be the manner that MPC implements the control moves, i.e., receding horizon optimization and receding horizon implementation.

In industrial applications and theoretical studies, in general MPC is based on the on-line optimization. By the optimization, a certain cost function is optimized, such that the future control moves are determined. This cost function involves the future behavior of a system, and usually is taken as minimizing the variance when the future output tracks the desired trajectory. However, it could be taken as a more general form, such as minimizing the energy of the control moves, etc. The future behavior involved in the cost function is determined by the model and future control strategies. But, the optimization in MPC differs largely from that in the traditional optimal control, i.e., the optimization in MPC does not utilize the globally time-invariant cost function and, rather, a receding, finite-horizon optimization strategy is adopted.

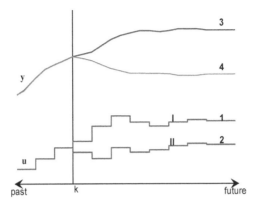

Figure 1.5.1: Prediction based on the model (1 - sequence of control moves I; 2 - sequence of control moves II; 3 - output corresponding to I; 4 - output corresponding to II).

At each sampling instant, the cost function often involves the finite length of future time; at the next sampling instant, the optimization horizon is moved forward (see Figure 1.5.2). Hence, at each sampling instant, MPC has a cost function for this instant; at a different sampling instant the relative form of the cost function can be the same, but the absolute form, i.e., the time window included, are different.

In MPC, the optimization is usually not determined off-line in a single optimization and, rather, it is performed repeatedly on-line. This is the meaning of receding horizon, and the intrinsic difference between MPC and the traditional optimal control. The limitation of this finite-horizon optimization is that, under ideal situations only the suboptimal solution for the global solution can be obtained. However, the receding horizon optimization can effectively incorporate the uncertainties incurred by model-plant mismatch, time-varying behavior and disturbances. With the effect of the uncertainties compensated, the new optimizations are always based on the real scenarios, and the control moves are optimal with an "in fact" manner.

For a real complicated industrial process, the uncertainties incurred by model-plant mismatch, time-varying behavior and disturbances are unavoidable and, hence, the receding finite-horizon optimization can be more powerful than the global one-time optimization.

Remark 1.5.2. Some off-line MPC algorithms will be given in the following chapters. In off-line MPC, the on-line optimization is not involved, rather, a set of optimization problems are solved off-line, with a sequence of control laws obtained. On-line, the real-time control law is selected according to the current state of the system. In spite of this, the control laws are implemented by receding horizon (i.e., at different time the control laws can be different), and each control law is determined by optimization.

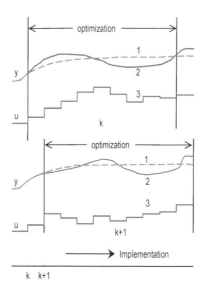

Figure 1.5.2: Receding horizon optimization (1 - reference trajectory; 2 - optimal predicted output; 3 - optimal control move).

Remark 1.5.3. Some MPC algorithms will be given in the following chapters which directly solve the infinite-horizon cost function. But, in order to obtain the optimal solution, the infinite-horizon optimization cannot be solved directly and, rather, it is transformed into the finite-horizon optimization. In some other MPC algorithms, the optimization problem is not solved at each sampling instant and, rather, at suitable times the optimization result in the previous sampling instant is inherited.

Remark 1.5.4. In a word, in MPC, all the ideas of on-line optimization, finite-horizon optimization and receding horizon optimization can be broken or temporarily broken. However, these "special" situations cannot include all the MPC algorithms and cannot deny the basic characteristics of MPC and, rather, they can be taken as the generalized forms complying with the basic characteristics. These "special" situations can be observed from another angle, i.e., the boundary between MPC and other control techniques can sometimes become rather blurry.

3. *While the optimal control rationale is adopted, MPC does not discard the feedback in the traditional control techniques.*

It is well known that feedback is essential and unique in overcoming disturbance, uncertainty and achieving closed-loop stability. Up to now, in MPC, the feedback is not discarded, but used more profoundly; the effect of feedback is never denied, but proven continuously.

Since its first proposal, industrial MPC has utilized feedback correction, and was concluded as one of the "three principles." The effect of feedback is realized in adaptive MPC by use of on-line refreshment of the model. In synthesis approaches, when the uncertainty is considered, feedback MPC (i.e., MPC based on closed-loop optimization, where a sequence of control laws are optimized) is better than open-loop MPC (i.e., MPC based on open-loop optimization, where a sequence of control moves are optimized) in performance (feasibility, optimality) (details are given in Chapter 9). Moreover, in synthesis approach, the local feedback control law is applied. More importantly, in the real applications, the "transparent control" is often applied, which sets up MPC based on PIDs, where PID is the feedback controller.

Further, it is noted that, without feedback, it is not effective for analyzing and studying MPC.

1.5.3 "Three principles" of industrial MPC

In order to distinguish synthesis approaches, one can call DMC, MAC, GPC etc. as classical MPC. Hence, classical MPC is coarsely referred to those MPC algorithms which were hot before the 1990s.

Here, industrial MPC refers to a part of classical MPC, which is the part of classical MPC that has been successfully applied in the industrial processes. Comparatively, for synthesis approaches, there are less reports for the industrial applications.

The key points of industrial MPC can be summarized as the "three principles," i.e., prediction model, receding horizon optimization and feedback correction. The "three principles" are the keys for successfully applying MPC in the real projects. It should be emphasized that, in industrial MPC, the on-line, finite-horizon optimization is preferred, and the effects of prediction model and cost function are more prominent.

In the following we talk about the feedback correction.

Industrial MPC is a kind of closed-loop control algorithm. When the receding horizon optimization is performed, the basis for optimization should comply with the real plant. However, the prediction model is only a coarse description of the real dynamics. Due to the unavoidable nonlinearity, time-varying behavior, model-plant mismatch and disturbance, the prediction based on the time-invariant model cannot be completely equivalent to the real situation, which needs additional prediction strategy to compensate for the deficiency in the model prediction, or the model needs to be refreshed on-line. The receding horizon optimization can only be advantageous when it is based on the feedback correction. For this reason, when a sequence of control moves are determined by optimization, in order to prevent the deviation of the control from the ideal status due to the model-plant mismatch and environmental disturbances, these control moves won't be implemented one by one. Rather, only the most current control move is implemented. At the next sampling instant, the real output is measured, the feedback correction is invoked to compensate the predictions (see Figure 1.5.3) or the prediction model, and the optimization is re-done.

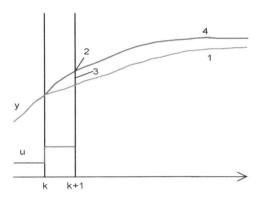

Figure 1.5.3: Error correction (1 - predicted output at k; 2 - measured output at $k+1$; 3 - prediction error; 4 - corrected predicted output at $k+1$).

The strategies for feedback correction are manifold. One can fix the model, and predict the future prediction error to compensate; one can also refresh the prediction model by on-line identification. Whichever strategy is adopted, MPC sets up its optimization based on the real plant, and tries to make an exact prediction of the future behavior along the optimization. Hence, the optimization in MPC is not only based on model, but also utilizes feedback information, which forms the actual closed-loop optimization.

Remark 1.5.5. Notice that, in the real applications, the feedback correction is very important. However, in the theoretical study of MPC, it is often supposed that the system and its model are equivalent (e.g., nominal stability of classical MPC), or that any possible dynamics of the real system can be included within the dynamics of the model (e.g., synthesis approach of robust MPC); then the feedback correction is not explicitly introduced. If we observe both from the side of theoretical studies and from the side of real applications, we can say that the key of MPC lies in its receding horizon optimization and receding horizon implementation. MPC without feedback correction is also called receding horizon control, which emphasizes the receding horizon nature. When one utilizes state space model to synthesis approaches, in general the feedback correction is not utilized. Rather, the state feedback strategy is often invoked so as to form "feedback MPC." Certainly, from the side of real control effect, this state feedback has the same effect with the feedback correction, hence is the feedback correction in the context of state space description.

1.6   Three typical optimal control problems of MPC

Due to the following reasons, the optimization problems of MPC are manifold (note that, sometimes one can classify MPC according to the optimization problem):

(i) the mathematical models that can be utilized are manifold;

(ii) the real systems are manifold;

(iii) there are large differences between the real applications and the theoretical investigations.

In the following, by taking the discrete-time state space model and the quadratic cost function, as an example, three categories of MPC optimization problems are given. Consider $x(k+1) = f(x(k), u(k))$, where $f(0,0) = 0$. The system is stabilizable. The state and input constraints are

$$x(k + i + 1) \in \mathcal{X}, u(k + i) \in \mathcal{U}, \ i \geq 0 \qquad (1.6.1)$$

satisfying $\{0\} \subset \mathcal{X} \subseteq \Re^n$ and $\{0\} \subset \mathcal{U} \subseteq \Re^m$.

In MPC, usually $x(k+i|k)$ is used to denote the prediction of x at a future time $k+i$, predicted at time k, and $x(k+i|k) = x(k+i)$, $i \leq 0$; $x^*(k+i|k)$, $i \geq 0$ are used to denote the optimal state predictions (i.e., predictions using the optimal solution of the MPC optimization problem).

Suppose the following state prediction is adopted:

$$x(k + i + 1|k) = f(x(k + i|k), u(k + i|k)), \ i \geq 0, \ x(k|k) = x(k). \quad (1.6.2)$$

The three categories are given in the following sections.

1.6.1 Infinite-horizon

The basic property of the infinite-horizon optimization is that the cost function is the sum of the positive-definite functions over an infinite time horizon. The cost function and the constraints are usually

$$J_\infty(x(k)) = \sum_{i=0}^{\infty} \left[\|x(k + i|k)\|_W^2 + \|u(k + i|k)\|_R^2 \right], \qquad (1.6.3)$$

$$\text{s.t. } (1.6.2), \ x(k + i + 1|k) \in \mathcal{X}, \ u(k + i|k) \in \mathcal{U}, \ i \geq 0, \qquad (1.6.4)$$

where $W \geq 0$, $R > 0$ are symmetric and, for any vector ϑ and non-negative matrix W, $\|\vartheta\|_W^2 \triangleq \vartheta^T W \vartheta$. At each time k, the decision variable (freedom for optimization) of (1.6.3)-(1.6.4) is

$$\tilde{u}(k) = \{u(k|k), u(k + 1|k), \cdots\}. \qquad (1.6.5)$$

Since an infinite number of decision variables are involved, the infinite-horizon optimization problem is generally not directly solvable.

1.6.2　Finite-horizon: classical MPC

The key property of classical MPC is that the cost function is the sum of the positive-definite functions over the finite time horizon, no off-line designed or on-line optimized terminal constraint set and terminal cost function, and no other kinds of artificial constraints are imposed. Usually the cost function and constraints are

$$J_N(x(k)) = \sum_{i=0}^{N-1} \left[\|x(k+i|k)\|_W^2 + \|u(k+i|k)\|_R^2 \right] + \|x(k+N|k)\|_W^2 ,$$

$$\tag{1.6.6}$$

$$\text{s.t. } (1.6.2),\ x(k+i+1|k) \in \mathcal{X},\ u(k+i|k) \in \mathcal{U},\ i \in \{0,1,\ldots,N-1\},$$

$$\tag{1.6.7}$$

or

$$J_{N,M}(x(k)) = \sum_{j=0}^{M-1} \|u(k+j|k)\|_R^2 + \sum_{i=0}^{N} \|x(k+i|k)\|_W^2 , \tag{1.6.8}$$

$$\text{s.t. } (1.6.2),\ \begin{cases} x(k+i|k) \in \mathcal{X},\ i \in \{1,\ldots,N\}, \\ u(k+j|k) \in \mathcal{U},\ j \in \{0,1,\ldots,M-1\} \\ u(k+s+M|k) = u(k+M-1|k), s \in \{0,1,\ldots,N-M-1\} \end{cases} ,$$

$$\tag{1.6.9}$$

where N is the prediction horizon, M control horizon, $M \leq N$. At each time k, the decision variables of (1.6.6)-(1.6.7) and (1.6.8)-(1.6.9) are, respectively,

$$\tilde{u}_N(k) = \{u(k|k), u(k+1|k), \cdots, u(k+N-1|k)\}, \tag{1.6.10}$$

$$\tilde{u}_M(k) = \{u(k|k), u(k+1|k), \cdots u(k+M-1|k)\}. \tag{1.6.11}$$

1.6.3　Finite-horizon: synthesis approaches

In the 1990s, there appeared some MPC algorithms by modifying the optimization problems (1.6.6)-(1.6.7) and (1.6.8)-(1.6.9), such that closed-loop stability can be guaranteed. Compared with classical MPC, the main property of this modified category is that, the off-line designed or on-line optimized terminal constraint set and terminal cost function are introduced into the optimization, such that the convergence property of the optimization algorithm is modified and the value function of the cost can monotonically decrease along the receding horizon optimization. Usually the cost function and constraints are

$$\bar{J}_N(x(k)) = \sum_{i=0}^{N-1} \left[\|x(k+i|k)\|_W^2 + \|u(k+i|k)\|_R^2 \right] + \|x(k+N|k)\|_{W_N}^2 ,$$

$$\tag{1.6.12}$$

$$\text{s.t. } (1.6.2),\ x(k+i+1|k) \in \mathcal{X},\ u(k+i|k) \in \mathcal{U},\ i \in \{0,1,\ldots,N-1\},$$
$$x(k+N|k) \in \mathcal{X}_f, \tag{1.6.13}$$

or

$$\bar{J}_{N,M}(x(k)) = \sum_{j=0}^{M-1} \|u(k+j|k)\|_R^2$$

$$+ \sum_{i=0}^{N-1} \|x(k+i|k)\|_W^2 + \|x(k+N|k)\|_{W_N}^2, \qquad (1.6.14)$$

$$\text{s.t. } (1.6.2), \begin{cases} x(k+i|k) \in \mathcal{X}, \ i \in \{1, \ldots, N\}, \\ x(k+N|k) \in \mathcal{X}_f \\ u(k+j|k) \in \mathcal{U}, \ j \in \{0, 1, \ldots, N-1\} \\ u(k+s+M|k) = Kx(k+s+M|k), \\ s \in \{0, 1, \ldots, N-M-1\} \end{cases}, \qquad (1.6.15)$$

where $\mathcal{X}_f$ is the terminal constraint set, K the local controller,

$$F(x(k+N|k)) = \|x(k+N|k)\|_{W_N}^2 \qquad (1.6.16)$$

the terminal cost function. At each time k, the decision variables of (1.6.12)-(1.6.13) and (1.6.14)-(1.6.15) are (1.6.10) and (1.6.11), respectively. By appropriately setting $\mathcal{X}_f$, K and W_N (called the three ingredients) we can obtain the "MPC algorithm with guaranteed stability." $F(\cdot)$ is used to form the infinite-horizon value function or approximate the infinite-horizon value function in a neighborhood of the origin, while $\mathcal{X}_f$ is usually selected as the subset of this neighborhood.

$\mathcal{X}_f$ is usually selected as a control invariant set. Refer to the following definitions.

Definition 1.6.1. Ω *is a positively invariant set of the autonomous system* $x(k+1) = f(x(k))$, *if* $x(k) \in \Omega$, $\forall k > 0$ *for any* $x(0) \in \Omega$.

Definition 1.6.2. *If there exists feedback law* $u(k) = g(x(k)) \in \mathcal{U}$, *such that* Ω *is the positively invariant set of the closed-loop system* $x(k+1) = f(x(k), g(x(k)))$, *then* Ω *is the control invariant set of the system* $x(k+1) = f(x(k), u(k))$.

By receding horizon solving (1.6.3)-(1.6.4), (1.6.6)-(1.6.7), (1.6.8)-(1.6.9), (1.6.12)-(1.6.13) or (1.6.14)-(1.6.15), the control move at time k is obtained,

$$u(k) = u(k|k).$$

Thus, various MPC algorithms are formed.

Certainly, in a concrete MPC algorithm, there may appear that

(i) the model adopted is not $x(k+1) = f(x(k), u(k))$;

(ii) the constraints handled are not (1.6.1);

(iii) the cost function adopted is different from (1.6.3), (1.6.6), (1.6.8), (1.6.12) and (1.6.14);

(iv) the future predictions on state/output are different from (1.6.2).

However, the above three categories of MPC problems, including the differences between finite-horizon and infinite-horizon, with or without terminal cost function and terminal constraint set, has a general meaning.

In this book, we call stability study on classical MPC as stability analysis, and that on synthesis approaches as stability synthesis.

1.7 Finite-horizon control: an example based on "three principles"

Suppose a nonlinear system is represented by the following model:

$$x(k + 1) = f(x(k), u(k)), \ y(k) = g(x(k)) \tag{1.7.1}$$

Based on this model, at time k, as long as the initial state $x(k)$ and the future control moves $u(k)$, $u(k + 1|k)$, $\cdots$ are known, the future output can be predicted as

$$\begin{aligned} x(k + i|k) =&f(x(k + i - 1|k), u(k + i - 1|k)), \ u(k|k) = u(k), \\ &\bar{y}(k + i|k) = g(x(k + i|k)), \ x(k|k) = x(k), \ i \in \{1, 2, \ldots\}. \end{aligned} \tag{1.7.2}$$

Based on the above recursive formula we can obtain

$$\bar{y}(k+i|k) = \phi_i(x(k), u(k), u(k+1|k), \cdots, u(k+i-1|k)), i \in \{1, 2, \ldots\} \tag{1.7.3}$$

where $\phi_i(\cdot)$ is composed of $f(\cdot)$ and $g(\cdot)$. Eq. (1.7.3) is the prediction model.

If there is mismatch between the model (1.7.1) and the real system, then based on the measured output, the output predictions can be compensated by error predictions. Denote the measured output at k as $y(k)$. Then, the error predictions can be constructed based on $\delta(k) = y(k) - \bar{y}(k|k - 1)$. The error predictions can be given based on the historical error information $\delta(k), \cdots, \delta(k - L)$,

$$\delta(k + i|k) = \varphi_i(\delta(k), \delta(k - 1), \cdots, \delta(k - L)), \tag{1.7.4}$$

where $\varphi_i(\cdot)$ is a linear or nonlinear function whose formula depends on the non-causal prediction method and L is the length of the used historical error information. By use of (1.7.4) to correct the model predictions, the closed-loop output predictions can be obtained as

$$y(k + i|k) = \bar{y}(k + i|k) + \delta(k + i|k). \tag{1.7.5}$$

Eq. (1.7.5) is the closed-loop prediction with feedback correction (1.7.4) based on the model (1.7.3).

At time k, the control objective is to find M control moves $u(k), u(k + 1|k), \cdots, u(k + M - 1|k)$ (suppose u is invariant after $k + M - 1$), such that the following cost function of output is minimized:

$$J(k) = F(\tilde{y}(k|k), \vec{\omega}(k)), \qquad (1.7.6)$$

where

$$\tilde{y}(k|k) = \begin{bmatrix} y(k+1|k) \\ \vdots \\ y(k+P|k) \end{bmatrix}, \quad \vec{\omega}(k) = \begin{bmatrix} \omega(k+1) \\ \vdots \\ \omega(k+P) \end{bmatrix},$$

$\omega(k + i)$ is the desired output at $k + i$, and M, P are control horizon and prediction horizon, $M \leq P$.

Thus, the on-line receding horizon optimization is, based on the closed-loop prediction (1.7.5), to find the control moves such that the cost function in (1.7.6) is minimized. If the optimal solution $u^*(k), u^*(k + 1|k), \cdots, u^*(k + M - 1|k)$ is found, then at time k, $u^*(k)$ is implemented. At the next sampling instant, the real output is measured, the prediction error is corrected, and the optimization is re-done. This is the general description of the nonlinear MPC based on the "three principles."

In general, even if the performance cost is quadratic, due to the system nonlinearity, we are still faced with a generally nonlinear optimization problem. Since the decision variables are submerged into the compound function, so that it is impossible to separate the decision variable, the closed-form solution to $u(k), u(k + 1|k), \cdots, u(k + M - 1|k)$ is in general unavailable (closed-form solution is the solution represented by formula, and is compared with the numerical solution). If problem (1.7.5)-(1.7.6) is taken as a nonlinear optimization (i.e., to find the numerical solution), in general an efficient algorithm is lacking. If we utilize the discrete maximum principle to write out a series of extremum conditions, then the computational burden involved will be very heavy and it is not admissible for a real-time control. Thus, although there is a general description for the optimization problem of MPC to a nonlinear system, there is intrinsic difficulty brought by the nonlinearity for solving the problem.

The above computational issue has always been the targeted issue of industrial MPC.

1.8   Infinite-horizon control: an example of dual-mode suboptimal control

Consider the time-invariant discrete-time nonlinear system described in the following state space equation:

$$x(k + 1) = f\left(x(k), u(k)\right), \ x(0) = x_0, \ k \geq 0. \qquad (1.8.1)$$

The state is measurable. The system input and state are constrained by

$$x(k) \in \mathcal{X}, \ u(k) \in \mathcal{U}, \ k \geq 0. \tag{1.8.2}$$

It is supposed that

(A1) $f : \mathfrak{R}^n \times \mathfrak{R}^m \to \mathfrak{R}^n$ is twice continuously differentiable and $f(0,0) = 0$ and, thus, $(x = 0, u = 0)$ is an equilibrium point of the system;

(A2) $\mathcal{X} \subseteq \mathfrak{R}^n, \mathcal{U} \subseteq \mathfrak{R}^m$ are compact, convex, and $\mathcal{X} \supset \{0\}, \mathcal{U} \supset \{0\}$;

(A3) by taking the Jacobean linearization of system at $(x = 0, u = 0)$, i.e.,

$$x(k+1) = Ax(k) + Bu(k), \ x(0) = x_0, k \geq 0, \tag{1.8.3}$$

the pair (A, B) is stabilizable.

The control objective is to regulate the state of system to the origin, at the same time satisfy both the state and control constraints and minimize the objective function

$$\Phi(x_0, u_0^\infty) = \sum_{i=0}^{\infty} \left[\|x(i)\|_W^2 + \|u(i)\|_R^2 \right]. \tag{1.8.4}$$

Suppose the pair $(A, W^{1/2})$ is observable. $u_0^\infty = \{u(0), u(1), u(2), \cdots\}$ are decision variables.

1.8.1　Three related control problems

Problem 1.1 Linear quadratic regulator (LQR):

$$\min_{u_0^\infty} \Phi(x_0, u_0^\infty), \text{ s.t. } (1.8.3). \tag{1.8.5}$$

Problem 1.1 was formulated and solved by Kalman, and the solution is the well-known linear feedback control law

$$u(k) = Kx(k). \tag{1.8.6}$$

The controller gain K is calculated by

$$K = -(R + B^T PB)^{-1} B^T PA \tag{1.8.7}$$

where P is obtained by solving the following discrete algebraic Riccati equation:

$$P = W + A^T PA - A^T PB(R + B^T PB)^{-1} B^T PA.$$

Problem 1.2 Nonlinear quadratic regulator (NLQR):

$$\min_{u_0^\infty} \Phi(x_0, u_0^\infty), \text{ s.t. } (1.8.1). \tag{1.8.8}$$

Problem 1.3 Constrained nonlinear quadratic regulator (CNLQR):

$$\min_{u_0^\infty} \Phi(x_0, u_0^\infty), \text{ s.t. } (1.8.1) - (1.8.2). \qquad (1.8.9)$$

CNLQR is a natural extension of NLQR and is more difficult but more practical than NLQR. In general, CNLQR and NLQR concern with infinite-horizon optimization and, hence, solution in the closed-form is usually impossible.

1.8.2 Suboptimal solution

A two-step suboptimal solution is proposed. In the first step a neighborhood of origin is constructed inside of which an inside mode controller is adopted in the form of (1.8.6). The neighborhood has to satisfy the following two conditions:

(A) the neighborhood is invariant for nonlinear system (1.8.1) controlled by (1.8.6);

(B) (1.8.2) should be satisfied in the neighborhood.

In the second step a finite-horizon optimization problem (FHOP) with additional terminal inequality constraints is solved to get an outside mode controller. The two controllers combine together forming an overall solution to suboptimal CNLQR. Notice that,

(a) the whole objective function $\Phi(x_0, u_0^\infty)$ is divided into $\Phi\left(x_0, u_0^{N-1}\right)$ and $\Phi(x(N), u_N^\infty)$ which are solved separately;

(b) the solution to $\Phi\left(x_0, u_0^{N-1}\right)$ is optimal and the solution to $\Phi(x(N), u_N^\infty)$ is suboptimal, so the overall solution is suboptimal.

This two-step type controller is also called a dual-mode controller. First consider the inside mode controller.

Lemma 1.8.1. *There exists a constant $\alpha \in (0, \infty)$ specifying a neighborhood Ω_α of the origin in the form of*

$$\Omega_\alpha \triangleq \left\{x \in \mathfrak{R}^n | x^T P x \leq \alpha\right\} \qquad (1.8.10)$$

such that

(i) Ω_α is control invariant with respect to control law (1.8.6) for (1.8.1);

(ii) $\forall x_0 \in \Omega_\alpha$, with (1.8.6) taken, $\lim_{k \to \infty} x(k) = 0$ and $\lim_{k \to \infty} u(k) = 0$.

Proof. (i) Since $\mathcal{X} \supset \{0\}, \mathcal{U} \supset \{0\}$, it is always possible to find a sufficiently small $\alpha_1 \in (0, \infty)$ that specifies a region in the form of (1.8.10) and satisfies

$x \in \mathcal{X}$, $Kx \in \mathcal{U}$, $\forall x \in \Omega_{\alpha_1}$. The deduction below will show that there exists $\alpha \in (0, \alpha_1]$ such that Ω_α is control invariant.

To achieve this aim, define Lyapunov function as

$$V(k) = x(k)^T P x(k),$$

and denote

$$\Theta(x) = f(x, Kx) - (A + BK)x.$$

For notational brevity, denote $x(k)$ as x. Thus,

$$
\begin{aligned}
&V(k+1) - V(k) \\
=&f(x, Kx)^T P f(x, Kx) - x^T P x \\
=&\left(\Theta(x) + (A + BK)x\right)^T P \left(\Theta(x) + (A + BK)x\right) - x^T P x \\
=&\Theta(x)^T P \Theta(x) + 2\Theta(x)^T P(A + BK)x \\
&+ x^T \left[A^T P A + 2K^T B^T P A + K^T B^T P B K - P\right] x \\
=&\Theta(x)^T P \Theta(x) + 2\Theta(x)^T P(A + BK)x + x^T \left[A^T P A - P\right] x \\
&- x^T \left[2A^T P B \left(R + B^T P B\right)^{-1} B^T P A - A^T P B \left(R + B^T P B\right)^{-1}\right. \\
&\times \left. B^T P B \left(R + B^T P B\right)^{-1} B^T P A\right] x \\
=&\Theta(x)^T P \Theta(x) + 2\Theta(x)^T P(A + BK)x + x^T \left[A^T P A - P\right] x \\
&- x^T \left[A^T P B \left(R + B^T P B\right)^{-1} B^T P A + A^T P B \left(R + B^T P B\right)^{-1}\right. \\
&\left. R \left(R + B^T P B\right)^{-1} B^T P A\right] x \\
=&\Theta(x)^T P \Theta(x) + 2\Theta(x)^T P(A + BK)x - x^T \left(W + K^T R K\right) x. \quad (1.8.11)
\end{aligned}
$$

Now take $\gamma > 0$ such that

$$\gamma < \lambda_{\min}\left(W + K^T R K\right)$$

and

$$V(k+1) - V(k) \leq -\gamma x^T x, \quad (1.8.12)$$

then

$$\Theta(x)^T P \Theta(x) + 2\Theta(x)^T P(A + BK)x \leq x^T \left(W + K^T R K\right) x - \gamma x^T x. \quad (1.8.13)$$

Define

$$L_\Theta = \sup_{x \in B_r} \frac{\|\Theta(x)\|}{\|x\|}$$

where $B_r = \{x| \|x\| \leq r\}$. L_Θ exists and is finite because f is twice continuously differentiable.

Then, for $\forall x \in B_r$, (1.8.13) is satisfied if

$$\{L_\Theta^2 \|P\| + 2L_\Theta \|P\| \|A + BK\|\} \|x\|^2 \le \{\lambda_{\min}(W + K^TRK) - \gamma\} \|x\|^2. \tag{1.8.14}$$

Since $\lambda_{\min}(W + K^TRK) - \gamma > 0$, and $L_\Theta \to 0$ as $r \to 0$, there exist suitable r and $\alpha \in (0, \alpha_1]$ such that (1.8.14) holds for $\forall x \in \Omega_\alpha \subseteq B_r$, which implies that (1.8.12) holds as well. Eq. (1.8.12) implies that the region Ω_α is control invariant with respect to $u = Kx$.

(ii) For $\forall x \in \Omega_\alpha$, (1.8.12) means that Ω_α is the region of attraction for asymptotic stability (a region of attraction is a set such that, when the initial state lies in this set, the closed-loop system has the specified stability property), so $\lim_{k \to \infty} x(k) = 0$ and $\lim_{k \to \infty} u(k) = 0$. $\qquad\square$

Remark 1.8.1. In the above deduction, one can directly choose

$$L_\Theta = \sup_{x \in \Omega_\alpha} \frac{\|\Theta(x)\|}{\|x\|}.$$

However, finding L_Θ is more difficult.

From the proof of Lemma 1.8.1, an applicable procedure to determine the region Ω_α is as follows:

Algorithm 1.1 The algorithm for determining Ω_α falls into the following steps:

Step 1. Solve Problem 1.1 to get a local linear state feedback gain matrix K.

Step 2. Find a suitable α_1 such that $x \in \mathcal{X}$, $Kx \in \mathcal{U}$ for all $x \in \Omega_{\alpha_1}$.

Step 3. Choose an arbitrary but suitable positive constant γ such that $\gamma < \lambda_{\min}(W + K^TRK)$.

Step 4. Choose an upper bound of L_Θ, L_Θ^u, such that L_Θ^u satisfies (1.8.14).

Step 5. Choose a suitable positive constant r such that $L_\Theta \le L_\Theta^u$.

Step 6. Choose a suitable $\alpha \in (0, \alpha_1]$ such that $\Omega_\alpha \subseteq B_r$.

Remark 1.8.2. Ω_α has an upper bound since it must guarantee the invariance of state and satisfaction of state and input constraints.

The inside mode controller is obtained by solving an LQR problem. The control law is denoted by

$$u_N^\infty = \{u_i^N, u_i^{N+1}, \cdots\}. \tag{1.8.15}$$

However, the special condition for applying the control sequence (1.8.15) is that the initial state $x(N)$ of the LQR problem should lie in Ω_α. If $x_0 \notin \Omega_\alpha$, then $x(N) \in \Omega_\alpha$ may be achieved by the following outside mode controller.

The outside mode control problem is formulated as a finite-horizon optimization problem with additional terminal state inequality constraints, coarsely,

$$\min_{u_0^{N-1}} \Phi(x(0), u_0^{N-1}), \text{ s.t. } (1.8.1), (1.8.2), x(N) \in \Omega_\alpha, \qquad (1.8.16)$$

where

$$\Phi\left(x(0), u_0^{N-1}\right) = \sum_{i=0}^{N-1} \left[\|x(i)\|_W^2 + \|u(i)\|_R^2 \right] + \|x(N)\|_W^2 \qquad (1.8.17)$$

with N being the optimization horizon. The selection of N should consider both the feasibility of (1.8.16) and the optimality of the whole CNLQR problem.

Remark 1.8.3. In general, the feasibility cannot be guaranteed for every initial state. If the original Problem 1.3 is feasible (this refers to the theoretical solution, which is difficult), but (1.8.16) is not feasible, i.e. the initial state x_0 is a feasible initial state with respect to Problem 1.3 but not to (1.8.16), then increasing N tends to retrieve feasibility.

Remark 1.8.4. The larger N has the advantage of optimality but results in more computational cost. Because the suboptimal control problem is solved off-line (not as MPC which do most computation on-line), the computational aspect is not so serious. So a large N may be chosen. If (1.8.16) is a complicated nonconvex optimization problem, its optimal solution is not easy to be found; in this case, increasing N at a special point $N = N_0$ may not have optimality advantage.

Remark 1.8.5. By properly selecting N, the artificial constraint $x(N) \in \Omega_\alpha$ can be automatically satisfied.

Suppose the solution to problem (1.8.16) is given as

$$u_0^{N-1} = \{u_o^*(0), u_o^*(1), \cdots, u_o^*(N-1)\}. \qquad (1.8.18)$$

The overall solution to the CNLQR problem is composed of the outside mode controller and the inside mode controller as

$$u_0^\infty = \{u_o^*(0), u_o^*(1), \cdots, u_o^*(N-1), u_i^N, u_i^{N+1}, \cdots\}. \qquad (1.8.19)$$

1.8.3   Feasibility and stability analysis

Usually, the closed-loop system by applying CNLQR cannot be globally stable. In the following we define a region in which (i.e., when the state lies in this region) the optimization problem (1.8.16) is feasible and the closed-loop system is stable.

Definition 1.8.1. $S_N(I,T)$ *is called an N-step stabilizable set contained in I for the system (1.8.1), if T is a control invariant subset of I and $S_N(I,T)$ contains all states in I for which there exists an admissible control sequence of length N which will drive the states of the system to T in N steps or less, while keeping the evolution of the state inside I, i.e.,*

$$S_N(I,T) \triangleq \{x_0 \in I : \exists u(0), u(1), \cdots, u(N-1) \in \mathcal{U}, \ \exists M \leq N \text{ such that } x(1),$$
$$x(2), \cdots, x(M-1) \in I \text{ and } x(M), x(M+1), \cdots, x(N) \in T, \ T \text{ is invariant}\}.$$

From Definition 1.8.1, it is easy to know that $S_i(I,T) \subseteq S_{i+1}(I,T)$ for every positive integer i.

Definition 1.8.2. *The feasible set $\Omega_F \subseteq \mathcal{X}$ is the set of initial state x_0 for which Problem 1.3 exists a feasible solution that results in a stable closed-loop system.*

Lemma 1.8.2. *Consider Problem (1.8.16) and Problem 1.3, then $S_N(\mathcal{X},\Omega_\alpha) \subseteq \Omega_F$ holds. Moreover, $S_N(\mathcal{X},\Omega_\alpha) \to \Omega_F$ as $N \to \infty$, that is, $S_\infty(\mathcal{X},\Omega_\alpha) = \Omega_F$.*

Proof. The initial state satisfies $x_0 \in \mathcal{X}$, so $\Omega_\alpha \subseteq \mathcal{X}$. In (1.8.16), the artificial constraint $x(N) \in \Omega_\alpha$ is added to the problem, so the feasible set must be smaller than that of Problem 1.3. As $N \to \infty$, since the original Problem 1.3 has asymptotic stability property and the constraint $x(N) \in \Omega_\alpha$ is not active, $S_N(\mathcal{X},\Omega_\alpha) \to \Omega_F$. □

Remark 1.8.6. Lemma 1.8.2 shows that in order for the suboptimal CNLQR problem to be solvable for some initial states, N may have to be chosen large. There may exist an integer j such that $S_j(\mathcal{X},\Omega_\alpha) = S_{j+1}(\mathcal{X},\Omega_\alpha)$. In this case, $S_\infty(\mathcal{X},\Omega_\alpha)$ is finite determined and j is the determinedness index.

Remark 1.8.7. If $S_\infty(\mathcal{X},\Omega_\alpha)$ is finite determined with determinedness index j, then as long as $N \geq j$, the feasible set of Problem 1.3 is equal to the feasible set of (1.8.16).

Theorem 1.8.1. *(Stability) For any $x_0 \in S_N(\Omega_F,\Omega_\alpha)$, (1.8.19) obtained from the dual-mode controller asymptotically stabilizes the nonlinear system (1.8.1).*

Proof. From Lemma 1.8.2, $\Omega_\alpha \in \Omega_F$ is trivial. As $x_0 \in S_N(\Omega_F,\Omega_\alpha)$, problem (1.8.16) is solvable, the control sequence (1.8.18) will drive states $x(N)$ into Ω_α. Inside Ω_α, the fact that (1.8.6) will drive the states to origin is guaranteed by Lemma 1.8.1. □

1.8.4   Numerical example

Consider a bilinear system represented by

$$\begin{cases} x_1(k+1) = -0.5x_2(k)0.5u(k) + 0.5u(k)x_1(k) \\ x_2(k+1) = x_1(k) + 0.5u(k) - 2u(k)x_2(k) \end{cases}.$$

The input and state are bounded by

$$\mathcal{U} = \left\{ u \in \mathfrak{R}^1 \mid -0.1 \le u \le 0.1 \right\}, \mathcal{X} = \left\{ (x_1, x_2) \in \mathfrak{R}^2 \middle| \begin{array}{c} -0.3 \le x_1 \le 0.3, \\ -0.3 \le x_2 \le 0.3 \end{array} \right\}.$$

The weighting matrices are $W = \begin{bmatrix} 1 & 0 \\ 0 & 1 \end{bmatrix}$ and $R = 1$. Linearize the system

at the origin, then $A = \begin{bmatrix} 0 & -0.5 \\ 1 & 0 \end{bmatrix}$, $B = \begin{bmatrix} 0.5 \\ 0.5 \end{bmatrix}$.

Make the following deductions according to Algorithm 1.1:

(i) Solve LQR problem, then

$$K = [-0.39 \ \ 0.2865], \ P = \begin{bmatrix} 2.0685 & 0.1434 \\ 0.1434 & 1.3582 \end{bmatrix}.$$

(ii) The choice of α_1 must satisfy $\Omega_{\alpha_1} \subseteq \Gamma_p$, where

$$\Gamma_p = \left\{ (x_1, x_2) \middle| \begin{array}{c} -0.1 \le -0.39 x_1 + 0.2865 x_2 \le 0.1, \\ -0.3 \le x_1 \le 0.3, -0.3 \le x_2 \le 0.3 \end{array} \right\}.$$

The largest α_1 can be chosen by optimization, but here only a feasible value is chosen. Let the long radius of the ellipsoid $x^T P x = \alpha_1$ be $0.2567/\sqrt{2}$. Then, $\alpha_1 = 0.0438$ is obtained. Ω_{α_1} lies in the shadowed region Γ_p in Figure 1.8.1.

(iii) Since $\lambda_{\min}\left(W + K^T R K\right) = 1$, choose $\gamma = 0.1$.

(iv) Since $\|P\| = 2.0937$ and $\|A + BK\| = 0.8624$, by (1.8.14), $L_{\ominus}^u = 0.22$ is obtained.

(v) Now choose $r = 0.13$. Then, $L_{\ominus} < L_{\ominus}^u$ for $x \in B_r$, where

$$\begin{cases} \Theta_1(x) = -0.19475 x_1^2 + 0.14325 x_1 x_2 \\ \Theta_2(x) = -0.573 x_2^2 + 0.779 x_1 x_2 \end{cases}.$$

(vi) Now choose $\beta = 0.0225$ such that $\Omega_\beta \subset B_r$, letting the long radius of the ellipsoid $x^T P x = \beta$ be 0.13. Finally, choose $\alpha = 0.0225$.

Then, choose $N = 10$. Use MATLAB Optimization Toolbox. The optimization is initialized as $[u(0), \cdots, u(9)] = [0, \cdots, 0]$. Figures 1.8.2-1.8.3 show the state responses and the control input signal with initial conditions $[u(-1), x_{10}, x_{20}] = [0, 0.25, 0.25]$, with those for samples $8 \sim 20$ magnified. During samples $1 \sim 10$, FHOP is performed. At sample 10, the states are driven to $(-0.00016, -0.00036)$ which lies in $\Omega_{0.0225}$. Then, controlled by the linear state feedback, the states are driven to the origin.

In this example, when the terminal inequality constraint is removed, the state responses are exactly the same (see Remark 1.8.5).

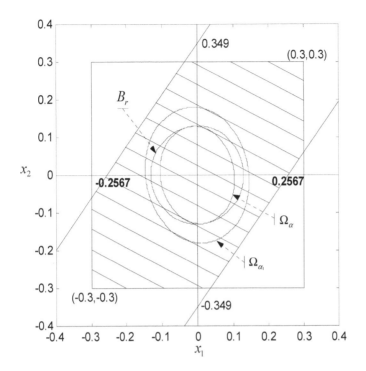

Figure 1.8.1: $\Gamma_p \supseteq \Omega_{\alpha_1} \supseteq B_r \supseteq \Omega_\alpha$

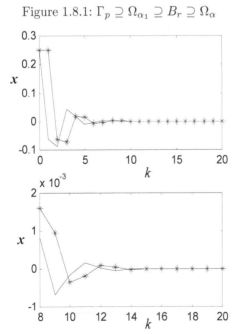

Figure 1.8.2: State responses of the closed-loop system (x_2 marked).

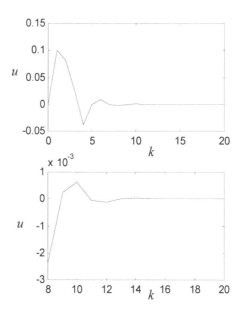

Figure 1.8.3: Control input signal.

1.9 Development from classical MPC to synthesis approaches

Currently, MPC algorithms adopted in the real projects mostly solve the finite-horizon optimization problem as in (1.7.6). In fact, the mature software adopts the linear models, so that the on-line optimization could be computationally admissible. For the classical MPC, it is always difficult to analyze closed-loop stability.

In order to overcome the analysis difficulty, from about the 1990s, people began to extensively research synthesis approaches. The so-called synthesis is the name from the side of stability. For classical MPC, by stability analysis, stability condition for the closed-loop system is obtained; for synthesis approach of MPC, by adding the three ingredients for stability, closed-loop stability can be guaranteed.

As has been explained in Remark 1.5.5, in synthesis approach of MPC, usually the feedback correction is not explicitly introduced. In synthesis approach of MPC, the uncertain model is utilized, and robust MPC is obtained. In industrial MPC, the linear model is often applied and, in theory, robust stability property in the existence of model-plant mismatch is analyzed. In the output/state prediction sets predicted by the uncertain models (often applied are polytopic description, bounded noise model), the real state/output evolutions are included, i.e., the real state evolution is always included by the state prediction set by the uncertain model.

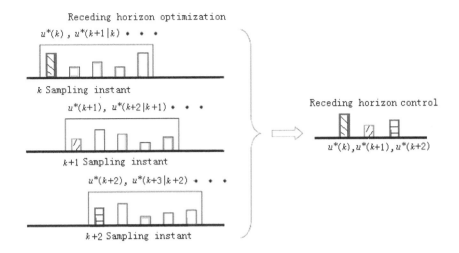

Figure 1.9.1: Receding horizon optimization in predictive control.

In the suboptimal controller of the last section, $\{u_o^*(0), u_o^*(1), \cdots, u_o^*(N-1)\}$ and K are calculated off-line. In the real applications, these data have been stored in the computer before implementation. At each sampling instant, it only needs to take the corresponding value from the computer memory. The most prominent difference between MPC (including classical MPC and synthesis approaches) and the traditional optimal control is that MPC adopts the receding horizon optimization (see Figure 1.9.1, which is inherited from [39]).

For the CNLQR proposed in the last section, if the receding horizon optimization is applied, i.e., the outside mode controller of CNLQR is solved at each sampling instant and the current control move $u^*(k)$ is implemented, then the closed-loop system is not necessarily asymptotically stable. In a synthesis approach of MPC, in order to achieve closed-loop stability, one needs to appropriately combine the terminal cost function, terminal constraint set, local controller (which are called the three ingredients of synthesis approach), for which the following conclusion is available (refer to [35]).

Lemma 1.9.1. *For system* (1.8.1)-(1.8.2), *suppose* (A1)-(A3) *hold and* $W > 0$, $R > 0$. *Then, there exist* K, $\beta > 0$ *and symmetric positive definite matrix* P *satisfying the following Lyapunov equation:*

$$(A + BK)^T P(A + BK) - P = -\beta P - W - K^T RK. \qquad (1.9.1)$$

Further, there exists a constant $\alpha \in (0, \infty)$, *such that the neighborhood of the origin* Ω_α *defined by* (1.8.10) *has the following properties:*

(i) *under the control law* $u(k) = Kx(k)$, Ω_α *is the control invariant set of* (1.8.1);

(ii) for $\forall x_0 \in \Omega_\alpha$, if $u(k) = Kx(k)$ is applied, then $\lim_{k \to \infty} x(k) = 0$ and $\lim_{k \to \infty} u(k) = 0$;

(iii) $\forall x(k) \in \Omega_\alpha$,

$$\|x(k)\|_P^2 \geq \sum_{i=0}^{\infty} [\|x(k+i|k)\|_W^2 + \|Kx(k+i|k)\|_R^2]. \qquad (1.9.2)$$

Proof. Since (A, B) is stabilizable, it is apparent that there exist K, $\beta > 0$ and symmetric positive definite matrix P satisfying (1.9.1). As in Lemma 1.8.1 we can obtain

$$V(k+1)-V(k) = \Theta(x)^T P\Theta(x)+2\Theta(x)^T P(A+BK)x-x^T(\beta P+W+K^T RK)x.$$

Similarly to Lemma 1.8.1, (i)-(ii) can be proved. Further, when α is sufficiently small, L_Θ is also sufficiently small. Hence, for sufficiently small α, when $x(k) \in \Omega_\alpha$, the following holds:

$$V(k+1) - V(k) \leq -x(k)^T(W + K^T RK)x(k). \qquad (1.9.3)$$

According to (1.9.3), for MPC,

$$V(k+i+1|k) - V(k+i|k) \leq -x(k+i|k)^T(W + K^T RK)x(k+i|k). \quad (1.9.4)$$

Summing (1.9.4) from $i = 0$ to $i = \infty$ obtains (1.9.2), where $V(\infty|k) = 0$. □

Simply speaking, let us

(I) choose K, $\beta > 0$ and symmetric positive definite matrix P according to (1.9.1) such that properties (i)-(iii) in Lemma 1.9.1 are satisfied;

(II) slightly modify the cost function $\Phi(x(0), u_0^{N-1})$ for (1.8.16) as

$$\Phi(x(k)) = \sum_{i=0}^{N-1} [\|x(k+i|k)\|_W^2 + \|u(k+i|k)\|_R^2] + \|x(k+N|k)\|_P^2;$$

(III) minimize, at each time k, $\Phi(x(k))$, at the same time satisfying the input/state constraints before the switching horizon N and satisfying $x(k+N|k) \in \Omega_\alpha$.

Then, the MPC corresponding to (I)-(III) is a synthesis approach, for which the closed-loop system is stable. The three ingredients are $\|x(k+N|k)\|_P^2$, Ω_α and K.

1. *Optimality and sub-optimality in the optimal control problem*

Optimality in the traditional optimal control refers to the minimization of a certain cost function. It is apparent the MPC also minimizes a cost function. If the cost function is not strictly minimized and, rather, a suboptimal solution is found, then it is called suboptimal control. The standard for distinguishing between MPC and the traditional optimal control does not lie in whether or not the cost function is minimized, but in whether or not the control law or control move is on-line refreshed.

Optimality and sub-optimality are two relative concepts. In synthesis approach of MPC, one usually utilizes the finite-horizon cost function to serve as the upper bound of the infinite-horizon cost function. This upper bound is usually conservative. Hence, corresponding to the infinite-horizon cost function, MPC only obtains the suboptimal solution; however, corresponding to the finite-horizon cost function, if the optimization problem is convex, MPC can find the optimal solution.

In MPC, the selection of the cost function is usually concerned with stability, i.e., the selection of the cost function should be advantageous for stability, and it is less important to consider optimality (optimality is a comparative property with respect to the optimum). The traditional optimal control pays more attention to optimality. If we select the infinite-horizon cost function as the sum of the positive definite functions over the infinite time horizon, then stability of the closed-loop system is equivalent to the boundedness of the summation function.

Both MPC and the traditional optimal control belong to the optimal control problems. However, it is not appropriate to say that the former includes the latter, or vice visa. In fact, they have different engineering backgrounds, large difference in the implementation, but in theory many equivalent aspects.

2. *Infinite-horizon optimal control is a bridge where a classical MPC transfers to a synthesis approach.*

This viewpoint has been illustrated in the former sections. CNLQR is originally an optimal control problem, but we can only give the suboptimal solution. The suboptimal solution is obtained by splitting the infinite-horizon cost function into two parts which are solved separately. Inside of the terminal constraint set Ω_α, the upper bound of the cost value is $\|x(k + N|k)\|_P^2$; when this upper bound is added on the cost function of finite-horizon optimal control problem for the suboptimal CNLQR, the upper bound of the infinite-horizon cost function is obtained; further, by use of this upper bound as the cost function of the predictive control, then the property that asymptotic stability is equivalent to boundedness of the summation function, possessed by infinite-horizon optimal control, is inherited.

Further details will be shown in the following chapters.

3. *Classical and synthesis approach of MPC are different forms, but emerged for pursuing the same target.*

In cases in which classical MPC is the version adopted in the engineering problems, why do we still investigate synthesis approaches? While synthesis approaches may be applied in other engineering areas, it is noted that the two categories of MPC are the inevitable results when the same kind of engineering problem is encountered, and the breakthrough can be made from different starters.

In order to control a system, we expect that our controller should be easy to implement, and closed-loop stability can be guaranteed. This is a paradox. Usually, given a system, if a simple model and a simple controller are applied, then a complicated closed-loop system can result. Classical MPC works this way, by adopting the easily obtainable impulse response and step response models, such that the implementation is simplified. However, it is difficult to analyze the closed-loop system. Given a system, if a more complex model and more complicated controller are utilized, then an easily analyzed closed-loop system can be obtained. Synthesis approach works this way, by adopting polytopic description, bounded noise model, the implementation of the controller relies on on-line solving a complicated optimization problem.

However, in cases in which the two categories are for the same kind of problem, the results of synthesis approaches can be invoked to guide the implementation of industrial MPC. For example, classical MPC cannot explicitly address the modeling uncertainty, but synthesis approach can. Hence, some results for uncertain systems with synthesis approaches can be applied to classical MPC.

Moreover, the existence of the large difference between synthesis approach and classical MPC is due to limited technical means in the areas of mathematical analysis, system identification, etc. In the continuous investigations and explorations, the two categories can solve the engineering problems from different angles and, even, combined and compensate for each other.

Chapter 2

Model algorithmic control (MAC)

The industrial production process is complex and the model we set up may be unsatisfactory. Even for the theoretically very complex modern control theory, its control effect may be unsatisfactory and even, to some extent, worse than the traditional PID control. In the 1970s, besides intensifying investigation on the modeling, identification, adaptive control, etc., people began to break the traditional ideas on the control, and tried to develop a new control algorithm having lower requirement on the model, involving convenient on-line computation and possessing better control performance. Under this situation, model algorithmic control (MAC), which is a kind of predictive control, was applied on the process control in France. Therefore, predictive control is not a theoretical product, but developed from engineering practice. Moreover, the development of the computer techniques also provides hardware and software for implementing the predictive control algorithm.

This chapter is mainly referred to in [57], [63].

2.1 Principle of MAC

MAC is also called model predictive heuristic control (MPHC). Its corresponding industrial software is IDCOM (Identification-Command). At its time, MPHC had the following unique properties:

(1) The multivariable process to be controlled is represented by its impulse responses which constitute the internal model (i.e., model stored in the computer memory). This model is used on-line for prediction, and its inputs and outputs are updated according to the actual state of the process. Although it could be identified on-line, the internal model is most of the time computed off-line. Usually the model is updated after

a long time period (e.g. one year).

(2) The strategy is fixed by means of a reference trajectory which defines the closed-loop behavior of the plant. This trajectory is initiated on the actual output of the process and tends to the desired setpoint. One of main tuning knobs of MPHC is the reference trajectory.

(3) Controls are not computed by a one-shot operator or controller but through a procedure which is heuristic in the general case. Future inputs are computed in such a way that, when applied to the fast time internal predictive model, they induce outputs as close as possible to the desired reference trajectory.

2.1.1   Impulse response model

The impulse response model did not just appear when the MPHC was proposed. Rather, this model appeared much earlier than the classical control theory. However, in the control theories before MPHC, the impulse response model was not thought to be convenient. Comparatively, the differential equation model, transfer function model and state space model are very suitable for theoretical analysis.

However, from the emergence of MPHC, there is no control algorithm (not adopting impulse response model and similar step response model) which can be more efficiently applied in the process industry. This phenomenon can be explained from several angles.

(I) From the side of system identification, the impulse response model (or step response model) is the easiest to obtain, the most original and the most accurate. When the transfer function model is identified by applying the input/output data, one also needs to first obtain the impulse response (or step response). The identification of the transfer function model cannot be finished with one calculation. Rather, it needs a number of iterations, such that the coefficients of identified model can converge to the consistent values. Whether or not the identified coefficients will converge to the true values depends on the identification algorithm and the characteristic of the identified system. If one wants to obtain the state space model, he/she usually should first obtain the input/output model. More importantly, when the orders of the input/output model are selected, there has to be a compromise, which makes a difference between the model and the real system; this difference can be even larger than when the impulse response model (or step response model) is adopted.

(II) For the implementation, the complex mathematical model obtained from principle analysis is often unnecessary. Moreover, when the principle is constructed, there have to be some assumptions. On the contrary, the impulse response model, as compared with the principle model, although adopting more model coefficients, preserves more system information.

(III) When the impulse response model (or step response model) is applied, the time for controller designing can be saved. MPHC, as well as DMC adopting the step response model, has simple algorithmic procedures, which can be easily accepted by the process engineers.

Denote the impulse response of the controlled process as $H(l)$. The process can be described by the following convolution model:

$$y(k) = \sum_{l=1}^{\infty} H(l)u(k-l). \qquad (2.1.1)$$

For the open-loop stable process, when l is increased, $H(l)$ tends to zero. Hence, we can use the finite convolution to approximate (2.1.1),

$$y(k) = \sum_{l=1}^{N} H(l)u(k-l). \qquad (2.1.2)$$

N is the modeling length or modeling horizon. The selection of N is closely related with the sampling interval T_s, i.e., NT_s should correspond to the response time of the controlled process. Denote

$$H = \begin{bmatrix} h_{11} & h_{12} & \cdots & h_{1m} \\ h_{21} & h_{22} & \cdots & h_{2m} \\ \vdots & \vdots & \ddots & \vdots \\ h_{r1} & h_{r2} & \cdots & h_{rm} \end{bmatrix}.$$

Then, according to (2.1.2), each output y_i ($i \in \{1,\ldots,r\}$) of the multi-variable system is the weighted summation of the m input data over the past N sampling instants, denoted as

$$y_i(k) = \sum_{j=1}^{m} \sum_{l=1}^{N} h_{ij}(l)u_j(k-l). \qquad (2.1.3)$$

2.1.2 Prediction model and feedback correction

By adopting the convolution model (2.1.2), the output prediction at the future time $k+j$ is

$$\bar{y}(k+j|k) = \sum_{l=1}^{N} H(l)u(k+j-l|k). \qquad (2.1.4)$$

By applying the difference between $y(k)$ and $\bar{y}(k|k)$ (defined as $\varepsilon(k) = y(k) - \bar{y}(k|k)$, or called real-time prediction error), the future output prediction $\bar{y}(k+$

$j|k)$ based on (2.1.4) can be corrected, which usually adopts the following form:

$$y(k + j|k) = \bar{y}(k + j|k) + f_j(y(k) - \bar{y}(k|k)) \qquad (2.1.5)$$

where f_j is the feedback correction coefficient.

If the setpoint value of the output is $y_s(k + j)$, then the prediction of the output tracking error (the output tracking error is defined as $e = y_s - y$) is

$$e(k + j|k) = y_s(k + j) - f_j y(k) - (\bar{y}(k + j|k) - f_j \bar{y}(k|k)). \qquad (2.1.6)$$

Applying (2.1.4) yields

$$\bar{y}(k + j|k) - f_j\bar{y}(k|k) \;=\; \sum_{l=1}^{j} H(l)u(k + j - l|k) + \sum_{l=1}^{N-j} H(l + j)u(k - l)$$
$$- f_j \sum_{l=1}^{N} H(l)u(k - l). \qquad (2.1.7)$$

In (2.1.7), the first item in the right is the effect on the tracking error by the current and future control moves, which is unknown at the current time; the latter two items are the effects of the historical control moves, which are known at the current time.

Substitute (2.1.7) into (2.1.6), we obtain the prediction on the tracking error

$$e(k + j|k) = e_0(k + j) - \sum_{l=1}^{j} H(l)u(k + j - l|k), \qquad (2.1.8)$$

where

$$e_0(k+j) = y_s(k+j) - f_j y(k) - \sum_{l=1}^{N-j} H(l+j)u(k-l) + f_j \sum_{l=1}^{N} H(l)u(k-l) \qquad (2.1.9)$$

is the prediction on the future tracking error based on the real measured output $y(k)$ and the historical control moves, when the current and future control moves are zeros.

Suppose the final desired output of the system is y_{ss}. Then $y_s(k + j)$ can be calculated in the following simple manner:

$$y_s(k + j) = a y_s(k + j - 1) + (1 - a)y_{ss}, \; j > 0 \qquad (2.1.10)$$

where $a > 0$ is a constant. $y_s(k + 1)$, $y_s(k + 2)$, $\cdots$ are reference trajectories.

2.1.3   Optimal control: case single input single output

In the real applications of MAC, one usually chooses $M < P$ (M is the control horizon, P the prediction horizon), and

$$u(k + i|k) = u(k + M - 1|k), \; i \in \{M, \dots, P - 1\}. \qquad (2.1.11)$$

First, let us consider the SISO system. The finite convolution model is

$$y(k) = \sum_{l=1}^{N} h_l u(k - l). \tag{2.1.12}$$

Applying (2.1.4), (2.1.11) and (2.1.12) easily yields

$$\tilde{y}(k|k) = G\tilde{u}(k|k) + G_{\mathrm{p}}\tilde{u}_{\mathrm{p}}(k) \tag{2.1.13}$$

where

$$\tilde{y}(k|k) = [\bar{y}(k+1|k), \bar{y}(k+2|k), \cdots, \bar{y}(k+P|k)]^T,$$
$$\tilde{u}(k|k) = [u(k|k), u(k+1|k), \cdots, u(k+M-1|k)]^T,$$
$$\tilde{u}_{\mathrm{p}}(k|k) = [u(k-1), u(k-2), \cdots, u(k-N+1)]^T,$$

$$G = \begin{bmatrix}
h_1 & 0 & \cdots & 0 & 0 \\
h_2 & h_1 & \cdots & 0 & 0 \\
\vdots & \vdots & \ddots & \vdots & \vdots \\
h_{M-1} & h_{M-2} & \cdots & h_1 & 0 \\
h_M & h_{M-1} & \cdots & h_2 & h_1 \\
h_{M+1} & h_M & \cdots & h_3 & (h_2 + h_1) \\
\vdots & \vdots & \ddots & \vdots & \vdots \\
h_P & h_{P-1} & \cdots & h_{P-M+2} & (h_{P-M+1} + \cdots + h_1)
\end{bmatrix},$$

$$G_{\mathrm{p}} = \begin{bmatrix}
h_2 & \cdots & h_{N-P+1} & h_{N-P+2} & \cdots & h_N \\
\vdots & \ddots & \vdots & \vdots & \cdot^{\cdot^{\cdot}} & 0 \\
h_P & \cdots & h_{N-1} & h_N & \cdot^{\cdot^{\cdot}} & \vdots \\
h_{P+1} & \cdots & h_N & 0 & \cdots & 0
\end{bmatrix}.$$

Notation "p" in the subscript denotes "past."

Further, applying (2.1.4)-(2.1.9) yields

$$\tilde{e}(k|k) = \tilde{e}_0(k) - G\tilde{u}(k|k), \tag{2.1.14}$$
$$\tilde{e}_0(k) = \tilde{y}_s(k) - G_{\mathrm{p}}\tilde{u}_{\mathrm{p}}(k) - \tilde{f}\varepsilon(k), \tag{2.1.15}$$

where

$$\tilde{f} = [f_1, f_2, \cdots, f_P]^T,$$
$$\tilde{e}(k|k) = [e(k+1|k), e(k+2|k), \cdots, e(k+P|k)]^T,$$
$$\tilde{e}_0(k) = [e_0(k+1), e_0(k+2), \cdots, e_0(k+P)]^T,$$
$$\tilde{y}_s(k) = [y_s(k+1), y_s(k+2), \cdots, y_s(k+P)]^T.$$

Suppose the criterion for optimizing $\tilde{u}(k|k)$ is to minimize the following cost function:

$$J(k) = \sum_{i=1}^{P} w_i e^2(k+i|k) + \sum_{j=1}^{M} r_j u^2(k+j-1|k) \tag{2.1.16}$$

where w_i and r_j are non-negative scalars. Then, when $G^T W G + R$ is nonsingular, by applying (2.1.14)-(2.1.15), the minimization of (2.1.16) yields

$$\tilde{u}(k|k) = (G^T W G + R)^{-1} G^T W \tilde{e}_0(k) \qquad (2.1.17)$$

where

$$W = \mathrm{diag}\{w_1, w_2, \cdots, w_P\}, \ \ R = \mathrm{diag}\{r_1, r_2, \cdots, r_M\}.$$

At each time k, implement the following control move:

$$u(k) = d^T(\tilde{y}_s(k) - G_p \tilde{u}_p(k) - \tilde{f}\varepsilon(k)) \qquad (2.1.18)$$

where

$$d^T = [1 \ \ 0 \ \cdots \ 0](G^T W G + R)^{-1} G^T W.$$

Algorithm 2.1 (Unconstrained MAC)

Step 0. Obtain $\{h_1, h_2, \cdots, h_N\}$. Calculate d^T. Choose $\tilde{f}$. Obtain $u(-N), u(-N+1), \cdots, u(-1)$.

Step 1. At each time $k \geq 0$,

Step 1.1. measure the output $y(k)$;

Step 1.2. determine $\tilde{y}_s(k)$;

Step 1.3. calculate $\varepsilon(k) = y(k) - \bar{y}(k|k)$, where $\bar{y}(k|k) = \sum_{l=1}^{N} h_l u(k - l)$;

Step 1.4. calculate $G_p \tilde{u}_p(k)$ in (2.1.13);

Step 1.5. calculate $u(k)$ by applying (2.1.18);

Step 1.6. implement $u(k)$.

Remark 2.1.1. Utilizing (2.1.5) for $j = 0$ yields

$$y(k|k) = \bar{y}(k|k) + f_0(y(k) - \bar{y}(k|k)).$$

Take $f_0 = 1$. Then the above formula yields $y(k|k) = y(k)$. This is the reason for applying $\varepsilon(k)$ in the feedback correction. The selection of f_j in (2.1.5) is artificial and, hence, f_j can be tuning parameter. In the earlier version of MPHC, $f_j = 1$.

2.1.4   Optimal control: case multi-input multi-output

For a system with m inputs and r outputs, the deduction of MAC control law is the generalization of that for the SISO system. The prediction of the output can be done in the following steps:

(i) *First step:* Suppose only the j-th input is nonzero, and the other inputs are zeros. Then considering the i-th output yields

$$\tilde{y}_{ij}(k|k) = G_{ij}\tilde{u}_j(k|k) + G_{ijp}\tilde{u}_{jp}(k), \qquad (2.1.19)$$

where

$$
\begin{aligned}
\tilde{y}_{ij}(k|k) &= [\bar{y}_{ij}(k+1|k), \bar{y}_{ij}(k+2|k), \cdots, \bar{y}_{ij}(k+P|k)]^T, \\
\tilde{u}_j(k|k) &= [u_j(k|k), u_j(k+1|k), \cdots, u_j(k+M-1|k)]^T, \\
\tilde{u}_{j,\mathrm{p}}(k|k) &= [u_j(k-1), u_j(k-2), \cdots, u_j(k-N+1)]^T,
\end{aligned}
$$

$$
G = \begin{bmatrix}
h_{ij}(1) & 0 & \cdots & 0 & 0 \\
h_{ij}(2) & h_{ij}(1) & \cdots & 0 & 0 \\
\vdots & \vdots & \ddots & \vdots & \vdots \\
h_{ij}(M-1) & h_{ij}(M-2) & \cdots & h_{ij}(1) & 0 \\
h_{ij}(M) & h_{ij}(M-1) & \cdots & h_{ij}(2) & h_{ij}(1) \\
h_{ij}(M+1) & h_{ij}(M) & \cdots & h_{ij}(3) & h_{ij}(2)+h_{ij}(1) \\
\vdots & \vdots & \ddots & \vdots & \vdots \\
h_{ij}(P) & h_{ij}(P-1) & \cdots & h_{ij}(P-M+2) & h_{ij}(P-M+1)+\cdots+h_{ij}(1)
\end{bmatrix},
$$

$$
G_{ij,\mathrm{p}} = \begin{bmatrix}
h_{ij}(2) & \cdots & h_{ij}(N-P+1) & h_{ij}(N-P+2) & \cdots & h_{ij}(N) \\
\vdots & \ddots & \vdots & \vdots & \cdots & 0 \\
h_{ij}(P) & \cdots & h_{ij}(N-1) & h_{ij}(N) & \cdots & \vdots \\
h_{ij}(P+1) & \cdots & h_{ij}(N) & 0 & \cdots & 0
\end{bmatrix}.
$$

(ii) *Second step:* Suppose all the inputs are not necessarily zeros. Then, considering the i-th output, applying the principle of superposition yields

$$
\tilde{y}_i(k|k) = \sum_{j=1}^{r} G_{ij}\tilde{u}_j(k|k) + \sum_{j=1}^{r} G_{ij\mathrm{p}}\tilde{u}_{j\mathrm{p}}(k), \tag{2.1.20}
$$

$$
\tilde{e}_i(k|k) = \tilde{e}_{i0}(k) + \sum_{j=1}^{r} G_{ij}\tilde{u}_j(k|k), \tag{2.1.21}
$$

$$
\tilde{e}_{i0}(k) = \tilde{y}_{is}(k) - \sum_{j=1}^{r} G_{ij\mathrm{p}}\tilde{u}_{j\mathrm{p}}(k) - \tilde{f}_i\varepsilon_i(k), \tag{2.1.22}
$$

where

$$
\begin{aligned}
\tilde{y}_i(k|k) &= [\bar{y}_i(k+1|k), \bar{y}_i(k+2|k), \cdots, \bar{y}_i(k+P|k)]^T, \\
\tilde{e}_i(k|k) &= [e_i(k+1|k), e_i(k+2|k), \cdots, e_i(k+P|k)]^T, \\
\tilde{e}_{i0}(k) &= [e_{i0}(k+1), e_{i0}(k+2), \cdots, e_{i0}(k+P)]^T, \\
\tilde{y}_{is}(k) &= [y_{is}(k+1), y_{is}(k+2), \cdots, y_{is}(k+P)]^T, \\
\tilde{f}_i &= [f_{i1}, f_{i2}, \cdots, f_{iP}]^T, \\
\varepsilon_i(k) &= y_i(k) - \bar{y}_i(k|k).
\end{aligned}
$$

(iii) *Third step:* Considering all the inputs and outputs yields

$$Y(k|k) = \tilde{G}U(k|k) + \tilde{G}_\mathrm{p}U_\mathrm{p}(k), \tag{2.1.23}$$

$$E(k|k) = E_0(k) - \tilde{G}U(k|k), \tag{2.1.24}$$

$$E_0(k) = Y_s(k) - \tilde{G}_\mathrm{p}U_\mathrm{p}(k) - \tilde{F}\Upsilon(k), \tag{2.1.25}$$

where

$$Y(k|k) = [\tilde{y}_1(k|k)^T, \tilde{y}_2(k|k)^T, \cdots, \tilde{y}_r(k|k)^T]^T,$$

$$U(k|k) = [\tilde{u}_1(k|k)^T, \tilde{u}_2(k|k)^T, \cdots, \tilde{u}_m(k|k)^T]^T,$$

$$U_\mathrm{p}(k) = [\tilde{u}_{1,\mathrm{p}}(k)^T, \tilde{u}_{2,\mathrm{p}}(k)^T, \cdots, \tilde{u}_{m,\mathrm{p}}(k)^T]^T,$$

$$E(k|k) = [\tilde{e}_1(k|k)^T, \tilde{e}_2(k|k)^T, \cdots, \tilde{e}_r(k|k)^T]^T,$$

$$E_0(k) = [\tilde{e}_{10}(k)^T, \tilde{e}_{20}(k)^T, \cdots, \tilde{e}_{r0}(k)^T]^T,$$

$$Y_s(k) = [\tilde{y}_{1s}(k)^T, \tilde{y}_{2s}(k)^T, \cdots, \tilde{y}_{rs}(k)^T]^T,$$

$$\Upsilon(k) = [\varepsilon_1(k), \varepsilon_2(k), \cdots, \varepsilon_r(k)]^T,$$

$$\tilde{F} = \begin{bmatrix} \tilde{f}_1 & 0 & \cdots & 0 \\ 0 & \tilde{f}_2 & \cdots & 0 \\ \vdots & \vdots & \ddots & \vdots \\ 0 & 0 & \cdots & \tilde{f}_r \end{bmatrix},$$

$$\tilde{G} = \begin{bmatrix} G_{11} & G_{12} & \cdots & G_{1m} \\ G_{21} & G_{22} & \cdots & G_{2m} \\ \vdots & \vdots & \ddots & \vdots \\ G_{r1} & G_{r2} & \cdots & G_{rm} \end{bmatrix},$$

$$\tilde{G}_\mathrm{p} = \begin{bmatrix} G_{11\mathrm{p}} & G_{12\mathrm{p}} & \cdots & G_{1m\mathrm{p}} \\ G_{21\mathrm{p}} & G_{22\mathrm{p}} & \cdots & G_{2m\mathrm{p}} \\ \vdots & \vdots & \ddots & \vdots \\ G_{r1\mathrm{p}} & G_{r2\mathrm{p}} & \cdots & G_{rm\mathrm{p}} \end{bmatrix}.$$

Suppose the criterion for optimization of $U(k|k)$ is the minimization of the following cost function:

$$J(k) = \|E(k|k)\|_{\tilde{W}}^2 + \|U(k|k)\|_{\tilde{R}}^2 \tag{2.1.26}$$

where $\tilde{W} \geq 0$ and $\tilde{R} \geq 0$ are symmetric matrices. Then, when $\tilde{G}^T\tilde{W}\tilde{G} + \tilde{R}$ is nonsingular, minimizing (2.1.26) yields

$$U(k|k) = (\tilde{G}^T\tilde{W}\tilde{G} + \tilde{R})^{-1}\tilde{G}^T\tilde{W}E_0(k). \tag{2.1.27}$$

Remark 2.1.2. If, in G, G_p of (2.1.13), all the h_l's are $r \times m$-dimensional matrices, then the closed-form solution to the unconstraint MIMO MAC can be represented as (2.1.17); thus, it is not required to use the deductions as in section 2.1.4. However, if for different inputs (outputs), different control horizons (prediction horizons) are utilized, then the expression by (2.1.27) is more convenient.

At each time k, implement the following control move:

$$u(k) = D(Y_s(k) - \tilde{G}_p U_p(k) - \tilde{F}\Upsilon(k)) \qquad (2.1.28)$$

where

$$D = L(\tilde{G}^T \tilde{W} \tilde{G} + \tilde{R})^{-1} \tilde{G}^T \tilde{W},$$

$$L = \begin{bmatrix} \theta & 0 & \cdots & 0 \\ 0 & \theta & \ddots & \vdots \\ \vdots & \ddots & \ddots & 0 \\ 0 & \cdots & 0 & \theta \end{bmatrix} \in \mathfrak{R}^{m \times mM}, \quad \theta = [1 \ 0 \ \cdots \ 0] \in \mathfrak{R}^M.$$

A simple method for selecting $\tilde{W}$ and $\tilde{R}$ is

$$\tilde{W} = \text{diag}\{W_1, W_2, \cdots, W_r\}, \quad \tilde{R} = \text{diag}\{R_1, R_2, \cdots, R_m\},$$
$$W_i = \text{diag}\{w_i(1), w_i(2), \cdots, w_i(P)\}, \quad i \in \{1, \ldots, r\},$$
$$R_j = \text{diag}\{r_j(1), r_j(2), \cdots, r_j(M)\}, \quad j \in \{1, \ldots, m\}.$$

Choosing $\tilde{R} > 0$ guarantees the nonsingularity of $\tilde{G}^T \tilde{W} \tilde{G} + \tilde{R}$.

2.2 Constraint handling

Before application of MPHC (design stage) and in the progress of application (run stage), users can tune the following parameters:

(i) sampling internal T_s;

(ii) prediction horizon P, control horizon M, weighting matrices $\tilde{W}$, $\tilde{R}$ or weighting coefficients w_i, r_j;

(iii) correction coefficients $\tilde{f}_i$ or f_i;

(iv) the modeling coefficients in the impulse response model;

(v) reference trajectory, such as parameter a;

(vi) constraints, including input magnitude constraint (upper limit, lower limit), the upper limit of the input increment, constraint on the intermediate variable (or combined variable), etc.

For (i)-(iii), one can refer to sections 5.1 and 6.2 of [63]. Although these sections in [63] are for DMC (see Chapter 3), most of the viewpoints are efficient for MAC. In this book, stability is mainly concerned with the results via mathematical proofs; when the experiential tuning methods are involved, the readers are referred to the related literature. In the real applications, the experiential tuning is very important.

In general, one does not need to often tune the modeling parameters of the impulse response model. The main reasons are as follows.

(I) MPHC has strong robustness, and in general can be suitable for the system changes.

(II) Testing the impulse response model, in general, will affect the real production process, i.e., some additional signal which is not desired by the real process will be added. Else, the obtained data can have low signal/noise ratio, and error is enlarged correspondingly.

(III) Although there is an appropriate identification tool, the hardware conditions of the real process can be insufficient (e.g., the measurements can be inaccurate). The operators can know the hardware problems in the controlled process. However, it is not easy for the identification tool to know the problems.

In the real applications, reference trajectory and constraints are the main parameters for tuning. In the following, let us discuss how to handle constraints in MPHC (take MIMO system as the example).

1. *Output magnitude constraint* $y_{i,\min} \leq y_i(k+l|k) \leq y_{i,\max}$

At each optimization cycle, the output prediction is $\tilde{G}_\mathrm{p} U_\mathrm{p}(k) + \tilde{F}\Upsilon(k) + \tilde{G}U(k|k)$. Hence, we can let the optimization problem satisfy the following constraint:

$$Y_{\min} \leq \tilde{G}_\mathrm{p} U_\mathrm{p}(k) + \tilde{F}\Upsilon(k) + \tilde{G}U(k|k) \leq Y_{\max} \qquad (2.2.1)$$

where

$$\begin{aligned}
Y_{\min} &= [\tilde{y}_{1,\min}^T, \tilde{y}_{2,\min}^T, \cdots, \tilde{y}_{r,\min}^T]^T, \\
\tilde{y}_{i,\min} &= [y_{i,\min}, y_{i,\min}, \cdots, y_{i,\min}]^T \in \Re^P, \\
Y_{\max} &= [\tilde{y}_{1,\max}^T, \tilde{y}_{2,\max}^T, \cdots, \tilde{y}_{r,\max}^T]^T, \\
\tilde{y}_{i,\max} &= [y_{i,\max}, y_{i,\max}, \cdots, y_{i,\max}]^T \in \Re^P.
\end{aligned}$$

2. *Input magnitude constraint* $u_{j,\min} \leq u_j(k+l|k) \leq u_{j,\max}$

We can let the optimization problem satisfy the following constraint:

$$U_{\min} \leq U(k|k) \leq U_{\max} \qquad (2.2.2)$$

where

$$
\begin{aligned}
U_{\min} &= [\tilde{u}_{1,\min}^T, \tilde{u}_{2,\min}^T, \cdots, \tilde{u}_{m,\min}^T]^T, \\
\tilde{u}_{j,\min} &= [u_{j,\min}, u_{j,\min}, \cdots, u_{j,\min}]^T \in \mathfrak{R}^M, \\
U_{\max} &= [\tilde{u}_{1,\max}^T, \tilde{u}_{2,\max}^T, \cdots, \tilde{u}_{m,\max}^T]^T, \\
\tilde{u}_{j,\max} &= [u_{j,\max}, u_{j,\max}, \cdots, u_{j,\max}]^T \in \mathfrak{R}^M.
\end{aligned}
$$

3. *Input rate constraint* $\Delta u_{j,\min} \le \Delta u_j(k+l|k) = u_j(k+l|k) - u_j(k+l-1|k) \le \Delta u_{j,\max}$

We can let the optimization problem satisfy the following constraint:

$$
\Delta U_{\min} \le BU(k|k) - \tilde{u}(k-1) \le \Delta U_{\max} \tag{2.2.3}
$$

where

$$
\begin{aligned}
\Delta U_{\min} &= [\Delta \tilde{u}_{1,\min}^T, \Delta \tilde{u}_{2,\min}^T, \cdots, \Delta \tilde{u}_{m,\min}^T]^T, \\
\Delta \tilde{u}_{j,\min} &= [\Delta u_{j,\min}, \Delta u_{j,\min}, \cdots, \Delta u_{j,\min}]^T \in \mathfrak{R}^M, \\
\Delta U_{\max} &= [\Delta \tilde{u}_{1,\max}^T, \Delta \tilde{u}_{2,\max}^T, \cdots, \Delta \tilde{u}_{m,\max}^T]^T, \\
\Delta \tilde{u}_{j,\max} &= [\Delta u_{j,\max}, \Delta u_{j,\max}, \cdots, \Delta u_{j,\max}]^T \in \mathfrak{R}^M, \\
B &= \text{diag}\{B_0, \cdots, B_0\} \ (m \text{ blocks}),
\end{aligned}
$$

$$
B_0 = \begin{bmatrix}
1 & 0 & 0 & \cdots & 0 \\
-1 & 1 & 0 & \ddots & \vdots \\
0 & -1 & 1 & \ddots & 0 \\
\vdots & \ddots & \ddots & \ddots & 0 \\
0 & \cdots & 0 & -1 & 1
\end{bmatrix} \in \mathfrak{R}^{M \times M},
$$

$$
\begin{aligned}
\tilde{u}(k-1) &= [\tilde{u}_1(k-1)^T, \tilde{u}_2(k-1)^T, \cdots, \tilde{u}_m(k-1)^T]^T, \\
\tilde{u}_j(k-1) &= [u_j(k-1), 0, \cdots, 0]^T \in \mathfrak{R}^M.
\end{aligned}
$$

Equations (2.2.1)-(2.2.3) can be written in a uniform manner as $CU(k|k) \le \bar{c}$, where C and $\bar{c}$ are known matrix and vector at time k. The optimization problem of MAC incorporating these constraints is

$$
\min_{U(k|k)} J(k) = \|E(k|k)\|_{\bar{W}}^2 + \|U(k|k)\|_{\bar{R}}^2, \quad \text{s.t. } CU(k|k) \le \bar{c}. \tag{2.2.4}
$$

An optimization problem, with quadratic cost function and linear equality and inequality constraints, is called a quadratic optimization problem.

2.3　The usual pattern for implementation of MPC

MPC (not restricted to MPHC) can be implemented in two different manners. One is the direct digital control (DDC), i.e., the output of the controller is directly acted on (transmitted to) the physical actuator. The other is the so-called "transparent control," i.e., the output of MPC is the setpoints of the PIDs, i.e., the decision variables of MPC are the setpoints of PID control loops. Usually, it is safer to adopt "transparent control" and in real applications "transparent control" is applied.

In the real applications of MPC, one usually meets with the following 4 hierarchical levels:

Level 0— control of ancillary systems (e.g., servo-valves) where PID controllers are quite efficient,

Level 1— predictive control of the multivariable process, satisfying physical constraints such as input saturation, limits on rate of input change, etc.,

Level 2— optimization of the setpoints of MPC, with minimization of cost-function ensuring quality and quantity of production,

Level 3— time and space scheduling of production (planning-operation research).

This 4-level hierarchical structure can answer a good number of readers. Firstly, implementing MPC does not mean substituting PIDs utterly with MPC. The effect of PIDs is still existing. This further explains why PID is still prominent in process control. Secondly, in real applications, the profits from Levels 0 and 1 can be overlooked, i.e., merely adopting the basic principles of MPC usually is not equivalent to gaining profits. Often, the profits are mainly gained from Level 2. The effect of Level 0 and 1 (especially Level 1) is to implement the results of Level 2. Since the profits are gained from Level 2, the significance of optimization problem in MPC is to enhance the control performance, not to gain profits. The selection of the performance cost in MPC is mainly for stability and fast responses, and there is usually no economic consideration (this can be different from the traditional optimal control).

So, *why not leave out Level 1 and directly optimize the setpoints of PID?* The plant to be handled by these 4 hierarchical levels usually has a large number of variables, and the number of inputs and number of outputs are not necessarily equal. Further, the number of variables can be changing. Moreover, the outputs to be controlled by MPC can be different from those of the PIDs. Since MPC can handle the multivariable control in a uniform manner other than PID's single-loop operation, the control effect of MPC is more forceful and enables the optimums of Level 2 to be implemented faster and more accurate. Note that MPC is based on optimization and is effective for handling constraints and time-delay, which are merits not possessed in PID. Still, if in Level 2, the setpoints of PID are directly optimized, then more complex and higher dimensional optimization problem can be involved, which is not practical.

Remark 2.3.1. There are other advantages by adopting transparent control, including overcoming the effect of model uncertainty, enlarging the region of attraction, etc. For processes with long settling time, if MPC based on the impulse response model or step response model is applied, then the sampling interval cannot be selected small by considering the computational burden (if the sampling interval is over small, then the model length has to be selected large). In this situation, by adopting transparent control, the disturbance rejection can be done by the fast-sampling PID controllers.

Due to the adoption of transparent control, multivariable control and multi-step prediction, stability of MPHC is not critical, i.e., stability of MPHC can be easily tuned. Moreover, MPHC has comparable robustness. In the real applications of MPHC, the control structure (the number of variables) often changes. More significantly, the plant to be controlled by MPHC is often very complex. Therefore, it is very difficult, and not likely to be effective for real applications, to analyze stability of MPHC theoretically.

Figure 2.3.1 shows the strategy to be considered in Level 2. This strategy adopts a static model for optimization. Hence, before running the optimization, one should check if the plant is in steady state. If the plant is settled, then refresh the parameters in the mathematical model for optimization, and run the optimization. Before implementing the optimization results, one should check if the plant lies on the original steady state.

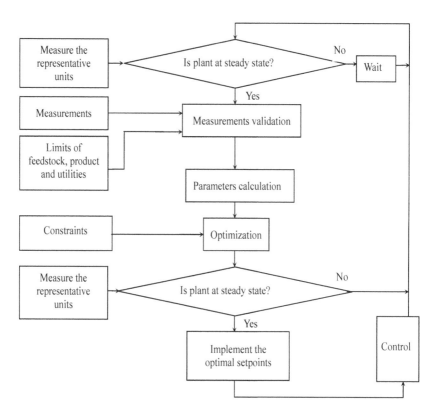

Figure 2.3.1: Static optimization of the setpoint values of MPC.

Chapter 3

Dynamic matrix control (DMC)

DMC has many similarities with MAC. It is an algorithm based on the step response model. Having an impulse response model is equivalent to having a step response model. However, DMC applies incremental algorithms, which is very effective in removing the steady-state error. Certainly, compared with DMC, MAC has its advantages, such as higher disturbance rejection capacity. In real applications, choosing between DMC and MAC depends on the precise situation. Up to now, DMC is the most widely accepted in the process industry.

This chapter mainly refers to in [50], [63].

3.1 Step response model and its identification

Suppose the system is at rest. For a linear time-invariant single-input single-output (SISO) system let the output change for a unit input change Δu be given by

$$\{0, s_1, s_2, \cdots, s_N, s_{N+1}, \cdots\}.$$

Here we suppose the system settles exactly after N steps. The step response $\{s_1, s_2, \cdots, s_N\}$ constitutes a complete model of the system, which allows us to compute the system output for any input sequence,

$$y(k) = \sum_{l=1}^{N} s_l \Delta u(k-l) + s_{N+1} u(k-N-1), \tag{3.1.1}$$

51

where $\Delta u(k-l) = u(k-l) - u(k-l-1)$. Note that, when $s_N = s_{N-1}$, (3.1.1) is equivalent to

$$y(k) = \sum_{l=1}^{N-1} s_l \Delta u(k-l) + s_N u(k-N). \qquad (3.1.2)$$

Step response model (3.1.1) can only be used in stable processes. For a MIMO process with m inputs and r outputs one obtains a series of step response coefficient matrices

$$S_l = \begin{bmatrix} s_{11l} & s_{12l} & \cdots & s_{1ml} \\ s_{21l} & s_{22l} & \cdots & s_{2ml} \\ \vdots & \vdots & \ddots & \vdots \\ s_{r1l} & s_{r2l} & \cdots & s_{rml} \end{bmatrix}$$

where s_{ijl} is the l-th step response coefficient relating j-th input to the i-th output.

The identification routines available in the Matlab MPC Toolbox are designed for multi-input single-output (MISO) systems. Based on a historical record of the output y_i and inputs $u_1, u_2, \cdots, u_m$,

$$\tilde{y}_i = \begin{bmatrix} y_i(1) \\ y_i(2) \\ y_i(3) \\ \vdots \end{bmatrix}, \quad \tilde{u} = \begin{bmatrix} u_1(1) & u_2(1) & \cdots & u_m(1) \\ u_1(2) & u_2(2) & \cdots & u_m(2) \\ u_1(3) & u_2(3) & \cdots & u_m(3) \\ \vdots & \vdots & & \vdots \end{bmatrix}$$

the step response coefficients

$$\begin{bmatrix} s_{i11} & s_{i21} & \cdots & s_{im1} \\ s_{i12} & s_{i22} & \cdots & s_{im2} \\ \vdots & \vdots & \ddots & \vdots \\ s_{i1l} & s_{i2l} & \cdots & s_{iml} \\ \vdots & \vdots & & \vdots \end{bmatrix}$$

are estimated.

For the estimation of the step response coefficients we write the SISO model in the form

$$\Delta y(k) = \sum_{l=1}^{N} h_l \Delta u(k-l) \qquad (3.1.3)$$

and firstly estimate h_l, where $\Delta y(k) = y(k) - y(k-1)$, $h_l = s_l - s_{l-1}$. s_l is given by

$$s_l = \sum_{j=1}^{l} h_j. \qquad (3.1.4)$$

For parameter estimation it is usually recommended to scale all the variables such that they are the same order of magnitude. This may be done via the MPC Toolbox functions "autosc" or "scal." Then the data has to be arranged into the form

$$Y = X\Theta \qquad (3.1.5)$$

where Y contains all the output information (for stable process, $\Delta y(k)$) and X all the input information ($\Delta u(k)$). Θ is a vector including all the parameters to be estimated (for stable process, h_l). The parameters Θ can be estimated via multivariable least square regression ("mlr" in Matlab) or partial least square regression ("plsr" in Matlab).

3.2 Principle of DMC

Consider the open-loop stable system. Giving the current and future control increments $\Delta u(k), \Delta u(k+1|k), \cdots, \Delta u(k+M-1|k)$, the future outputs $y(k+1|k), y(k+2|k), \cdots, y(k+P|k)$ can be predicted. $M \leq P \leq N$. The current and future control increments are obtained by solving the optimization problem. Although in total M control increments are obtained, only the first ($\Delta u(k)$) is implemented. At the next sampling instant, based on the new measurement, the control time horizon is moved forward with one sampling interval, and the same optimization as in the previous sampling instant is repeated.

3.2.1 Case single input single output

At time k, by utilizing (3.1.1), the output prediction for the future P sampling intervals are

$$\bar{y}(k+1|k) = y_0(k+1|k-1) + s_1 \Delta u(k)$$

$$\vdots$$

$$\bar{y}(k+M|k) = y_0(k+M|k-1) + s_M \Delta u(k) + s_{M-1} \Delta u(k+1|k) + \cdots$$
$$+ s_1 \Delta u(k+M-1|k)$$

$$\bar{y}(k+M+1|k) = y_0(k+M+1|k-1) + s_{M+1} \Delta u(k) + s_M \Delta u(k+1|k) + \cdots$$
$$+ s_2 \Delta u(k+M-1|k)$$

$$\vdots$$

$$\bar{y}(k+P|k) = y_0(k+P|k-1) + s_P \Delta u(k) + s_{P-1} \Delta u(k+1|k) + \cdots$$
$$+ s_{P-M+1} \Delta u(k+M-1|k)$$

where

$$y_0(k+i|k-1) = \sum_{j=i+1}^{N} s_j \Delta u(k+i-j) + s_{N+1}u(k+i-N-1)$$

$$= \sum_{j=1}^{N-i} s_{i+j} \Delta u(k-j) + s_{N+1}u(k+i-N-1)$$

$$= s_{i+1}u(k-1) + \sum_{j=2}^{N-i}(s_{i+j}-s_{i+j-1})u(k-j), \; i \in \{1,2,\ldots,P\}$$

$$(3.2.1)$$

is the output prediction by assuming that the current and future control moves keep invariant.

Denote

$$\bar{\varepsilon}(k) = y(k) - y_0(k|k-1) \tag{3.2.2}$$

where

$$y_0(k|k-1) = s_1 u(k-1) + \sum_{j=2}^{N}(s_j - s_{j-1})u(k-j). \tag{3.2.3}$$

We can use (3.2.2) to correct the future output predictions. Denote

$$y_0(k+i|k) = y_0(k+i|k-1) + f_i\bar{\varepsilon}(k), \; i \in \{1,2,\ldots,P\}, \tag{3.2.4}$$
$$y(k+i|k) = \bar{y}(k+i|k) + f_i\bar{\varepsilon}(k), \; i \in \{1,2,\ldots,P\}. \tag{3.2.5}$$

Write the output predictions corrected via (3.2.4)-(3.2.5) as the following vector form

$$\tilde{y}(k|k) = \tilde{y}_0(k|k) + A\Delta\tilde{u}(k|k), \tag{3.2.6}$$

where

$$\tilde{y}(k|k) = [y(k+1|k), y(k+2|k), \cdots, y(k+P|k)]^T,$$
$$\tilde{y}_0(k|k) = [y_0(k+1|k), y_0(k+2|k), \cdots, y_0(k+P|k)]^T,$$
$$\Delta\tilde{u}(k|k) = [\Delta u(k), \Delta u(k+1|k), \cdots, \Delta u(k+M-1|k)]^T,$$

$$A = \begin{bmatrix} s_1 & 0 & \cdots & 0 \\ s_2 & s_1 & \cdots & 0 \\ \vdots & \vdots & \ddots & \vdots \\ s_M & s_{M-1} & \cdots & s_1 \\ \vdots & \vdots & \ddots & \vdots \\ s_P & s_{P-1} & \cdots & s_{P-M+1} \end{bmatrix}.$$

Suppose the criterion for optimizing $\Delta\tilde{u}(k|k)$ is to minimize the following cost function:

$$J(k) = \sum_{i=1}^{P} w_i e^2(k+i|k) + \sum_{j=1}^{M} r_j \Delta u^2(k+j-1|k), \tag{3.2.7}$$

where w_i and r_j are non-negative scalars; $e(k+i|k) = y_s(k+i) - y(k+i|k)$ is the tracking error; $y_s(k+i)$ is the setpoint value of the future output. The second item in the cost function (3.2.7) is to restrict the magnitude of the control increment, so as to prevent the system from exceeding its limits or oscillating.

When $A^T W A + R$ is nonsingular, by applying (3.2.6), the minimization of (3.2.7) yields

$$\Delta \tilde{u}(k|k) = (A^T W A + R)^{-1} A^T W \tilde{e}_0(k) \qquad (3.2.8)$$

where

$$
\begin{aligned}
W &= \mathrm{diag}\{w_1, w_2, \cdots, w_P\}, R = \mathrm{diag}\{r_1, r_2, \cdots, r_M\}, \\
\tilde{e}_0(k) &= \tilde{y}_s(k) - \tilde{y}_0(k|k), \\
\tilde{e}_0(k) &= [e_0(k+1), e_0(k+2), \cdots, e_0(k+P)]^T, \\
\tilde{y}_s(k) &= [y_s(k+1), y_s(k+2), \cdots, y_s(k+P)]^T,
\end{aligned}
$$

e_0 is the prediction on the tracking error based on the measured output $y(k)$ and historical control moves, when the current and future control moves keep invariant.

At each time k, implement the following control moves:

$$\Delta u(k) = d^T (\tilde{y}_s(k) - \tilde{y}_0(k|k)) \qquad (3.2.9)$$

where

$$d^T = [1 \ 0 \ \cdots \ 0](A^T W A + R)^{-1} A^T W.$$

In fact, by applying (3.2.1) we can obtain the following vector form:

$$\tilde{y}_0(k|k-1) = A_{\mathrm{p}} \tilde{u}_{\mathrm{p}}(k) \qquad (3.2.10)$$

where

$$\tilde{y}_0(k|k-1) = [y_0(k+1|k-1), y_0(k+2|k-1), \cdots, y_0(k+P|k-1)]^T,$$

$$
A_{\mathrm{p}} =
\begin{bmatrix}
s_2 & s_3 - s_2 & \cdots & s_{N-P+1} - s_{N-P} & s_{N-P+2} - s_{N-P+1} & \cdots & s_N - s_{N-1} \\
\vdots & \vdots & \ddots & \vdots & \vdots & \cdots & 0 \\
s_P & s_{P+1} - s_P & \cdots & s_{N-1} - s_{N-2} & s_N - s_{N-1} & \cdots & \vdots \\
s_{P+1} & s_{P+2} - s_{P+1} & \cdots & s_N - s_{N-1} & 0 & \cdots & 0
\end{bmatrix},
$$

$$\tilde{u}_{\mathrm{p}}(k) = [u(k-1), u(k-2), \cdots, u(k-N+1)]^T.$$

Then applying (3.2.4)-(3.2.5) yields

$$
\begin{aligned}
\tilde{e}(k|k) &= \tilde{e}_0(k) - A\Delta \tilde{u}(k|k), \qquad &(3.2.11) \\
\tilde{e}_0(k) &= \tilde{y}_s(k) - A_{\mathrm{p}} \tilde{u}_{\mathrm{p}}(k) - \tilde{f}\tilde{e}(k), \qquad &(3.2.12)
\end{aligned}
$$

where

$$\tilde{e}(k|k) = [e(k+1|k), e(k+2|k), \cdots, e(k+P|k)]^T,$$
$$\tilde{f} = [f_1, f_2, \cdots, f_P]^T.$$

Thus, at each time k, implement the following control move:

$$\Delta u(k) = d^T(\tilde{y}_s(k) - A_p \tilde{u}_p(k) - \tilde{f}\bar{\varepsilon}(k)). \qquad (3.2.13)$$

Equations (3.2.13) and (3.2.9) are equivalent.

The above method can be summarized as follows.

Algorithm 3.1 (Type-I unconstrained DMC)

Step 0. Obtain $\{s_1, s_2, \cdots, s_N\}$. Calculate d^T. Choose $\tilde{f}$. Obtain $u(-N), u(-N+1), \cdots, u(-1)$.

Step 1. At each time $k \geq 0$,

Step 1.1. measure $y(k)$;

Step 1.2. determine $\tilde{y}_s(k)$ (the method for MAC in Chapter 2 can be adopted);

Step 1.3. calculate $\bar{\varepsilon}(k)$ via (3.2.2)-(3.2.3);

Step 1.4. calculate $A_p \tilde{u}_p(k)$ in (3.2.10);

Step 1.5. calculate $\Delta u(k)$ via (3.2.13);

Step 1.6. implement $\Delta u(k)$.

Remark 3.2.1. Use (3.2.4) for $i = 0$. Then applying (3.2.2) yields

$$y_0(k|k) = y_0(k|k-1) + f_0[y(k) - y_0(k|k-1)].$$

If we choose $f_0 = 1$, then $y_0(k|k) = y(k)$. This is the reason for choosing (3.2.2) as the feedback correction. In (3.2.4), the selection of f_i depends on the concrete situations. Hence, f_i can be tuning parameter. In Matlab MPC Toolbox, $f_i = 1$.

3.2.2   Case single input single output: alternative procedure of deduction

In some literature (e.g., in [63]), DMC is introduced from a different angle. Suppose $y_0(k+i|k)$ is the output prediction when the current and future control moves keep invariant. Then, it is shown that, when the current and future moves are changed, the output predictions in future P sampling instants are

$$y(k+1|k) = y_0(k+1|k) + s_1 \Delta u(k)$$

$$\vdots$$

$$y(k+M|k) = y_0(k+M|k) + s_M \Delta u(k) + s_{M-1} \Delta u(k+1|k) + \cdots \\ + s_1 \Delta u(k+M-1|k)$$

$$y(k+M+1|k) = y_0(k+M+1|k) + s_{M+1} \Delta u(k) + s_M \Delta u(k+1|k) + \cdots \\ + s_2 \Delta u(k+M-1|k)$$

$$\vdots$$

$$y(k+P|k) = y_0(k+P|k) + s_P \Delta u(k) + s_{P-1} \Delta u(k+1|k) + \cdots \\ + s_{P-M+1} \Delta u(k+M-1|k)$$

where $y_0(k+i|k)$ will be explained later.

Writing the output predictions in a vector form, we directly obtain (3.2.6). Suppose the criterion for optimizing $\Delta \tilde{u}(k|k)$ is to minimize the cost function (3.2.7). Then, when $A^T W A + R$ is nonsingular, it is shown that minimization of (3.2.7) yields (3.2.8).

In the following, we show how to calculate $\tilde{y}_0(k|k)$ at each sampling instant.

First, at the initial time $k = 0$, suppose the system is in steady-state. When DMC is in startup, choose $y_0(i|0) = y(0)$ $(i = 1, 2, \cdots, P+1)$.

At the time $k \geq 0$, implement (3.2.9). Consider the time $k+1$. For this, note that

$$\tilde{y}_0(k+1|k) = [y_0(k+2|k), y_0(k+3|k), \cdots, y_0(k+P+1|k)]^T.$$

According to its definition, $\tilde{y}_0(k+1|k)$ is the output prediction when the control moves at $k+1$ and future time instants keep invariant. Denote

$$\bar{y}_0(k+2|k+1) = y_0(k+2|k) + s_2 \Delta u(k),$$

$$\vdots$$

$$\bar{y}_0(k+M+1|k+1) = y_0(k+M+1|k) + s_{M+1} \Delta u(k),$$

$$\vdots$$

$$\bar{y}_0(k+P+1|k+1) = y_0(k+P+1|k) + s_{P+1} \Delta u(k).$$

$\bar{y}_0(k+1+i|k+1)$, $i \in \{1, 2, \ldots, P\}$ can be the basis for constructing $\tilde{y}_0(k+1|k+1)$.

Also denote

$$\varepsilon(k+1) = y(k+1) - y(k+1|k).$$

Since $\varepsilon(k+1)$ is the effect on the output by the un-modeled uncertainties, it can be used to predict the future prediction error, so as to compensate the predictions based on the model. In summary, we can use the following to predict $y_0(k+i|k+1)$:

$$y_0(k+2|k+1) =\bar{y}_0(k+2|k+1) + f_1\varepsilon(k+1),$$

$$\vdots$$

$$y_0(k+M+1|k+1) =\bar{y}_0(k+M+1|k+1) + f_M\varepsilon(k+1),$$

$$\vdots$$

$$y_0(k+P+1|k+1) =\bar{y}_0(k+P+1|k+1) + f_P\varepsilon(k+1).$$

By summarizing the above deductions, at each time $k > 0$, $\tilde{y}_0(k|k)$ can be calculated by

$$\tilde{y}_0(k|k) = \tilde{y}_0(k|k-1) + A_1\Delta u(k-1) + \tilde{f}\varepsilon(k), \qquad (3.2.14)$$

where

$$\varepsilon(k) =y(k) - y(k|k-1), \qquad (3.2.15)$$
$$y(k|k-1) =y_0(k|k-1) + s_1\Delta u(k-1), \qquad (3.2.16)$$
$$A_1 =[\begin{array}{cccc} s_2 & s_3 & \cdots & s_{P+1} \end{array}]^T.$$

The above method can be summarized as follows.
Algorithm 3.2 (Type-II unconstrained DMC)

Step 0. Obtain $\{s_1, s_2, \cdots, s_N\}$. Calculate d^T. Choose $\tilde{f}$.

Step 1. At $k = 0$,

    Step 1.1. measure $y(0)$;
    Step 1.2. determine $\tilde{y}_s(0)$;
    Step 1.3. choose $y_0(i|0) = y(0)$, $i \in \{1, 2, \ldots, P\}$ and construct $\tilde{y}_0(0|0)$;
    Step 1.4. use (3.2.9) to calculate $\Delta u(0)$;
    Step 1.5. implement $\Delta u(0)$.

Step 2. At each time $k > 0$,

    Step 2.1. measure $y(k)$;
    Step 2.2. determine $\tilde{y}_s(k)$;
    Step 2.3. use (3.2.15) to calculate $\varepsilon(k)$;
    Step 2.4. use (3.2.14) to calculate $\tilde{y}_0(k|k)$;
    Step 2.5. use (3.2.9) to calculate $\Delta u(k)$;
    Step 2.6. implement $\Delta u(k)$.

Remark 3.2.2. In (3.2.14), $\tilde{y}_0(k|k)$ includes $i \in \{1, 2, \ldots, P\}$. Hence, (3.2.14) can be expressed as

$$y_0(k+i|k) = y_0(k+i|k-1) + s_{i+1}\Delta u(k-1) + f_i\varepsilon(k), i \in \{1, 2, \ldots, P\}.$$

By utilizing the above formula for $i = 0$, applying (3.2.15)-(3.2.16) yields

$$y_0(k|k) = y_0(k|k-1) + s_1\Delta u(k-1) + f_0[y(k) - y_0(k|k-1) - s_1\Delta u(k-1)].$$

If we take $f_0 = 1$, then $y_0(k|k) = y(k)$. This is the reason for choosing (3.2.15) as the feedback correction. In (3.2.14), the selection of f_i depends on the concrete situation. Hence, f_i can be tuning parameter.

Remark 3.2.3. In Algorithm 3.1, the feedback correction has the following form:

$$y_0(k+i|k) = y_0(k+i|k-1) + f_i[y(k) - y_0(k|k-1)], i \in \{1, 2, \ldots, P\}.$$

In Algorithm 3.2, the feedback correction has the following form:

$$y_0(k+i|k) = y_0(k+i|k-1) + s_{i+1}\Delta u(k-1) + f_i[y(k) - y_0(k|k-1) \\ - s_1\Delta u(k-1)], i \in \{1, 2, \ldots, P\}.$$

Hence, even for $f_i = 1$, the two algorithms are different.

Remark 3.2.4. By considering Remarks 3.2.1, 3.2.2 and 3.2.3, it is shown that the feedback correction has the heuristic property, i.e., it is not set arbitrarily but can be different for different designers.

Remark 3.2.5. For prediction errors, there is no causal description. Hence, the error prediction is artificial. The deduction of the predictive control law is strict except that the prediction error is artificial.

3.2.3 Case multi-input multi-output

DMC of system with m input and r output is a generalization of SISO case. In the following, based on Algorithm 3.2, MIMO DMC is given.

(i) *First step:* Suppose only the j-th input is changed and other inputs keep invariant. Then considering the i-th output yields

$$\tilde{y}_{ij}(k|k) = \tilde{y}_{ij0}(k|k) + A_{ij}\Delta\tilde{u}_j(k|k) \tag{3.2.17}$$

where

$$\tilde{y}_{ij}(k|k) = [y_{ij}(k+1|k), y_{ij}(k+2|k), \cdots, y_{ij}(k+P|k)]^T,$$
$$\tilde{y}_{ij0}(k|k) = [y_{ij0}(k+1|k), y_{ij0}(k+2|k), \cdots, y_{ij0}(k+P|k)]^T,$$
$$\Delta\tilde{u}_j(k|k) = [\Delta u_j(k), \Delta u_j(k+1|k), \cdots, \Delta u_j(k+M-1|k)]^T,$$

$$A_{ij} = \begin{bmatrix} s_{ij1} & 0 & \cdots & 0 \\ s_{ij2} & s_{ij1} & \cdots & 0 \\ \vdots & \vdots & \ddots & \vdots \\ s_{ijM} & s_{ij,M-1} & \cdots & s_{ij,1} \\ \vdots & \vdots & \ddots & \vdots \\ s_{ijP} & s_{ij,P-1} & \cdots & s_{ij,P-M+1} \end{bmatrix}.$$

(ii) *Second step:* Suppose all the input can be changed. Then considering the i-th output and applying the superposition principle yields

$$\tilde{y}_i(k|k) = \tilde{y}_{i0}(k|k) + \sum_{j=1}^{r} A_{ij}\Delta\tilde{u}_j(k|k) \qquad (3.2.18)$$

where

$$\tilde{y}_i(k|k) = [y_i(k+1|k), y_i(k+2|k), \cdots, y_i(k+P|k)]^T,$$
$$\tilde{y}_{i0}(k|k) = [y_{i0}(k+1|k), y_{i0}(k+2|k), \cdots, y_{i0}(k+P|k)]^T.$$

(iii) *Third step:* Considering all the inputs and outputs yields

$$Y(k|k) = Y_0(k|k) + \tilde{A}\Delta U(k|k), \qquad (3.2.19)$$

where

$$Y(k|k) = [\tilde{y}_1(k|k)^T, \tilde{y}_2(k|k)^T, \cdots, \tilde{y}_r(k|k)^T]^T,$$
$$Y_0(k|k) = [\tilde{y}_{10}(k|k)^T, \tilde{y}_{20}(k|k)^T, \cdots, \tilde{y}_{r0}(k|k)^T]^T,$$
$$\Delta U(k|k) = [\Delta\tilde{u}_1(k|k)^T, \Delta\tilde{u}_2(k|k)^T, \cdots, \Delta\tilde{u}_m(k|k)^T]^T,$$

$$\tilde{A} = \begin{bmatrix} A_{11} & A_{12} & \cdots & A_{1m} \\ A_{21} & A_{22} & \cdots & A_{2m} \\ \vdots & \vdots & \ddots & \vdots \\ A_{r1} & A_{r2} & \cdots & A_{rm} \end{bmatrix}.$$

Suppose the criterion for optimizing $\Delta U(k|k)$ is to minimize the following cost function:

$$J(k) = \|E(k|k)\|_{\tilde{W}}^2 + \|\Delta U(k|k)\|_{\tilde{R}}^2, \qquad (3.2.20)$$

where $\tilde{W} \geq 0$ and $\tilde{R} \geq 0$ are symmetric matrices,

$$E(k|k) = [\tilde{e}_1(k|k)^T, \tilde{e}_2(k|k)^T, \cdots, \tilde{e}_r(k|k)^T]^T,$$
$$\tilde{e}_i(k|k) = [e_i(k+1|k), e_i(k+2|k), \cdots, e_i(k+P|k)]^T,$$
$$e_i(k+l|k) = y_{is}(k+l) - y_i(k+l|k).$$

Then, when $\tilde{A}^T\tilde{W}\tilde{A} + \tilde{R}$ is nonsingular, minimization of (3.2.20) yields

$$\Delta U(k|k) = (\tilde{A}^T\tilde{W}\tilde{A} + \tilde{R})^{-1}\tilde{A}^T\tilde{W}E_0(k) \qquad (3.2.21)$$

where

$$E_0(k) = [\tilde{e}_{10}(k)^T, \tilde{e}_{20}(k)^T, \cdots, \tilde{e}_{r0}(k)^T]^T,$$
$$\tilde{e}_{i0}(k) = [e_{i0}(k+1), e_{i0}(k+2), \cdots, e_{i0}(k+P)]^T,$$
$$e_{i0}(k+l) = y_{is}(k+l) - y_{i0}(k+l|k).$$

$y_{is}(k+l)$ is the setpoint value at the future time $k+l$ for the i-th output; $y_{i0}(k+l|k)$ is the prediction on the i-th output at future time $k+l$, when the control moves for the time k and future sampling instants are kept invariant.

At each time k, implement the following control move:

$$\Delta u(k) = DE_0(k) \qquad (3.2.22)$$

where

$$D = L(\tilde{A}^T \tilde{W} \tilde{A} + \tilde{R})^{-1} \tilde{A}^T \tilde{W},$$

$$L = \begin{bmatrix} \theta & 0 & \cdots & 0 \\ 0 & \theta & \cdots & 0 \\ \vdots & \vdots & \ddots & \vdots \\ 0 & 0 & \cdots & \theta \end{bmatrix} \in \mathfrak{R}^{m \times mM}, \ \theta = [\ 1 \ \ 0 \ \ \cdots \ \ 0\ ] \in \mathfrak{R}^M.$$

A simple selection of $\tilde{W}$ and $\tilde{R}$ is

$$\tilde{W} = \text{diag}\{W_1, W_2, \cdots, W_r\}, \ \tilde{R} = \text{diag}\{R_1, R_2, \cdots, R_m\},$$
$$W_i = \text{diag}\{w_{i1}, w_{i2}, \cdots, w_{iP}\}, \ i \in \{1, \dots, r\},$$
$$R_j = \text{diag}\{r_{j1}, r_{j2}, \cdots, r_{jM}\}, \ j \in \{1, \dots, m\}.$$

Taking $\tilde{R} > 0$ guarantees the nonsingularity of $\tilde{A}^T \tilde{W} \tilde{A} + \tilde{R}$.

(iv) *Fourth step:* At the initial time $k = 0$, suppose the system is in the steady-state. For startup of DMC we can take $y_{i0}(l|0) = y_i(0)$ ($l = 1, 2, \cdots, P+1$).

For each time $k > 0$, $\tilde{y}_0(k|k)$ can be calculated as

$$Y_0(k|k) = Y_0(k|k-1) + \tilde{A}_1 \Delta U(k-1) + \tilde{F} \Upsilon(k) \qquad (3.2.23)$$

where

$$Y_0(k|k-1) = [\tilde{y}_{10}(k|k-1)^T, \tilde{y}_{20}(k|k-1)^T, \cdots, \tilde{y}_{r0}(k|k-1)^T]^T,$$
$$\tilde{y}_{i0}(k|k-1) = [y_{i0}(k+1|k-1), y_{i0}(k+2|k-1), \cdots, y_{i0}(k+P|k-1)]^T,$$
$$\Delta U(k-1) = [\Delta u_1(k-1), \Delta u_2(k-1), \cdots, \Delta u_m(k-1)]^T,$$

$$\tilde{A}_1 = \begin{bmatrix} A_{111} & A_{121} & \cdots & A_{1m1} \\ A_{211} & A_{221} & \cdots & A_{2m1} \\ \vdots & \vdots & \ddots & \vdots \\ A_{r11} & A_{r21} & \cdots & A_{rm1} \end{bmatrix},$$

$$A_{ij1} = [\ s_{ij2} \ \ s_{ij3} \ \ \cdots \ \ s_{ij,P+1}\ ]^T,$$

$$\tilde{F} = \begin{bmatrix} \tilde{f}_1 & 0 & \cdots & 0 \\ 0 & \tilde{f}_2 & \cdots & 0 \\ \vdots & \vdots & \ddots & \vdots \\ 0 & 0 & \cdots & \tilde{f}_r \end{bmatrix},$$

$$\tilde{f}_i = [f_{i1}, f_{i2}, \cdots, f_{iP}]^T,$$

$$\Upsilon(k) = [\varepsilon_1(k), \varepsilon_2(k), \cdots, \varepsilon_r(k)]^T,$$

$$\varepsilon_i(k) = y_i(k) - y_i(k|k-1),$$

$$y_i(k|k-1) = y_{i0}(k|k-1) + \sum_{j=1}^{r} s_{ij1}\Delta u_j(k-1). \tag{3.2.24}$$

Remark 3.2.6. If, in A of (3.2.6) and A_p of (3.2.10), all the s_l's are $r \times m$-dimensional matrices, then the closed-form solution to DMC for the unconstrained MIMO systems can be expressed as (3.2.8), rather than adopting the deduction manner as in section 3.2.3. However, if different inputs (outputs) adopt different control horizons (prediction horizons), the expression in (3.2.23) will be more convenient.

3.2.4 Remarks on Matlab MPC Toolbox

In DMC provided in Matlab MPC Toolbox, the predicted process outputs $y(k+1|k), y(k+2|k), \cdots, y(k+P|k)$ depend on the current measurement $y(k)$ and assumptions we make about the unmeasurable disturbances and measurement noise affecting the outputs.

For unconstrained plant, the linear time-invariant feedback control law can be obtained (which can be solved by "mpccon" in Matlab):

$$\Delta u(k) = K_{\text{DMC}} E_0(k). \tag{3.2.25}$$

For open-loop stable plants, nominal stability of the closed-loop system depends only on K_{DMC} which in turn is affected by P, M, W_i, R_j, etc. No precise conditions on P, M, W_i, R_j exist which guarantee closed-loop stability. In general, decreasing M relative to P makes the control action less aggressive and tends to stabilize a system. For $M = 1$, nominal stability of the closed-loop system is guaranteed for any finite P and time invariant W_i and R_j. More commonly, R_j is used as tuning parameter. Increasing R_j always has the effect of making the control action less aggressive. We can obtain the state-space description of the closed-loop system with the command "mpccl" and then determine the pole locations with "smpcpole." The closed-loop system is stable if all the poles are inside or on the unit-circle.

The algorithm in Matlab MPC Toolbox can track the step-change setpoint value without steady-state error. For Matlab MPC toolbox one can refer to [50].

3.3 Constraint handling

Before DMC is implemented (design stage) and in the real application of DMC (run stage), users can tune the following parameters:

(i) sampling interval T_s,

(ii) prediction horizon P, control horizon M, weighting matrices $\tilde{W}$, $\tilde{R}$ or weighting coefficients w_i, r_j,

(iii) the correction coefficients $\tilde{f}_i$ or f_i,

(iv) modeling coefficients of the step response model,

(v) constraints, including input magnitude constraint (upper limit, lower limit), the upper bound of the control increment, the constraints on the intermediate variable (combined variable), etc.

In general, it does not need to often tune the modeling coefficients of the step response model; the main reason is the same as that for the impulse response model (refer to section 2.2). The constraints in (v) are usually hard constraints (i.e., constraints not violable). In the following we discuss how to handle the constraint in DMC (take MIMO system as an example).

1. *Output magnitude constraint* $y_{i,\min} \leq y_i(k+l|k) \leq y_{i,\max}$

At each optimization cycle, the output prediction is (3.2.19). Hence, we can let the optimization problem satisfy the following constraint:

$$Y_{\min} \leq Y_0(k|k) + \tilde{A}\Delta U(k|k) \leq Y_{\max} \tag{3.3.1}$$

where

$$Y_{\min} = [\tilde{y}_{1,\min}^T, \tilde{y}_{2,\min}^T, \cdots, \tilde{y}_{r,\min}^T]^T,$$
$$\tilde{y}_{i,\min} = [y_{i,\min}, y_{i,\min}, \cdots, y_{i,\min}]^T \in \Re^P,$$
$$Y_{\max} = [\tilde{y}_{1,\max}^T, \tilde{y}_{2,\max}^T, \cdots, \tilde{y}_{r,\max}^T]^T,$$
$$\tilde{y}_{i,\max} = [y_{i,\max}, y_{i,\max}, \cdots, y_{i,\max}]^T \in \Re^P.$$

2. *The input increment constraint* $\Delta u_{j,\min} \leq \Delta u_j(k+l|k) = u_j(k+l|k) - u_j(k+l-1|k) \leq \Delta u_{j,\max}$

We can let the optimization problem satisfy the following constraint:

$$\Delta U_{\min} \leq \Delta U(k|k) \leq \Delta U_{\max} \tag{3.3.2}$$

where

$$\Delta U_{\min} = [\Delta\tilde{u}_{1,\min}^T, \Delta\tilde{u}_{2,\min}^T, \cdots, \Delta\tilde{u}_{m,\min}T]^T,$$
$$\Delta\tilde{u}_{j,\min} = [\Delta u_{j,\min}, \Delta u_{j,\min}, \cdots, \Delta u_{j,\min}]^T \in \Re^M,$$
$$\Delta U_{\max} = [\Delta\tilde{u}_{1,\max}^T, \Delta\tilde{u}_{2,\max}^T, \cdots, \Delta\tilde{u}_{m,\max}^T]^T,$$
$$\Delta\tilde{u}_{j,\max} = [\Delta u_{j,\max}, \Delta u_{j,\max}, \cdots, \Delta u_{j,\max}]^T \in \Re^M.$$

3. *The input magnitude constraint $u_{j,\min} \leq u_j(k+l|k) \leq u_{j,\max}$*
We can let the optimization problem satisfy the following constraint:

$$U_{\min} \leq B\Delta U(k|k) + \tilde{u}(k-1) \leq U_{\max} \qquad (3.3.3)$$

where

$$U_{\min} = [\tilde{u}_{1,\min}^T, \tilde{u}_{2,\min}^T, \cdots, \tilde{u}_{m,\min}T]^T,$$

$$\tilde{u}_{j,\min} = [u_{j,\min}, u_{j,\min}, \cdots, u_{j,\min}]^T \in \mathfrak{R}^M,$$

$$U_{\max} = [\tilde{u}_{1,\max}^T, \tilde{u}_{2,\max}^T, \cdots, \tilde{u}_{m,\max}T]^T,$$

$$\tilde{u}_{j,\max} = [u_{j,\max}, u_{j,\max}, \cdots, u_{j,\max}]^T \in \mathfrak{R}^M,$$

$$B = \mathrm{diag}\{B_0, \cdots, B_0\} \ (m \text{ blocks}),$$

$$B_0 = \begin{bmatrix} 1 & 0 & \cdots & 0 \\ 1 & 1 & \ddots & \vdots \\ \vdots & \ddots & \ddots & 0 \\ 1 & \cdots & 1 & 1 \end{bmatrix} \in \mathfrak{R}^{M \times M},$$

$$\tilde{u}(k-1) = [\tilde{u}_1(k-1)^T, \tilde{u}_2(k-1)^T, \cdots, \tilde{u}_m(k-1)^T]^T,$$

$$\tilde{u}_j(k-1) = [u_j(k-1), u_j(k-1), \cdots, u_j(k-1)]^T \in \mathfrak{R}^M.$$

Equations (3.3.1)-(3.3.3) can be written in a uniform form as $C\Delta U(k|k) \leq \bar{c}$, where C and $\bar{c}$ are matrix and vector known at time k. DMC optimization problem considering these constraints can be written as

$$\min_{\Delta U(k|k)} J(k) = \|E(k|k)\|_W^2 + \|\Delta U(k|k)\|_R^2, \text{ s.t. } C\Delta U(k|k) \leq \bar{c}. \qquad (3.3.4)$$

Problem (3.3.4) is a quadratic optimization problem. The feedback law solution to the constrained quadratic optimization problem is, in general, nonlinear.

In Matlab MPC Toolbox, for DMC optimization problem for constrained systems one can adopt "cmpc."

Chapter 4

Generalized predictive control (GPC)

In the 1980s, the adaptive control techniques, such as minimum variance adaptive control, had been widely recognized and developed in the process control area. However, these adaptive control algorithms rely on models with high accuracy which, to a large extent, limits the applicability in the complex industrial processes. GPC was developed along the investigation of adaptive control. GPC not only inherits the advantages of adaptive control for its applicability in stochastic systems, on-line identification etc., but preserves the advantages of predictive control for its receding-horizon optimization, lower requirement on the modeling accuracy, etc. GPC has the following characteristics:

(1) It relies on the traditional parametric models. Hence, there are fewer parameters in the system model. Note that MAC and DMC both apply non-parametric models, i.e., impulse response model and step response model, respectively.

(2) It was developed along the investigation of adaptive control. It inherits the advantage of adaptive control but is more robust.

(3) The techniques of multi-step prediction, dynamic optimization and feedback correction are applied. Hence, the control effect is better and more suitable for industrial processes.

Due to the above advantages, GPC has been widely acknowledged both in control theory academia and in process control studies, which makes GPC the most active MPC algorithm.

Section 4.1 is referred to in [63], [6]. Section 4.3 is referred to in [18].

65

4.1   Principle of GPC

4.1.1   Prediction model

Consider the following SISO CARIMA model:

$$A(z^{-1})y(k) = B(z^{-1})u(k-1) + \frac{C(z^{-1})\xi(k)}{\Delta} \qquad (4.1.1)$$

where

$$
\begin{aligned}
A(z^{-1}) &= 1 + a_1 z^{-1} + \cdots + a_{n_a} z^{-n_a}, \ \ \deg A(z^{-1}) = n_a, \\
B(z^{-1}) &= b_0 + b_1 z^{-1} + \cdots + b_{n_b} z^{-n_b}, \ \ \deg B(z^{-1}) = n_b, \\
C(z^{-1}) &= c_0 + c_1 z^{-1} + \cdots + c_{n_c} z^{-n_c}, \ \ \deg c(z^{-1}) = n_c;
\end{aligned}
$$

z^{-1} is the backward shift operator, i.e., $z^{-1}y(k) = y(k-1)$, $z^{-1}u(k) = u(k-1)$; $\Delta = 1 - z^{-1}$ is the difference operator; $\{\xi(k)\}$ is the white noise sequence with zero mean value. For systems with q samples time delay, $b_0 \sim b_{q-1} = 0$, $n_b \geq q$.

Suppose $C(z^{-1}) = 1$. Thus, (4.1.1) actually utilizes the impulse transfer function to describe the controlled plant. The transfer function from the input u to the output y is

$$G(z^{-1}) = \frac{z^{-1}B(z^{-1})}{A(z^{-1})}. \qquad (4.1.2)$$

In order to deduce the prediction $y(k+j|k)$ via (4.1.1), let us first introduce the Diophantine equation

$$1 = E_j(z^{-1})A(z^{-1})\Delta + z^{-j}F_j(z^{-1}), \qquad (4.1.3)$$

where $E_j(z^{-1})$, $F_j(z^{-1})$ are polynomials uniquely determined by $A(z^{-1})$ and length j,

$$
\begin{aligned}
E_j(z^{-1}) &= e_{j,0} + e_{j,1}z^{-1} + \cdots + e_{j,j-1}z^{-(j-1)}, \\
F_j(z^{-1}) &= f_{j,0} + f_{j,1}z^{-1} + \cdots + f_{j,n_a}z^{-n_a}.
\end{aligned}
$$

Multiplying (4.1.1) by $E_j(z^{-1})\Delta z^j$, and utilizing (4.1.3), we can write out the output prediction at time $k + j$,

$$y(k+j|k) = E_j(z^{-1})B(z^{-1})\Delta u(k+j-1|k) + F_j(z^{-1})y(k) + E_j(z^{-1})\xi(k+j). \qquad (4.1.4)$$

Pay attention to the formulations of $E_j(z^{-1})$, $F_j(z^{-1})$, it is known that $E_j(z^{-1})B(z^{-1})\Delta u(k+j-1|k)$ has relation with $\{u(k+j-1|k), u(k+j-2|k), \cdots\}$, $F_j(z^{-1})y(k)$ has relation with $\{y(k), y(k-1), \cdots\}$ and $E_j(z^{-1})\xi(k+j)$ has relation with $\{\xi(k+j), \cdots, \xi(k+2), \xi(k+1)\}$.

Since, at time k, the future noises $\xi(k+i)$, $i \in \{1, \ldots, j\}$ are unknown, the most suitable predicted value of $y(k+j)$ can be represented by the following:

$$\bar{y}(k+j|k) = E_j(z^{-1})B(z^{-1})\Delta u(k+j-1|k) + F_j(z^{-1})y(k). \qquad (4.1.5)$$

In (4.1.5), denote $G_j(z^{-1}) = E_j(z^{-1})B(z^{-1})$. Combining with (4.1.3) yields

$$G_j(z^{-1}) = \frac{B(z^{-1})}{A(z^{-1})\Delta}[1 - z^{-j}F_j(z^{-1})]. \qquad (4.1.6)$$

Let us introduce another Diophantine equation

$$G_j(z^{-1}) = E_j(z^{-1})B(z^{-1}) = \tilde{G}_j(z^{-1}) + z^{-(j-1)}H_j(z^{-1}),$$

where

$$\begin{aligned}\tilde{G}_j(z^{-1}) &= g_{j,0} + g_{j,1}z^{-1} + \cdots + g_{j,j-1}z^{-(j-1)},\\ H_j(z^{-1}) &= h_{j,1}z^{-1} + h_{j,2}z^{-2} + \cdots + h_{j,n_b}z^{-n_b}.\end{aligned}$$

Then, applying (4.1.4)-(4.1.5) yields

$$\bar{y}(k+j|k) = \tilde{G}_j(z^{-1})\Delta u(k+j-1|k) + H_j(z^{-1})\Delta u(k) + F_j(z^{-1})y(k), \qquad (4.1.7)$$

$$y(k+j|k) = \bar{y}(k+j|k) + E_j(z^{-1})\xi(k+j). \qquad (4.1.8)$$

All the equations (4.1.4), (4.1.5), (4.1.7) and (4.1.8) can be the prediction models of GPC. Thus, the future output can be predicted by applying the known input, output and the future input.

4.1.2 Solution to the Diophantine equation

In order to predict the future output by applying (4.1.4) or (4.1.5), one has to first know $E_j(z^{-1})$, $F_j(z^{-1})$. For different $j \in \{1, 2, \ldots\}$, this amounts to solve a set of Diophantine equations (4.1.3) in parallel, which involves a heavy computational burden. For this reason, [6] gave an iterative algorithm for calculating $E_j(z^{-1})$, $F_j(z^{-1})$.

First, according to (4.1.3),

$$\begin{aligned}1 &= E_j(z^{-1})A(z^{-1})\Delta + z^{-j}F_j(z^{-1}),\\ 1 &= E_{j+1}(z^{-1})A(z^{-1})\Delta + z^{-(j+1)}F_{j+1}(z^{-1}).\end{aligned}$$

Making subtractions from both sides of the above two equations yields

$$A(z^{-1})\Delta[E_{j+1}(z^{-1}) - E_j(z^{-1})] + z^{-j}[z^{-1}F_{j+1}(z^{-1}) - F_j(z^{-1})] = 0.$$

Denote

$$\begin{aligned}\tilde{A}(z^{-1}) &= A(z^{-1})\Delta = 1 + \tilde{a}_1 z^{-1} + \cdots + \tilde{a}_{n_a}z^{-n_a} + \tilde{a}_{n_a+1}z^{-(n_a+1)}\\ &= 1 + (a_1 - 1)z^{-1} + \cdots + (a_{n_a} - a_{n_a-1})z^{-n_a} - a_{n_a}z^{-(n_a+1)},\\ E_{j+1}(z^{-1}) &- E_j(z^{-1}) = \tilde{E}(z^{-1}) + e_{j+1,j}z^{-j}.\end{aligned}$$

Then,

$$\tilde{A}(z^{-1})\tilde{E}(z^{-1}) + z^{-j}[z^{-1}F_{j+1}(z^{-1}) - F_j(z^{-1}) + \tilde{A}(z^{-1})e_{j+1,j}] = 0. \quad (4.1.9)$$

A necessary condition for (4.1.9) to be consistently satisfied is that all the items in $\tilde{A}(z^{-1})\tilde{E}(z^{-1})$ with order less than j should be equal to zeros. Since the coefficient for the first item in $\tilde{A}(z^{-1})$ is 1, it is easy to obtain that the necessary condition for consistently satisfying (4.1.9) is

$$\tilde{E}(z^{-1}) = 0. \quad (4.1.10)$$

Further, the necessary and sufficient condition for consistently satisfying (4.1.9) is that (4.1.10) and the following equation holds:

$$F_{j+1}(z^{-1}) = z[F_j(z^{-1}) - \tilde{A}(z^{-1})e_{j+1,j}]. \quad (4.1.11)$$

Comparing the items with the same orders in both sides of (4.1.11), we obtain

$$\begin{aligned}
e_{j+1,j} &= f_{j,0}, \\
f_{j+1,i} &= f_{j,i+1} - \tilde{a}_{i+1}e_{j+1,j} = f_{j,i+1} - \tilde{a}_{i+1}f_{j,0}, \ i \in \{0, \ldots, n_a - 1\}, \\
f_{j+1,n_a} &= -\tilde{a}_{n_a+1}e_{j+1,j} = -\tilde{a}_{n_a+1}f_{j,0}.
\end{aligned}$$

This formulation for deducing the coefficients of $F_j(z^{-1})$ can be written in the vector form

$$f_{j+1} = \tilde{A}f_j,$$

where

$$\begin{aligned}
f_{j+1} &= [f_{j+1,0}, \cdots, f_{j+1,n_a}]^T, \\
f_j &= [f_{j,0}, \cdots, f_{j,n_a}]^T,
\end{aligned}$$

$$\tilde{A} = \begin{bmatrix}
1 - a_1 & 1 & 0 & \cdots & 0 \\
a_1 - a_2 & 0 & 1 & \ddots & \vdots \\
\vdots & \vdots & \ddots & \ddots & 0 \\
a_{n_a-1} - a_{n_a} & 0 & \cdots & 0 & 1 \\
a_{n_a} & 0 & \cdots & 0 & 0
\end{bmatrix}.$$

Moreover, the iterative formula for coefficients of $E_j(z^{-1})$ is

$$E_{j+1}(z^{-1}) = E_j(z^{-1}) + e_{j+1,j}z^{-j} = E_j(z^{-1}) + f_{j,0}z^{-j}.$$

When $j = 1$, equation (4.1.3) is

$$1 = E_1(z^{-1})\tilde{A}(z^{-1}) + z^{-1}F_1(z^{-1}).$$

Hence, we should choose $E_1(z^{-1}) = 1$, $F_1(z^{-1}) = z[1 - \tilde{A}(z^{-1})]$ as the initial values of $E_j(z^{-1})$, $F_j(z^{-1})$. Thus, $E_{j+1}(z^{-1})$ and $F_{j+1}(z^{-1})$ can be iteratively calculated by

$$\left. \begin{array}{l} f_{j+1} = \tilde{A} f_i, \ f_0 = [1, 0 \cdots 0]^T, \\ E_{j+1}(z^{-1}) = E_j(z^{-1}) + f_{j,0} z^{-j}, \ E_0 = 0. \end{array} \right\} \quad (4.1.12)$$

From (4.1.10), it is shown that $e_{j,i}, i < j$ is not related with j. Hence, $e_i \triangleq e_{j,i}, i < j$.

Remark 4.1.1. In (4.1.3), we can directly choose $E_j(z^{-1}) = e_0 + e_1 z^{-1} + \cdots + e_{j-1} z^{-(j-1)}$. According to (4.1.12), simplifying $e_{j,j-1}$ as e_{j-1} does not affect the results.

Consider the second Diophantine equation. From (4.1.6) we know that the first j items of $G_j(z^{-1})$ have no relation with j; the first j coefficients of $G_j(z^{-1})$ are the unit impulse response values, which are denoted as $g_1, \cdots, g_j$. Thus,

$$G_j(z^{-1}) = E_j(z^{-1}) B(z^{-1}) = g_1 + g_2 z^{-1} + \cdots + g_j z^{-(j-1)} + z^{-(j-1)} H_j(z^{-1}).$$

Therefore,

$$g_{j,i} = g_{i+1}, \ i < j.$$

Since $G_j(z^{-1})$ is the convolution of $E_j(z^{-1})$ and $B(z^{-1})$, it is easy to calculate $G_j(z^{-1})$ and, hence, the coefficients of $\tilde{G}_j(z^{-1})$ and $H_j(z^{-1})$ can be easily obtained.

4.1.3 Receding horizon optimization

In GPC, the cost function at time k has the following form:

$$\min J(k) = \mathrm{E} \left\{ \sum_{j=N_1}^{N_2} [y(k+j|k) - y_s(k+j)]^2 + \sum_{j=1}^{N_u} \lambda(j)[\Delta u(k+j-1|k)]^2 \right\}$$

$$(4.1.13)$$

where $\mathrm{E}\{\bullet\}$ represents the mathematical expectation; y_s is the desired value of the output; N_1 and N_2 are the starting and ending instant for the optimization horizon; N_u is the control horizon, i.e., the control moves after N_u steps keep invariant,

$$u(k+j-1|k) = u(k+N_u-1|k), \ j > N_u;$$

$\lambda(j)$ is the control weighting coefficient. For simplicity, in general, $\lambda(j)$ is supposed to be a constant λ.

In the cost function (4.1.13), N_1 should be larger than number of delayed intervals, N_2 should be as large when the plant dynamics is sufficiently represented (i.e., as large when the effect of the current control increment is included). Since the multi-step prediction and optimization is adopted, even when the delay is not estimated correctly or the delay is changed, one can still achieve reasonable control from the overall optimization. This is the important reason why GPC has robustness with respect to the modeling inaccuracy.

The above cost function is quite similar to that for DMC in Chapter 3, except that the stochastic systems are considered. In the cost function for DMC in Chapter 3, if we set the weighting coefficients w_i's before N_1 as zeros, and those at and after N_1 as 1, then we obtain the same cost function. Hence, for simplifying the notations, in the following discussion we suppose $N_1 = 1$ and take $N_2 = N$. If it is desired, then we can obtain different starting instant for the optimization horizon by setting the coefficients w_i as in DMC (of course, in the sequel, some results for $N_1 > 1$ will be given).

In (4.1.13), the desired values for the output can be selected as in the reference trajectory of MAC, i.e.,

$$y_s(k) = y(k), \ y_s(k+j) = \alpha y_s(k+j-1) + (1-\alpha)\omega, \ 0 < \alpha < 1, j \in \{1,\ldots,N\} \tag{4.1.14}$$

where $\alpha \in [0,1)$ is called soften factor and ω is the output setpoint.

Applying the prediction model (4.1.5) yields

$$\begin{aligned}
\bar{y}(k+1|k) &= G_1(z^{-1})\Delta u(k) + F_1(z^{-1})y(k) = g_{1,0}\Delta u(k|k) + f_1(k) \\
\bar{y}(k+2|k) &= G_2(z^{-1})\Delta u(k+1|k) + F_2(z^{-1})y(k) \\
&= g_{2,0}\Delta u(k+1|k) + g_{2,1}\Delta u(k|k) + f_2(k) \\
&\vdots \\
\bar{y}(k+N|k) &= G_N(z^{-1})\Delta u(k+N-1|k) + F_N(z^{-1})y(k) \\
&= g_{N,0}\Delta u(k+N-1|k) + \cdots + g_{N,N-N_u}\Delta u(k+N_u-1|k) \\
&\quad + \cdots + g_{N,N-1}\Delta u(k|k) + f_N(k) \\
&= g_{N,N-N_u}\Delta u(k+N_u-1|k) + \cdots + g_{N,N-1}\Delta u(k) + f_N(k),
\end{aligned}$$

where

$$\left.\begin{aligned}
f_1(k) &= [G_1(z^{-1}) - g_{1,0}]\Delta u(k) + F_1(z^{-1})y(k) \\
f_2(k) &= z[G_2(z^{-1}) - z^{-1}g_{2,1} - g_{2,0}]\Delta u(k) + F_2(z^{-1})y(k) \\
&\vdots \\
f_N(k) &= z^{N-1}[G_N(z^{-1}) - z^{-(N-1)}g_{N,N-1} - \cdots - g_{N,0}] \\
&\quad \Delta u(k) + F_N(z^{-1})y(k)
\end{aligned}\right\} \tag{4.1.15}$$

can be calculated by applying $\{y(\tau), \tau \leq k\}$ and $\{u(\tau), \tau < k\}$ which are known at time k.

Denote

$$\vec{y}(k|k) = [\bar{y}(k+1|k), \cdots, \bar{y}(k+N|k)]^T,$$
$$\Delta\tilde{u}(k|k) = [\Delta u(k|k), \cdots, \Delta u(k+N_u-1|k)]^T,$$
$$\overleftarrow{f}(k) = [f_1(k), \cdots, f_N(k)]^T.$$

Notice that $g_{j,i} = g_{i+1}(i < j)$ is the step response coefficient. Then

$$\vec{y}(k|k) = G\Delta\tilde{u}(k|k) + \overleftarrow{f}(k) \qquad (4.1.16)$$

where

$$G = \begin{bmatrix} g_1 & 0 & \cdots & & 0 \\ g_2 & g_1 & \ddots & & \vdots \\ \vdots & \vdots & \ddots & & 0 \\ g_{N_u} & g_{N_u-1} & \cdots & & g_1 \\ \vdots & \vdots & \ddots & & \vdots \\ g_N & g_{N-1} & \cdots & & g_{N-N_u+1} \end{bmatrix}.$$

By using $\bar{y}(k+j|k)$ to substitute $y(k+j|k)$ in (4.1.13), the cost function can be written in the vector form,

$$J(k) = [\vec{y}(k|k) - \vec{\omega}(k)]^T[\vec{y}(k|k) - \vec{\omega}(k)] + \lambda\Delta\tilde{u}(k|k)^T\Delta\tilde{u}(k|k),$$

where

$$\vec{\omega}(k) = [y_s(k+1), \cdots, y_s(k+N)]^T.$$

Thus, when $\lambda I + G^T G$ is nonsingular, the optimal solution to cost function (4.1.13) is obtained as

$$\Delta\tilde{u}(k|k) = (\lambda I + G^T G)^{-1}G^T[\vec{\omega}(k) - \overleftarrow{f}(k)]. \qquad (4.1.17)$$

The real-time optimal control moves is given by

$$u(k) = u(k-1) + d^T[\vec{\omega}(k) - \overleftarrow{f}(k)], \qquad (4.1.18)$$

where d^T is the first row of $(\lambda I + G^T G)^{-1}G^T$.

One can further utilize (4.1.8) to write the output prediction in the following vector form:

$$\tilde{y}(k|k) = G\Delta\tilde{u}(k|k) + F(z^{-1})y(k) + H(z^{-1})\Delta u(k) + \tilde{\varepsilon}(k),$$

where

$$\tilde{y}(k|k) = [y(k+1|k), \cdots, y(k+N|k)]^T,$$
$$F(z^{-1}) = [F_1(z^{-1}), \cdots, F_N(z^{-1})]^T,$$
$$H(z^{-1}) = [H_1(z^{-1}), \cdots, H_N(z^{-1})]^T,$$
$$\tilde{\varepsilon}(k) = [E_1(z^{-1})\xi(k+1), \cdots, E_N(z^{-1})\xi(k+N)]^T.$$

Thus, the cost function is written as the vector form

$$J(k) = \mathrm{E}\{[\tilde{y}(k|k) - \vec{\omega}(k)]^T [\tilde{y}(k|k) - \vec{\omega}(k)] + \lambda \Delta \tilde{u}(k|k)^T \Delta \tilde{u}(k|k)\}.$$

Thus, when $\lambda I + G^T G$ is nonsingular, the optimal control law is

$$\Delta \tilde{u}(k|k) = (\lambda I + G^T G)^{-1} G^T \left[\vec{\omega}(k) - F(z^{-1})y(k) - H(z^{-1})\Delta u(k)\right].$$

Since the mathematical expectation is adopted, $\tilde{\varepsilon}(k)$ does not appear in the above control law. The real-time optimal control move is given by

$$u(k) = u(k-1) + d^T \left[\vec{\omega}(k) - F(z^{-1})y(k) - H(z^{-1})\Delta u(k)\right]. \qquad (4.1.19)$$

4.1.4   On-line identification and feedback correction

In GPC, the modeling coefficients are on-line estimated continuously based on the real-time input/output data, and the control law is modified correspondingly. In DMC and MAC, it amounts to utilization of a time-invariant prediction model, combined with an additional error prediction model, so as to make an accurate prediction on the future output. In GPC, the error prediction model is not considered, and the model is on-line modified so as to make an accurate prediction.

Remark 4.1.2. In DMC and MAC, since the non-parametric models are applied, and the error prediction is invoked, the inherent heuristic nature exists. Here, the so-called "heuristic" means not depending on the strict theoretical deduction, which is like what we have pointed out in the last two chapters that the feedback correction depends on the users. If input/output model is selected, we can still apply the heuristic feedback correction as in DMC and MAC, i.e., use the current prediction error to modify the future output predictions (some literature did it this way, although in this book we have not done it this way when we introduce GPC).

Let us write (4.1.1) as

$$A(z^{-1})\Delta y(k) = B(z^{-1})\Delta u(k-1) + \xi(k).$$

Then,

$$\Delta y(k) = -A_1(z^{-1})\Delta y(k) + B(z^{-1})\Delta u(k-1) + \xi(k),$$

where $A_1(z^{-1}) = A(z^{-1}) - 1$. Denote the modeling parameters and data as the vector forms,

$$\theta = [a_1 \cdots a_{n_a} \; b_0 \cdots b_{n_b}]^T,$$
$$\varphi(k) = [-\Delta y(k-1) \cdots - \Delta y(k-n_a) \; \Delta u(k-1) \cdots \Delta u(k-n_b-1)]^T.$$

Then,

$$\Delta y(k) = \varphi(k)^T \theta + \xi(k).$$

Here, we can utilize the iterative least square method with fading memory to estimate the parameter vector:

$$
\left.
\begin{aligned}
\hat{\theta}(k) &= \hat{\theta}(k-1) + K(k)[\Delta y(k) - \varphi(k)^T \hat{\theta}(k-1)] \\
K(k) &= P(k-1)\varphi(k)[\varphi(k)^T P(k-1)\varphi(k) + \mu]^{-1} \\
P(k) &= \frac{1}{\mu}[I - K(k)\varphi(k)^T]P(k-1)
\end{aligned}
\right\}
\qquad (4.1.20)
$$

where $0 < \mu < 1$ is the forgetting factor usually chosen as $0.95 < \mu < 1$; $K(k)$ is the weighting factor; $P(k)$ is the positive-definite covariance matrix. In the startup of the controller, it needs to set the initial values of the parameter vector θ and covariance matrix P. Usually, we can set $\hat{\theta}(-1) = 0$, $P(-1) = \alpha^2 I$ where α is a sufficiently large positive scalar. At each control step, first setup the data vector, and then calculate $K(k)$, $\hat{\theta}(k)$ and $P(k)$ by applying (4.1.20).

After the parameters in $A(z^{-1})$, $B(z^{-1})$ are obtained by identification, d^T and $\overleftarrow{f}(k)$ in the control law (4.1.18) can be re-calculated and the optimal control move can be computed.

Algorithm 4.1 (Adaptive GPC)
The on-line implementation of GPC falls into the following steps:

Step 1. Based on the newly obtained input/output data, use the iterative formula (4.1.20) to estimate the modeling parameters, so as to obtain $A(z^{-1})$, $B(z^{-1})$.

Step 2. Based on the obtained $A(z^{-1})$, iteratively calculate $E_j(z^{-1})$, $F_j(z^{-1})$ according to (4.1.12).

Step 3. Based on $B(z^{-1})$, $E_j(z^{-1})$, $F_j(z^{-1})$, calculate the elements g_i's of G, and calculate $f_i(k)$ according to (4.1.15).

Step 4. Re-compute d^T, and calculate $u(k)$ according to (4.1.18). Implement $u(k)$ to the plant. This step involves the inversion of a $N_u \times N_u$-dimensional matrix and, hence, the on-line computational burden should be considered in selecting N_u.

4.2 Some basic properties

In the last section, we chose $N_1 = 1$. For $N_1 \neq 1$, define

$$
\begin{aligned}
F(z^{-1}) &= [F_{N_1}(z^{-1}), F_{N_1+1}(z^{-1}), \cdots, F_{N_2}(z^{-1})]^T, \\
H(z^{-1}) &= [H_{N_1}(z^{-1}), H_{N_1+1}(z^{-1}), \cdots, H_{N_2}(z^{-1})]^T, \\
\overleftarrow{f}(k) &= [f_{N_1}(k), f_{N_1+1}(k), \cdots, f_{N_2}(k)]^T, \\
\bar{\omega}(k) &= [y_s(k+N_1), \cdots, y_s(k+N_2)]^T, \\
d^T &= [1, 0, \cdots, 0](\lambda I + G^T G)^{-1} G^T,
\end{aligned}
$$

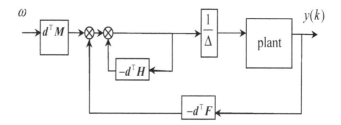

Figure 4.2.1: The block diagram of GPC (z^{-1} omitted).

$$G = \begin{bmatrix} g_{N1} & g_{N1-1} & \cdots & g_{N1-Nu+1} \\ g_{N1+1} & g_{N1} & \cdots & g_{N1-Nu+2} \\ \vdots & \vdots & \ddots & \vdots \\ g_{N2} & g_{N2-1} & \cdots & g_{N2-Nu+1} \end{bmatrix}, \ g_j = 0, \ \forall j \le 0.$$

Then, when $\lambda I + G^T G$ is nonsingular, the real-time control law of GPC is $\Delta u(k) = d^T[\vec{\omega}(k) - \overleftarrow{f}(k)]$, where $\overleftarrow{f}(k)$ is a vector composed of the past input, past output and the current output.

Lemma 4.2.1. *(The control structure of GPC) If we take $\vec{\omega}(k) = [\omega, \omega, \cdots, \omega]^T$ (i.e., the output setpoint is not softened), then the block diagram of GPC is shown in Figure 4.2.1, where $M = [1, 1, \cdots, 1]^T$.*

Proof. Adopt $\Delta u(k) = d^T[\vec{\omega}(k) - \overleftarrow{f}(k)]$, and consider the structure of (4.1.19). It is easy to obtain Figure 4.2.1. □

Of course, if $\vec{\omega}(k) = [\omega, \omega, \cdots, \omega]^T$ is not taken, i.e., the soften technique is adopted, then the corresponding block diagram can also be obtained. The structure in Figure 4.2.1 is for future use.

Lemma 4.2.2. *(The internal model control structure of GPC) If we take $\vec{\omega}(k) = [\omega, \omega, \cdots, \omega]^T$, then the internal model control structure is shown in Figure 4.2.2.*

Proof. This can be easily obtained by transformation of the block diagram. □

The so-called internal model means "inside model" which coarsely indicates that the controller contains the model, or, the computer stores the system model.
Define

$$\Delta \overleftarrow{u}(k) = [\Delta u(k-1), \Delta u(k-2), \cdots, \Delta u(k-n_b)]^T,$$
$$\overleftarrow{y}(k) = [y(k), y(k-1), \cdots, y(k-n_a)]^T.$$

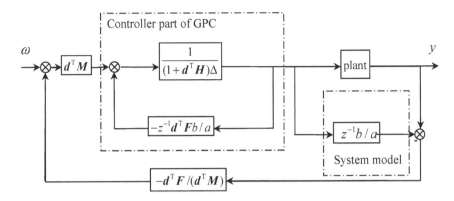

Figure 4.2.2: The internal model control structure of GPC (z^{-1} omitted).

Then the future output predictions $\vec{y}(k|k) = [\bar{y}(k+N_1|k), \bar{y}(k+N_1+1|k), \cdots, \bar{y}(k+N_2|k)]^T$ can be represented as

$$\vec{y}(k|k) = G\Delta\tilde{u}(k|k) + H\Delta\overleftarrow{u}(k) + F\overleftarrow{y}(k), \qquad (4.2.1)$$

where

$$H = \begin{bmatrix} h_{N_1,1} & h_{N_1,2} & \cdots & h_{N_1,nb} \\ h_{N_1+1,1} & h_{N_1+1,2} & \cdots & h_{N_1+1,nb} \\ \vdots & \vdots & \ddots & \vdots \\ h_{N_2,1} & h_{N_2,2} & \cdots & h_{N_2,nb} \end{bmatrix},$$

$$F = \begin{bmatrix} f_{N_1,0} & f_{N_1,1} & \cdots & f_{N_1,na} \\ f_{N_1+1,0} & f_{N_1+1,1} & \cdots & f_{N_1+1,na} \\ \vdots & \vdots & \ddots & \vdots \\ f_{N_2,0} & f_{N_2,1} & \cdots & f_{N_2,na} \end{bmatrix},$$

i.e., when $\lambda I + G^T G$ is nonsingular, the optimal control move is

$$\Delta u(k) = d^T \left[\vec{\omega}(k) - H\Delta\overleftarrow{u}(k) - F\overleftarrow{y}(k) \right].$$

Theorem 4.2.1. *(The optimal cost value) Suppose $\vec{\omega}(k) = 0$. Then the optimum of GPC cost function is*

$$J^*(k) = \overleftarrow{f}(k)^T \left[I - G(\lambda I + G^T G)^{-1} G^T \right] \overleftarrow{f}(k), \qquad (4.2.2)$$

$$J^*(k) = \lambda \overleftarrow{f}(k)^T (\lambda I + GG^T)^{-1} \overleftarrow{f}(k), \quad \lambda \neq 0, \qquad (4.2.3)$$

where $\overleftarrow{f}(k) = H\Delta\overleftarrow{u}(k) + F\overleftarrow{y}(k)$.

Proof. Substitute (4.1.16)-(4.1.17) into the cost function. By invoking an appropriate simplification, it yields (4.2.2). By using the following matrix inversion formula:

$$(Q + MTS)^{-1} = Q^{-1} - Q^{-1}M\left(SQ^{-1}M + T^{-1}\right)^{-1}SQ^{-1} \qquad (4.2.4)$$

where Q, M, T, S are matrices satisfying the corresponding requirement of nonsingularity, and applying (4.2.2), (4.2.3) can be obtained. □

4.3 Stability results not related to the concrete model coefficients

This section is mainly for integrity, by showing the relationship between the classical GPC and the special case of synthesis approach (Kleinman's controller). It should be noted that constructing the relationship of industrial MPC with synthesis approach of MPC is not easy. The content in this section is limited (only suitable to SISO linear time-invariant unconstrained systems). The content in this section is relatively independent and readers can omit this section. However, readers can review some knowledge about linear control theory in this section.

4.3.1 Transformation to the linear quadratic control problem

Consider the model in (4.1.1), $C(z^{-1}) = 1$, which can be transformed into

$$\tilde{A}(z^{-1})y(k) = \tilde{B}(z^{-1})\Delta u(k) + \xi(k) \qquad (4.3.1)$$

where $\tilde{A}(z^{-1}) = 1 + \tilde{a}_1 z^{-1} + \cdots + \tilde{a}_{n_A} z^{-n_A}$, $\tilde{B}(z^{-1}) = \tilde{b}_1 z^{-1} + \tilde{b}_2 z^{-2} + \cdots + \tilde{b}_{n_B} z^{-n_B}$, $n_A = n_a + 1$, $n_B = n_b + 1$. Suppose $\tilde{a}_{n_A} \neq 0$, $\tilde{b}_{n_B} \neq 0$ and $\left(\tilde{A}(z^{-1}), \tilde{B}(z^{-1})\right)$ is an irreducible pair. Take $\vec{\omega} = [\omega, \omega, \cdots, \omega]^T$.

In order to transform GPC into a receding horizon LQ problem, we do not consider $\xi(k)$ here (since noise does not affect stability), and (4.3.1) is transformed into the following state-space model (the observable canonical and minimal realization model)

$$x(k+1) = Ax(k) + B\Delta u(k), \quad y(k) = Cx(k), \qquad (4.3.2)$$

where $x \in \mathfrak{R}^n$, $n = \max\{n_A, n_B\}$, $A = \begin{bmatrix} -\tilde{\alpha}^T & -\tilde{a}_n \\ I_{n-1} & 0 \end{bmatrix}$, $B = [1 \ 0 \ \cdots \ 0]^T$, $C = \begin{bmatrix} \tilde{b}_1 & \tilde{b}_2 & \cdots & \tilde{b}_n \end{bmatrix}$, I_{n-1} a $n-1$-ordered identity matrix, $\tilde{\alpha}^T = \begin{bmatrix} \tilde{a}_1 & \tilde{a}_2 & \cdots & \tilde{a}_{n-1} \end{bmatrix}$. When $i > n_A$, $\tilde{a}_i = 0$; when $i > n_B$, $\tilde{b}_i = 0$; when $n_A < n_B$, A is singular.

Since stability is considered, assuming $\omega = 0$ is without loss of generality. Take

$$Q_i = \begin{cases} C^T C, & N_1 \leq i \leq N_2 \\ 0, & i < N_1 \end{cases}, \quad \lambda_j = \begin{cases} \lambda, & 1 \leq j \leq N_u \\ \infty, & j > N_u \end{cases}.$$

Then (4.1.13) can be equivalently transformed into the objective function of LQ problem:

$$J(k) = x(k+N_2)^T C^T C x(k+N_2) + \sum_{i=0}^{N_2-1} \left[x(k+i)^T Q_i x(k+i) + \lambda_{i+1} \Delta u(k+i)^2 \right].$$

$$(4.3.3)$$

By the standard solution of LQ problem, the control law can be obtained as

$$\Delta u(k) = -\left(\lambda + B^T P_1 B \right)^{-1} B^T P_1 A x(k). \tag{4.3.4}$$

This is the control law obtained by taking GPC as LQ problem, and is called GPC's LQ control law, where P_1 can be obtained by Riccati iteration formula:

$$P_i = Q_i + A^T P_{i+1} A - A^T P_{i+1} B \left(\lambda_{i+1} + B^T P_{i+1} B \right)^{-1} B^T P_{i+1} A,$$
$$i = N_2 - 1, \dots, 2, 1, \quad P_{N_2} = C^T C. \tag{4.3.5}$$

Stability equivalence between the control law (4.3.4) and GPC's routine control law (4.1.18) is referred to [41].

Lemma 4.3.1. *(Special case of Riccati iteration) When* $\lambda_{j+1} = \lambda$, $Q_j = 0$, $1 \leq j \leq i$, *applying (4.3.5) yields*

$$P_1 = \left(A^T \right)^i \left\{ P_{i+1} - P_{i+1} J_{i+1} \left(J_{i+1}^T P_{i+1} J_{i+1} + \lambda I \right)^{-1} J_{i+1}^T P_{i+1} \right\} A^i,$$

$$(4.3.6)$$

where $J_{i+1} = \begin{bmatrix} B & AB & \cdots & A^{i-1}B \end{bmatrix}$.

Proof. (By induction) See [8]. □

4.3.2 Tool for stability proof: Kleinman's controller

Kleinman *et al.* have pointed out that, for systems represented by n-dimensional state space equation

$$x(k+1) = Ax(k) + B\Delta u(k), \tag{4.3.7}$$

the control law (called Kleinman's controller)

$$\Delta u(k) = -\gamma^{-1} B^T \left(A^T \right)^N \left[\sum_{h=m}^{N} A^h B \gamma^{-1} B^T \left(A^T \right)^h \right]^{-1} A^{N+1} x(k), \tag{4.3.8}$$

where $\gamma > 0$, has the following stability properties (see [36], [42], [64]).

Lemma 4.3.2. *(Stability of Kleinman's controller) If the system is completely controllable and A is nonsingular, then the control law (4.3.8) stabilizes the system (4.3.7) iff $N - m \geq n-1$, and is a deadbeat controller iff $N - m = n-1$.*

Lemma 4.3.3. *(Stability of Kleinman's controller) If the system is completely controllable and A is singular, then the control law (4.3.8) stabilizes the system (4.3.7) iff $N \geq n-1$, $m = 0$, and is a deadbeat controller iff $N = n-1$, $m = 0$.*

The above two lemmas have close relation with the notion of controllability and, hence, their proofs are omitted. Kleinman's controller was proposed early in the 1970s and, in the 1990s, it was included as a special case of synthesis approach.

In the following, we deduce extended Kleiman's controller for singular systems. For this reason, we make a nonsingular transformation for (4.3.2) (including system (4.3.7)) which leads to

$$\bar{x}(k+1) = \bar{A}\bar{x}(k) + \bar{B}\Delta u(k), \ y(k) = \bar{C}\bar{x}(k),$$

where $\bar{A} = \begin{bmatrix} A_0 & 0 \\ 0 & A_1 \end{bmatrix}$, $\bar{B} = \begin{bmatrix} B_0 \\ B_1 \end{bmatrix}$, $\bar{C} = [C_0, \ C_1]$, A_0 nonsingular,

$A_1 = \begin{bmatrix} 0 & 0 \\ I_{p-1} & 0 \end{bmatrix} \in \Re^{p \times p}$, $C_1 = \begin{bmatrix} 0 & \cdots & 0 & 1 \end{bmatrix}$, p the number of zero eigenvalues in A, I_{p-1} a $p-1$-ordered identity matrix.

For the above transformed system, Kleinman's controller is designed only for the subsystem $\{A_0, B_0\}$, and the following control law is constructed:

$$\Delta u(k) = \left[-\gamma^{-1}B_0^T \left(A_0^T\right)^N \left[\sum_{h=m}^{N} A_0^h B_0 \gamma^{-1} B_0^T \left(A_0^T\right)^h\right]^{-1} A_0^{N+1} \quad 0 \right] \bar{x}(k),$$
(4.3.9)

which is called extended Kleinman's controller. Substitute (4.3.9) into $\bar{x}(k+1) = \bar{A}\bar{x}(k) + \bar{B}\Delta u(k)$, and note that (A_0, B_0) is completely controllable if (A, B) is completely controllable. Then the following conclusion can be obtained by virtue of Lemma 4.3.2.

Theorem 4.3.1. *(Stability of extended Kleinman's controller) If the system is completely controllable and A is singular, then the control law (4.3.9) stabilizes the systems (4.3.7) iff $N - m \geq n - p - 1$, and is a deadbeat controller iff $N - m = n - p - 1$.*

Based on Lemmas 4.3.1-4.3.3 and Theorem 4.3.1, we will discuss closed-loop stability of GPC under sufficiently small λ for the following four cases:

(A) $n_A \geq n_B$, $N_1 \geq N_u$,

(B) $n_A \leq n_B$, $N_1 \leq N_u$,

(C) $n_A \leq n_B$, $N_1 \geq N_u$,

(D) $n_A \geq n_B$, $N_1 \leq N_u$.

Applying (4.3.6), under certain conditions the control law (4.3.4) can be transformed into the form

$$\Delta u(k) = -B^T \left(A^T\right)^N \left[\lambda P_{N+1}^{-1} + \sum_{h=0}^{N} A^h B B^T \left(A^T\right)^h\right]^{-1} A^{N+1} x(k). \quad (4.3.10)$$

Then, when λP_{N+1}^{-1} tends to be sufficiently small, stability conclusion of GPC can be obtained by virtue of that for Kleinman's controller (4.3.8). Moreover, for singular A, if A, B, C are substituted by $\bar{A}$, $\bar{B}$, $\bar{C}$ in Riccati iteration, then under certain conditions the control law (4.3.4) can be transformed into the following form, by applying (4.3.6),

$$\Delta u(k) = \left[\, -B_0^T \left(A_0^T\right)^N \left[\lambda P_{0,N+1}^{-1} + \sum_{h=0}^{N} A_0^h B_0 B_0^T \left(A_0^T\right)^h\right]^{-1} A_0^{N+1} \quad 0\,\right] \bar{x}(k)$$

$$(4.3.11)$$

where the result of Riccati iteration is $P_{N+1} = \begin{bmatrix} P_{0,N+1} & 0 \\ 0 & 0 \end{bmatrix}$ with $P_{0,N+1}$ a matrix having the same dimension with A_0. Then, when $\lambda P_{0,N+1}^{-1}$ tends to be sufficiently small, stability conclusion of GPC can be obtained by virtue of that for extended Kleinman's controller (4.3.9).

Concretely speaking, by adding corresponding conditions and applying (4.3.6) and matrix inversion formula, we can transform (4.3.4) into the form of (4.3.10) under cases (A), (B) and (D), and into the form (4.3.11) under case (C). During the deduction, we will also demonstrate that P_{N+1} (or $P_{0,N+1}$) is nonsingular for both $\lambda = 0$ and $\lambda > 0$. Thus, when λ tends to be sufficiently small, so does λP_{N+1}^{-1} (or $\lambda P_{0,N+1}^{-1}$), and (4.3.10) or (4.3.11) can be sufficiently close to Kleinman's controller (4.3.8) or (4.3.9).

4.3.3 GPC law resembling Kleinman's controller

Lemma 4.3.4. *When $n_A \geq n_B$, choose $N_1 \geq N_u$, $N_2 - N_1 \geq n - 1$. Then,*

(i) for $\lambda \geq 0$, P_{N_u} is nonsingular;

(ii) for $\lambda > 0$, GPC control law (4.3.4) can be transformed into

$$\Delta u(k) = -B^T \left(A^T\right)^{N_u-1} \left[\lambda P_{N_u}^{-1} + \sum_{h=0}^{N_u-1} A^h B B^T \left(A^T\right)^h\right]^{-1} A^{N_u} x(k).$$

$$(4.3.12)$$

Proof. In feedback control law (4.3.4), P_1 can be obtained by Riccati iteration (4.3.5); we discuss this in stages.

(s1) When $N_1 \leq i < N_2$, iteration becomes $P_i = C^T C + A^T P_{i+1} A$. Since $N_2 - N_1 \geq n - 1$ and the system is observable, P_{N_1} is nonsingular.

(s2) When $N_u \leq i < N_1$, iteration becomes $P_i = A^T P_{i+1} A$. Since $n_A \geq n_B$, A, and consequently P_{N_u}, is nonsingular. Moreover, P_{N_u} is not affected by λ.

(s3) When $1 \leq i < N_u$, iteration becomes

$$P_i = A^T P_{i+1} A - A^T P_{i+1} B \left(\lambda + B^T P_{i+1} B \right)^{-1} B^T P_{i+1} A.$$

Utilizing (4.3.6) we obtain

$$P_1 = \left(A^T \right)^{N_u - 1} \left\{ P_{N_u} - P_{N_u} J_{N_u} \left(J_{N_u}^T P_{N_u} J_{N_u} + \lambda I \right)^{-1} J_{N_u}^T P_{N_u} \right\} A^{N_u - 1}. \tag{4.3.13}$$

Applying the matrix inversion formula (4.2.4) we obtain

$$P_1 = \left(A^T \right)^{N_u - 1} \left(P_{N_u}^{-1} + \lambda^{-1} J_{N_u} J_{N_u}^T \right)^{-1} A^{N_u - 1};$$

substituting P_1 into (4.3.4) and applying (4.2.4) once again we obtain (4.3.12). $\square$

Lemma 4.3.5. *When $n_A \leq n_B$, choose $N_u \geq N_1$, $N_2 - N_u \geq n - 1$. Then,*

(i) for $\lambda \geq 0$, P_{N_1} is nonsingular;

(ii) for $\lambda > 0$, GPC control law (4.3.4) can be transformed into

$$\Delta u(k) = -B^T \left(A^T \right)^{N_1 - 1} \left[\lambda P_{N_1}^{-1} + \sum_{h=0}^{N_1 - 1} A^h B B^T \left(A^T \right)^h \right]^{-1} A^{N_1} x(k). \tag{4.3.14}$$

Proof. The proof is similar to that of Lemma 4.3.4 and is omitted here. Details are in [8]. $\square$

Lemma 4.3.6. *When $n_A \leq n_B$, let $p = n_B - n_A$ and choose $N_1 \geq N_u$, $N_2 - N_1 \geq n - p - 1$, $N_2 - N_u \geq n - 1$. Then,*

(a)-(i) for $N_1 - N_u \geq p$ and $\lambda \geq 0$, $P_{N_u} = \begin{bmatrix} P_{0,N_u} & 0 \\ 0 & 0 \end{bmatrix}$ where $P_{0,N_u} \in \mathfrak{R}^{(n-p) \times (n-p)}$ is nonsingular;

(a)-(ii) for $N_1 - N_u \geq p$ and $\lambda > 0$, GPC control law (4.3.4) can be transformed into

$$\Delta u(k) = - \left[B_0^T \left(A_0^T \right)^{N_u - 1} \left[\lambda P_{0,N_u}^{-1} + \sum_{h=0}^{N_u - 1} A_0^h B_0 B_0^T \left(A_0^T \right)^h \right]^{-1} A_0^{N_u} 0 \right]$$
$$\times \bar{x}(k); \tag{4.3.15}$$

(b)-(i) for $N_1 - N_u < p$ and $\lambda \geq 0$, $P_{N_1-p} = \begin{bmatrix} P_{0,N_1-p} & 0 \\ 0 & 0 \end{bmatrix}$ where

$P_{0,N_1-p} \in \Re^{(n-p)\times(n-p)}$ is nonsingular;

(b)-(ii) for $N_1 - N_u < p$ and $\lambda > 0$, GPC control law (4.3.4) can be transformed into

$$\Delta u(k) = - \left[B_0^T \left(A_0^T \right)^{N_1-p-1} \left[\lambda P_{0,N_1-p}^{-1} + \sum_{h=0}^{N_1-p-1} A_0^h B_0 B_0^T \left(A_0^T \right)^h \right]^{-1} \right.$$
$$\left. \times A_0^{N_1-p} \right] \bar{x}(k). \tag{4.3.16}$$

Proof. Take the transformation as illustrated before Theorem 4.3.1. For the transformed system, P_1 can be obtained by Riccati iteration.

(a)-(s1) When $N_u \leq i < N_2$, since $N_2 - N_1 \geq n - p - 1$ and $N_1 - N_u \geq p$ (note that these two conditions induce $N_2 - N_u \geq n - 1$), applying the special form of $(\bar{A}, \bar{C})$ (for details refer to [8]) we can obtain $P_{N_u} = \begin{bmatrix} P_{0,N_u} & 0 \\ 0 & 0 \end{bmatrix}$, where $P_{0,N_u} \in \Re^{(n-p)\times(n-p)}$ is nonsingular and P_{0,N_u} is not affected by λ.

(a)-(s2) When $1 \leq i < N_u$, the iteration becomes

$$P_i = \bar{A}^T P_{i+1} \bar{A} - \bar{A}^T P_{i+1} \bar{B} \left(\lambda + \bar{B}^T P_{i+1} \bar{B} \right)^{-1} \bar{B}^T P_{i+1} \bar{A}. \tag{4.3.17}$$

It is not difficult to see that the iteration corresponds only to $\{A_0, B_0\}$. At the end of the iteration, $P_1 = \begin{bmatrix} P_{0,1} & 0 \\ 0 & 0 \end{bmatrix}$, hence, the control law (4.3.4) becomes

$$\Delta u(k) = - \left(\lambda + B_0^T P_{0,1} B_0 \right)^{-1} \left[B_0^T P_{0,1} A_0 \quad 0 \right] \bar{x}(k), \tag{4.3.18}$$

where $P_{0,1}$ can be obtained, by analogy to (4.3.13), as

$$P_{0,1} = \left(A_0^T \right)^{N_u-1} \left\{ P_{0,N_u} - P_{0,N_u} J_{0,N_u} \left(J_{0,N_u}^T P_{0,N_u} J_{0,N_u} + \lambda I \right)^{-1} J_{0,N_u}^T P_{0,N_u} \right\}$$
$$\times A_0^{N_u-1}. \tag{4.3.19}$$

The matrices in (4.3.19), except that they correspond to $\{A_0, B_0\}$, have the same meanings as those in the proof of Lemma 4.3.4. By (4.3.18), (4.3.19) and deduction similar to (s3) in Lemma 4.3.4 we can obtain (4.3.15).

(b) The proof is similar to (a) and is omitted here. $\qquad\square$

Lemma 4.3.7. *When $n_A \geq n_B$, let $q = n_A - n_B$, $N^0 = \min\{ q, N_u - N_1 \}$ and choose $N_u \geq N_1$, $N_2 - N_u \geq n - q - 1$, $N_2 - N_1 \geq n - 1$. Then,*

(i) for $\lambda \geq 0$, if P_{N_1} is taken as the initial value to calculate $P^*_{N_1+N^0}$ via

$$P^*_i = A^T P^*_{i+1} A - A^T P^*_{i+1} B \left(\lambda + B^T P^*_{i+1} B\right)^{-1} B^T P^*_{i+1} A,$$

$$P^*_{N_1} = P_{N_1}, \ i = N_1, N_1 + 1, \ldots, N_1 + N^0 - 1, \qquad (4.3.20)$$

then $P^*_{N_1+N^0}$ is nonsingular;

(ii) for $\lambda > 0$, GPC control law (4.3.4) can be transformed into

$$\Delta u(k) = - B \left(A^T\right)^{N_1+N^0-1} \left[\lambda P^{*-1}_{N_1+N^0} + \sum_{h=0}^{N_1+N^0-1} A^h BB^T \left(A^T\right)^h\right]^{-1}$$

$$\times A^{N_1+N^0} x(k). \qquad (4.3.21)$$

Proof. The Riccati iteration is again directed to the original system.

(s1) First we prove that, for $\lambda > 0$, $P^*_{N_1+N^0}$ is nonsingular.

When $N_u \leq i < N_2$, by analogy to Lemma 4.3.4 we obtain $\mathrm{rank} P_{N_u} = \min\{n, N_2 - N_u + 1\}$.

When $N_1 \leq i < N_u$, since the system is observable and $N_2 - N_1 \geq n - 1$, P_{N_1} is nonsingular.

When $N_1 \leq i < N_1 + N^0$, calculate $P^*_{N_1+N^0}$ by (4.3.20) with P_{N_1} as initial iteration value. Since $\mathrm{rank} P^*_{N_1+N^0} \geq \mathrm{rank} P_{N_1}$, $P^*_{N_1+N^0}$ is nonsingular.

(s2) Second we prove that, for $\lambda = 0$, $P^*_{N_1+N^0}$ is nonsingular.

When $N_u \leq i < N_2$, since $N_2 - N_u \geq n - q - 1$, by analogy to Lemma 4.3.4 we obtain $\mathrm{rank} P_{N_u} \geq n - q$.

When $N_1 \leq i < N_u$, since $N_2 - N_1 \geq n - 1$, applying the last q zero elements of C (for details refer to [8]) we obtain $P_{N_1} = \begin{bmatrix} P_{0,N_1} & 0 \\ 0 & 0 \end{bmatrix}$, where $P_{0,N_1} \in \Re^{(n-N^0)\times(n-N^0)}$ is nonsingular.

When $N_1 \leq i < N_1 + N^0$, calculate $P^*_{N_1+N^0}$ by (4.3.20) with P_{N_1} as the initial value, then $P^*_{N_1+N^0}$ is nonsingular (for details refer to [8]).

(s3) When $1 \leq i < N_1 + N^0$, (4.3.20) is applied for iteration while $P^*_{N_1+N^0}$ is taken as initial value, that is

$$P^*_i = A^T P^*_{i+1} A - A^T P^*_{i+1} B \left(\lambda + B^T P^*_{i+1} B\right)^{-1} B^T P^*_{i+1} A, \ i$$

$$= N_1 + N^0 - 1, \ldots 2, 1. \qquad (4.3.22)$$

The iterated matrices will satisfy $P^*_j = P_j$, $1 \leq j \leq N_1$. Then (4.3.6) is also applicable. By proof analogous to Lemma 4.3.4 we obtain (4.3.21). $\qquad \square$

4.3.4 Stability based on Kleinman's controller

Based on Lemmas 4.3.4-4.3.7, we will discuss how GPC's state feedback control law approaches Kleinman's controller or its extended form when λ tends to be sufficiently small.

Theorem 4.3.2. *There exists a sufficiently small λ_0 such that for any $0 < \lambda < \lambda_0$ the closed-loop system of GPC is stable if the following condition is satisfied:*

$$N_1 \geq n_B, \ N_u \geq n_A, \ N_2 - N_u \geq n_B - 1, \ N_2 - N_1 \geq n_A - 1. \quad (4.3.23)$$

Proof. Condition (4.3.23) is combined by the following four conditions:

(i) $n_A \geq n_B$, $N_1 \geq N_u \geq n_A$, $N_2 - N_1 \geq n_A - 1$;

(ii) $n_A \leq n_B$, $N_u \geq N_1 \geq n_B$, $N_2 - N_u \geq n_B - 1$;

(iii) $n_A \leq n_B$, $N_1 \geq N_u \geq n_A$, $N_1 \geq n_B$, $N_2 - N_1 \geq n_A - 1$, $N_2 - N_u \geq n_B - 1$;

(iv) $n_A \geq n_B$, $N_u \geq N_1 \geq n_B$, $N_u \geq n_A$, $N_2 - N_1 \geq n_A - 1$, $N_2 - N_u \geq n_B - 1$.

They correspond to the cases (A)-(D) mentioned in section 4.3.2, respectively, and will be discussed one by one in the following.

(A) According to Lemma 4.3.4, GPC control law (4.3.4) has the form (4.3.12)) as $n_A \geq n_B$, $N_1 \geq N_u$, $N_2 - N_1 \geq n_A - 1$, $\lambda > 0$. Furthermore, since P_{N_u} is nonsingular for $\lambda \geq 0$, when λ tends to be sufficiently small, (4.3.12) tends to Kleinman's controller

$$\Delta u(k) = -B^T \left(A^T\right)^{N_u-1} \left[\sum_{h=0}^{N_u-1} A^h B B^T \left(A^T\right)^h\right]^{-1} A^{N_u} x(k). \quad (4.3.24)$$

Thus, by Lemma 4.3.2, the closed-loop system is stable when $N_u - 1 \geq n_A - 1$, i.e., when $N_u \geq n_A$. Combining these conditions yields condition (i).

(B) According to Lemma 4.3.5, GPC control law (4.3.4) has the form (4.3.14) as $n_A \leq n_B$, $N_u \geq N_1$, $N_2 - N_u \geq n_B - 1$, $\lambda > 0$. Furthermore, since P_{N_1} is nonsingular for $\lambda \geq 0$, when λ tends to be sufficiently small, (4.3.14) tends to Kleinman's controller

$$\Delta u(k) = -B^T \left(A^T\right)^{N_1-1} \left[\sum_{h=0}^{N_1-1} A^h B B^T \left(A^T\right)^h\right]^{-1} A^{N_1} x(k). \quad (4.3.25)$$

Thus, by Lemma 4.3.3, the closed-loop system is stable when $N_1 - 1 \geq n_B - 1$, i.e., when $N_1 \geq n_B$. Combining these conditions leads to condition (ii).

(C) According to Lemma 4.3.6, when $n_A \leq n_B$, $N_1 \geq N_u$, $N_2 - N_1 \geq n_A - 1$, $N_2 - N_u \geq n_B - 1$, $\lambda > 0$,

(a) if $N_1 - N_u \geq n_B - n_A$, GPC control law (4.3.4) has the form (4.3.15). Furthermore, since P_{0,N_u} is nonsingular for $\lambda \geq 0$, when λ tends to be sufficiently small, (4.3.15) tends to the extended Kleinman's controller

$$\Delta u(k) = - \left[\ B_0^T \left(A_0^T\right)^{N_u-1} \left[\sum_{h=0}^{N_u-1} A_0^h B_0 B_0^T \left(A_0^T\right)^h\right]^{-1} A_0^{N_u} \quad 0 \ \right] \bar{x}(k). \quad (4.3.26)$$

Thus, by Theorem 4.3.1, the closed-loop system is stable when $N_u - 1 \geq n_A - 1$, i.e., when $N_u \geq n_A$.

(b) if $N_1 - N_u < n_B - n_A$, similarly (4.3.16) tends to the extended Kleinman's controller

$$\Delta u(k) = - \left[B_0^T \left(A_0^T \right)^{N_1-p-1} \left[\sum_{h=0}^{N_1-p-1} A_0^h B_0 B_0^T \left(A_0^T \right)^h \right]^{-1} A_0^{N_1-p} \quad 0 \right] \bar{x}(k).$$

$$(4.3.27)$$

Thus, by Theorem 4.3.1, the closed-loop system is stable when $N_1 - p - 1 \geq n_B - p - 1$, i.e., when $N_1 \geq n_B$. It is not difficult to verify that combining the conditions in both (a) and (b) yields condition (iii).

(D) According to Lemma 4.3.7, GPC control law (4.3.4) has the form (4.3.21) as $n_A \geq n_B$, $N_u \geq N_1$, $N_2 - N_u \geq n_B - 1$, $N_2 - N_1 \geq n_A - 1$, $\lambda > 0$. Furthermore, since $P^*_{N_1+N^0}$ is nonsingular for $\lambda \geq 0$, when λ tends to be sufficiently small, (4.3.21) tends to Kleinman's controller

$$\Delta u(k) = -B \left(A^T \right)^{N_1+N^0-1} \left[\sum_{h=0}^{N_1+N^0-1} A^h BB^T \left(A^T \right)^h \right]^{-1} A^{N_1+N^0} x(k).$$

$$(4.3.28)$$

Thus, by Lemma 4.3.2, the closed-loop system is stable when $N_1 + N^0 - 1 \geq n_A - 1$, i.e., when $\min \{ N_1 + n_A - n_B, N_u \} \geq n_A$. It is not difficult to verify that combining all these conditions gives condition (iv). $\square$

Furthermore, by applying the dead beat control properties in Lemmas 4.3.2, 4.3.3 and Theorem 4.3.1 and through deduction analogous to Theorem 4.3.2, the following dead beat property of GPC can be obtained.

Theorem 4.3.3. *Suppose $\xi(k) = 0$. GPC is a dead beat controller if either of the following two conditions is satisfied :*

(i) $\lambda = 0$, $N_u = n_A$, $N_1 \geq n_B$, $N_2 - N_1 \geq n_A - 1$;

(ii) $\lambda = 0$, $N_u \geq n_A$, $N_1 = n_B$, $N_2 - N_u \geq n_B - 1$.

Remark 4.3.1. Theorem 4.3.2 investigates closed-loop stability of GPC under sufficiently small λ, while Theorem 4.3.3 the deadbeat property of GPC under $\lambda = 0$. If we fix other conditions in Theorem 4.3.3 but $\lambda > 0$ be sufficiently small, then GPC does not have deadbeat property any more, but the closed-loop system is stable, that is, this results in a part of Theorem 4.3.2. However Theorem 4.3.2 cannot cover Theorem 4.3.3 by taking $\lambda = 0$ because, when $\lambda = 0$, the conditions in Theorem 4.3.2 cannot guarantee the solvability of GPC control law. Therefore, Theorem 4.3.3 can be taken as deadbeat conclusion deduced on the basis of Theorem 4.3.2, letting $\lambda = 0$ and considering solvability conditions. Theorem 4.3.3 considers the necessity of solvability and, hence, cannot be simply covered by Theorem 4.3.2. These two theorems establish the overall equivalence relationship between Kleinman's controller (including the extended) and GPC from closed-loop stability to the deadbeat property. Using Kleinman's controller to investigate stability and deadbeat property relates only to system order and not to concrete model coefficients.

4.4 Cases of multivariable systems and constrained systems

4.4.1 Multivariable GPC

For a system with m inputs and r outputs, GPC law is a generalization of the SISO case. Then, each possible pair of input and output must adopt the Diophantine equation as in (4.1.3) and definition as in (4.1.6). Suppose the order of the model and the horizons are the same as the SISO case. Based on (4.2.1), the output prediction can be given according to the following steps.

(i) *First step:* Suppose only the j-th input is changed and other inputs keep invariant. Then considering the i-th output yields

$$\vec{y}_{ij}(k|k) = G_{ij}\Delta\tilde{u}_j(k|k) + H_{ij}\Delta\overleftarrow{u}_j(k) + \sum_{l=1}^{r} F_{il}\overleftarrow{y}_l(k), \qquad (4.4.1)$$

where

$$\vec{y}_{ij}(k|k) = [\bar{y}_{ij}(k+N_1|k), \bar{y}_{ij}(k+N_1+1|k), \cdots, \bar{y}_{ij}(k+N_2|k)]^T,$$
$$\Delta\tilde{u}_j(k|k) = [\Delta u_j(k|k), \cdots, \Delta u_j(k+N_u-1|k)]^T,$$
$$\Delta\overleftarrow{u}_j(k) = [\Delta u_j(k-1), \Delta u_j(k-2), \cdots, \Delta u_j(k-n_b)]^T,$$
$$\overleftarrow{y}_l(k) = [y_l(k), y_l(k-1), \cdots, y_l(k-n_a)]^T,$$

$$G_{ij} = \begin{bmatrix} g_{ij,N_1} & g_{ij,N_1-1} & \cdots & g_{ij,N_1-N_u+1} \\ g_{ij,N_1+1} & g_{ij,N_1} & \cdots & g_{ij,N_1-N_u+2} \\ \vdots & \vdots & \ddots & \vdots \\ g_{ij,N_2} & g_{ij,N_2-1} & \cdots & g_{ij,N_2-N_u+1} \end{bmatrix}, \quad g_{ij,l} = 0, \ \forall l \leq 0,$$

$$H_{ij} = \begin{bmatrix} h_{ij,N_1,1} & h_{ij,N_1,2} & \cdots & h_{ij,N_1,nb} \\ h_{ij,N_1+1,1} & h_{ij,N_1+1,2} & \cdots & h_{ij,N_1+1,nb} \\ \vdots & \vdots & \ddots & \vdots \\ h_{ij,N_2,1} & h_{ij,N_2,2} & \cdots & h_{ij,N_2,nb} \end{bmatrix},$$

$$F_{il} = \begin{bmatrix} f_{il,N_1,0} & f_{il,N_1,1} & \cdots & f_{il,N_1,na} \\ f_{il,N_1+1,0} & f_{il,N_1+1,1} & \cdots & f_{il,N_1+1,na} \\ \vdots & \vdots & \ddots & \vdots \\ f_{il,N_2,0} & f_{il,N_2,1} & \cdots & f_{il,N_2,na} \end{bmatrix}.$$

(ii) *Second step:* Suppose all the control inputs may be changed. Then, considering the i-th output, applying the superposition principle yields

$$\vec{y}_i(k|k) = \sum_{j=1}^{m} G_{ij}\Delta \tilde{u}_j(k|k) + \sum_{j=1}^{m} H_{ij}\Delta \overleftarrow{u}_j(k) + \sum_{l=1}^{r} F_{il}\overleftarrow{y}_l(k), \qquad (4.4.2)$$

where

$$\vec{y}_i(k|k) = [\bar{y}_i(k+N_1|k), \bar{y}_i(k+N_1+1|k), \cdots, \bar{y}_i(k+N_2|k)]^T.$$

(iii) *Third step:* Considering all the inputs and outputs yields

$$Y(k|k) = \tilde{G}\Delta U(k|k) + \tilde{H}\Delta \overleftarrow{U}(k) + \tilde{F}\overleftarrow{Y}(k), \qquad (4.4.3)$$

where

$$Y(k|k) = [\vec{y}_1(k|k)^T, \vec{y}_2(k|k)^T, \cdots, \vec{y}_r(k|k)^T]^T,$$
$$\Delta U(k|k) = [\Delta \tilde{u}_1(k|k)^T, \Delta \tilde{u}_2(k|k)^T, \cdots, \Delta \tilde{u}_m(k|k)^T]^T,$$
$$\Delta \overleftarrow{U}(k) = [\Delta \overleftarrow{u}_1(k)^T, \Delta \overleftarrow{u}_2(k)^T, \cdots, \Delta \overleftarrow{u}_m(k)^T]^T,$$
$$\overleftarrow{Y}(k) = [\overleftarrow{y}_1(k)^T, \overleftarrow{y}_2(k)^T, \cdots, \overleftarrow{y}_r(k)^T]^T,$$

$$\tilde{G} = \begin{bmatrix} G_{11} & G_{12} & \cdots & G_{1m} \\ G_{21} & G_{22} & \cdots & G_{2m} \\ \vdots & \vdots & \ddots & \vdots \\ G_{r1} & G_{r2} & \cdots & G_{rm} \end{bmatrix},$$

$$\tilde{H} = \begin{bmatrix} H_{11} & H_{12} & \cdots & H_{1m} \\ H_{21} & H_{22} & \cdots & H_{2m} \\ \vdots & \vdots & \ddots & \vdots \\ H_{r1} & H_{r2} & \cdots & H_{rm} \end{bmatrix},$$

$$\tilde{F} = \begin{bmatrix} F_{11} & F_{12} & \cdots & H_{1r} \\ F_{21} & F_{22} & \cdots & F_{2r} \\ \vdots & \vdots & \ddots & \vdots \\ F_{r1} & F_{r2} & \cdots & F_{rr} \end{bmatrix}.$$

Remark 4.4.1. If, in (4.2.1), the elements g, h, f in G, H, F are all $r \times m$-dimensional matrices, then the closed-form solution to the unconstrained MIMO GPC can be written in the form of $\Delta u(k) = (\lambda I + G^T G)^{-1} G^T \left[\vec{\omega}(k) - H\Delta \overleftarrow{u}(k) - F\overleftarrow{y}(k) \right]$, rather than as that in section 4.4.1. However, if for different inputs (outputs), different control horizons (prediction horizons) are adopted, and/or different input/output models have different orders, then the expression in (4.4.3) will be more convenient.

Suppose the criterion for optimization of $\Delta U(k|k)$ is to minimize the following cost function:

$$J(k) = \|Y(k|k) - Y_s(k)\|_{\tilde{W}}^2 + \|\Delta U(k|k)\|_{\tilde{\Lambda}}^2 \qquad (4.4.4)$$

where $\tilde{W} \geq 0$ and $\tilde{\Lambda} \geq 0$ are symmetric matrices;

$$Y_s(k) = [\vec{y}_{1s}(k)^T, \vec{y}_{2s}(k)^T, \cdots, \vec{y}_{rs}(k)^T]^T,$$
$$\vec{y}_{is}(k) = [y_{is}(k+N_1), y_{is}(k+N_1+1), \cdots, y_{is}(k+N_2)]^T;$$

$y_{is}(k+l)$ is the setpoint value for the i-th output at future time $k+l$.

Then, when $\tilde{G}^T \tilde{W} \tilde{G} + \tilde{\Lambda}$ is nonsingular, minimization of (4.4.4) yields

$$\Delta U(k|k) = (\tilde{G}^T \tilde{W} \tilde{G} + \tilde{\Lambda})^{-1} \tilde{G}^T \tilde{W} [Y_s(k) - \tilde{H} \Delta \overleftarrow{U}(k) - \tilde{F} \overleftarrow{Y}(k)]. \quad (4.4.5)$$

At each time k, implement the following control move:

$$\Delta u(k) = D[Y_s(k) - \tilde{H} \Delta \overleftarrow{U}(k) - \tilde{F} \overleftarrow{Y}(k)], \quad (4.4.6)$$

where

$$D = L(\tilde{G}^T \tilde{W} \tilde{G} + \tilde{\Lambda})^{-1} \tilde{G}^T \tilde{W},$$

$$L = \begin{bmatrix} \theta & 0 & \cdots & 0 \\ 0 & \theta & \cdots & 0 \\ \vdots & \vdots & \ddots & \vdots \\ 0 & 0 & \cdots & \theta \end{bmatrix} \in \mathfrak{R}^{m \times mN_u}, \quad \theta = [\, 1 \quad 0 \quad \cdots \quad 0 \,] \in \mathfrak{R}^{N_u}.$$

A simple selection of $\tilde{W}$ and $\tilde{\Lambda}$ is

$$\tilde{W} = \mathrm{diag}\{W_1, W_2, \cdots, W_r\}, \quad \tilde{\Lambda} = \mathrm{diag}\{\Lambda_1, \Lambda_2, \cdots, \Lambda_m\},$$
$$W_i = \mathrm{diag}\{w_i(N_1), w_i(N_1+1), \cdots, w_i(N_2)\}, \quad i \in \{1, \ldots, r\},$$
$$\Lambda_j = \mathrm{diag}\{\lambda_j(1), \lambda_j(2), \cdots, \lambda_j(N_u)\}, \quad j \in \{1, \ldots, m\}.$$

Taking $\tilde{\Lambda} > 0$ guarantees nonsingularity of $\tilde{G}^T \tilde{W} \tilde{G} + \tilde{\Lambda}$.

4.4.2   Constraint handling

In the following, we discuss how to handle constraint in GPC (take MIMO system as an example).

1. *Output magnitude constraint* $y_{i,\min} \leq y_i(k+l|k) \leq y_{i,\max}$

At each optimization instant, the output prediction is (4.4.3). Hence, we can let the optimization problem satisfy the following constraint:

$$Y_{\min} \leq \tilde{G}\Delta U(k|k) + \tilde{H} \Delta \overleftarrow{U}(k) + \tilde{F} \overleftarrow{Y}(k) \leq Y_{\max}, \quad (4.4.7)$$

where

$$Y_{\min} = [\tilde{y}_{1,\min}^T, \tilde{y}_{2,\min}^T, \cdots, \tilde{y}_{r,\min}^T]^T,$$
$$\tilde{y}_{i,\min} = [y_{i,\min}, y_{i,\min}, \cdots, y_{i,\min}]^T \in \mathfrak{R}^{N_2-N_1+1},$$
$$Y_{\max} = [\tilde{y}_{1,\max}^T, \tilde{y}_{2,\max}^T, \cdots, \tilde{y}_{r,\max}^T]^T,$$
$$\tilde{y}_{i,\max} = [y_{i,\max}, y_{i,\max}, \cdots, y_{i,\max}]^T \in \mathfrak{R}^{N_2-N_1+1}.$$

2. *Input increment constraints* $\Delta u_{j,\min} \leq \Delta u_j(k+l|k) = u_j(k+l|k) - u_j(k+l-1|k) \leq \Delta u_{j,\max}$

We can let the optimization problem satisfy the following constraint:

$$\Delta U_{\min} \leq \Delta U(k|k) \leq \Delta U_{\max}, \tag{4.4.8}$$

where

$$
\begin{aligned}
\Delta U_{\min} &= [\Delta \tilde{u}_{1,\min}^T, \Delta \tilde{u}_{2,\min}^T, \cdots, \Delta \tilde{u}_{m,\min}^T]^T, \\
\Delta \tilde{u}_{j,\min} &= [\Delta u_{j,\min}, \Delta u_{j,\min}, \cdots, \Delta u_{j,\min}]^T \in \mathfrak{R}^{N_u}, \\
\Delta U_{\max} &= [\Delta \tilde{u}_{1,\max}^T, \Delta \tilde{u}_{2,\max}^T, \cdots, \Delta \tilde{u}_{m,\max}^T]^T, \\
\Delta \tilde{u}_{j,\max} &= [\Delta u_{j,\max}, \Delta u_{j,\max}, \cdots, \Delta u_{j,\max}]^T \in \mathfrak{R}^{N_u}.
\end{aligned}
$$

3. *Input magnitude constraint* $u_{j,\min} \leq u_j(k+l|k) \leq u_{j,\max}$

We can let the optimization problem satisfy the following constraint:

$$U_{\min} \leq B\Delta U(k|k) + \tilde{u}(k-1) \leq U_{\max}, \tag{4.4.9}$$

where

$$
\begin{aligned}
U_{\min} &= [\tilde{u}_{1,\min}^T, \tilde{u}_{2,\min}^T, \cdots, \tilde{u}_{m,\min}^T]^T, \\
\tilde{u}_{j,\min} &= [u_{j,\min}, u_{j,\min}, \cdots, u_{j,\min}]^T \in \mathfrak{R}^{N_u}, \\
U_{\max} &= [\tilde{u}_{1,\max}^T, \tilde{u}_{2,\max}^T, \cdots, \tilde{u}_{m,\max}^T]^T, \\
\tilde{u}_{j,\max} &= [u_{j,\max}, u_{j,\max}, \cdots, u_{j,\max}]^T \in \mathfrak{R}^{N_u}, \\
B &= \mathrm{diag}\{B_0, \cdots, B_0\} \ (m \text{ blocks}),
\end{aligned}
$$

$$
B_0 = \begin{bmatrix} 1 & 0 & \cdots & 0 \\ 1 & 1 & \ddots & \vdots \\ \vdots & \ddots & \ddots & 0 \\ 1 & \cdots & 1 & 1 \end{bmatrix} \in \mathfrak{R}^{N_u \times N_u},
$$

$$
\begin{aligned}
\tilde{u}(k-1) &= [\tilde{u}_1(k-1)^T, \tilde{u}_2(k-1)^T, \cdots, \tilde{u}_m(k-1)^T]^T, \\
\tilde{u}_j(k-1) &= [u_j(k-1), u_j(k-1), \cdots, u_j(k-1)]^T \in \mathfrak{R}^{N_u}.
\end{aligned}
$$

Equations (4.4.7)-(4.4.9) can be written in the uniform manner as $\tilde{C}\Delta U(k|k) \leq \tilde{c}$, where $\tilde{C}$ and $\tilde{c}$ are matrix and vector known at time k. GPC optimization problem considering these constraints can be written as

$$\min_{\Delta U(k|k)} J(k) = \|Y(k|k) - Y_s(k)\|_{\tilde{W}}^2 + \|\Delta U(k|k)\|_{\tilde{\Lambda}}^2, \ \text{s.t. } \tilde{C}\Delta U(k|k) \leq \tilde{c}. \tag{4.4.10}$$

Problem (4.4.10) is a quadratic optimization problem which is in the same form as DMC.

4.5 GPC with terminal equality constraint

GPC with terminal equality constraint is a special synthesis approach of MPC. The main results in this section will be restricted to SISO linear deterministic time-invariant unconstrained systems. The content of this section is independent. Readers can omit this section. However, some linear control techniques can be seen in this section.

Stability was not guaranteed in the routine GPC. This has been overcome since the 1990s with new versions of GPC. One idea is that stability of the closed-loop system could be guaranteed if in the last part of the prediction horizon the future outputs are constrained at the desired setpoint and the prediction horizon is properly selected. The obtained predictive control is the predictive control with terminal equality constraint, or SIORHC (stabilizing input/output receding horizon control; see [51]) or CRHPC (constrained receding horizon predictive control; see [7]).

Consider the model the same as in section 4.3. $\xi(k) = 0$. At sampling time k the objective function of GPC with terminal equality constraint is

$$J = \sum_{i=N_0}^{N_1-1} q_i y(k+i|k)^2 + \sum_{j=1}^{N_u} \lambda_j \Delta u^2(k+j-1|k), \qquad (4.5.1)$$

$$s.t. \quad y(k+l|k) = 0, \quad l \in \{N_1, \dots, N_2\}, \qquad (4.5.2)$$

$$\Delta u(k+l-1|k) = 0, \quad l \in \{N_u+1, \dots, N_2\} \qquad (4.5.3)$$

where $q_i \geq 0$ and $\lambda_j \geq 0$ are the weighting coefficients, N_0, N_1 and N_1, N_2 are the starting and end points of the prediction horizon and constraint horizon respectively, and N_u is the control horizon. Other notations: I_i is i-ordered identity matrix,

$$W_o = [C^T \ A^T C^T \ \cdots \ (A^T)^{n-1} C^T]^T,$$

$$W_i = [A^{i-1}B \ \cdots \ AB \ B],$$

$$\Delta U_i(k) = [\Delta u(k) \ \cdots \ \Delta u(k+i-1|k)]^T.$$

In deducing the deadbeat properties of GPC with terminal equality constraints, we apply the following procedure:

(a) Substitute $x(k+N_1|k)$ by $x(k)$ and $\Delta U_i(k)$, where $i = N_u$, n_A or N_1.

(b) Express (4.5.2) by $x(k+N_1|k)$, but if $N_1 < N_u$, then express (4.5.2) by $x(k+N_1|k)$ and $[\Delta u(k+N_1|k) \ \Delta u(k+N_1+1|k) \ \cdots \ \Delta u(k+N_u-1|k)]$.

(c) Solve $\Delta u(k)$ as Ackermann's formula for deadbeat control.

Lemma 4.5.1. *Consider the completely controllable single input system $x(k+1) = Ax(k) + Bu(k)$. By adopting the following controller (called Ackermann's formula):*

$$u(k) = -[0 \ \cdots \ 0 \ 1][B \ AB \ A^{n-1}B]^{-1} A^n x(k),$$

the closed-loop system is deadbeat stable.

Actually, Ackermann's formula has close relation with the notion of controllability and, hence, more details for this formula are omitted here.

Lemma 4.5.2. *Under the following conditions the closed-loop system of GPC with terminal equality constraint is deadbeat stable:*

$$n_A < n_B, \ N_u = n_A, \ N_1 \geq n_B, \ N_2 - N_1 \geq n_A - 1. \tag{4.5.4}$$

Proof. Firstly, since $N_1 > N_u$,

$$x(k + N_1|k) = A^{N_1} x(k) + A^{N_1 - N_u} W_{N_u} \Delta U_{N_u}(k). \tag{4.5.5}$$

Take a nonsingular linear transformation to (4.3.2), we can obtain

$$\bar{x}(k + 1) = \bar{A}\bar{x}(k) + \bar{B}\Delta u(k), \ y(k) = \bar{C}\bar{x}(k)$$

where $\bar{x} = [x_0^T \ x_1^T]^T$, $\bar{A} = \text{block-diag}\{A_0, \ A_1\}$, $\bar{B} = [B_0^T \ B_1^T]^T$ and $\bar{C} = [C_0 \ C_1]$, with $A_0 \in \Re^{n_A \times n_A}$ nonsingular, all the eigenvalues of A_1 zero. Denote $n_B = n_A + p$, then $A_1 \in \Re^{p \times p}$. Since $N_1 \geq n_B$ and $N_u = n_A$, $A_1^h = 0$ $\forall h \geq N_1 - N_u$. Then (4.5.5) becomes

$$\begin{bmatrix} x_0(k + N_1|k) \\ x_1(k + N_1|k) \end{bmatrix} = \begin{bmatrix} A_0^{N_1} & 0 \\ 0 & 0 \end{bmatrix} \begin{bmatrix} x_0(k) \\ x_1(k) \end{bmatrix} + \begin{bmatrix} A_0^{N_1 - 1} B_0 & \cdots & A_0^{N_1 - N_u} B_0 \\ 0 & \cdots & 0 \end{bmatrix}$$
$$\times \Delta U_{N_u}(k). \tag{4.5.6}$$

According to (4.5.6), $x_1(k+N_1|k) = 0$ is automatically satisfied. Therefore, considering deadbeat control of (4.3.2) is equivalent to considering deadbeat control of its subsystem $\{A_0, \ B_0, \ C_0\}$. Further, consider $N_2 - N_1 = n_A - 1$, then (4.5.2) becomes

$$\begin{bmatrix} C_0 & C_1 \\ C_0 A_0 & C_1 A_1 \\ \vdots & \vdots \\ C_0 A_0^{n_A - 1} & C_1 A_1^{n_A - 1} \end{bmatrix} \begin{bmatrix} x_0(k + N_1|k) \\ 0 \end{bmatrix} = 0. \tag{4.5.7}$$

Since $(A_0, \ C_0)$ is observable, imposing (4.5.7) is equivalent to letting $x_0(k + N_1|k) = 0$. Then (4.5.6) becomes

$$0 = A_0^{N_u} x_0(k) + W_{0, N_u} \Delta U_{N_u}(k) \tag{4.5.8}$$

where $W_{0,j} = [A_0^{j-1} B_0 \ \cdots \ A_0 B_0 \ B_0]$, $\forall j \geq 1$. By applying (4.5.8), the optimal control law of GPC with terminal equality constraint is given by:

$$\Delta u(k) = - \begin{bmatrix} 0 & \cdots & 0 & 1 \end{bmatrix} \begin{bmatrix} B_0 & A_0 B_0 & \cdots & A_0^{N_u - 1} B_0 \end{bmatrix}^{-1} A_0^{N_u} x_0(k). \tag{4.5.9}$$

Since $N_u = n_A$, (4.5.9) is Ackermann's formula for deadbeat control of $\{A_0, \ B_0, \ C_0\}$.   □

Lemma 4.5.3. *Under the following conditions the closed-loop system of GPC with terminal equality constraint is deadbeat stable:*

$$n_A < n_B, \quad N_u \geq n_A, \quad N_1 = n_B, \quad N_2 - N_u \geq n_B - 1. \qquad (4.5.10)$$

Proof. (a) $N_1 \geq N_u$. For $N_u = n_A$, the conclusion follows from Lemma 4.5.2. For $N_u > n_A$, take a nonsingular transformation the same as in Lemma 4.5.2, then

$$
\begin{bmatrix} x_0(k+N_1|k) \\ x_1(k+N_1|k) \end{bmatrix} = \begin{bmatrix} A_0^{N_1} & 0 \\ 0 & 0 \end{bmatrix} \begin{bmatrix} x_0(k) \\ x_1(k) \end{bmatrix}
$$
$$
+ \begin{bmatrix} A_0^{N_1-1} & \cdots & A_0^p B_0 & A_0^{p-1} B_0 & \cdots & A_0^{N_1-N_u} B_0 \\ 0 & \cdots & 0 & A_1^{p-1} B_1 & \cdots & A_1^{N_1-N_u} B_1 \end{bmatrix} \Delta U_{N_u}(k).
$$
$$(4.5.11)$$

Suppose $A_1 = \begin{bmatrix} 0 & I_{p-1} \\ 0 & 0 \end{bmatrix}$ and $B_1 = [0 \cdots 0 \; 1]^T$, then
$[A_1^{p-1} B_1 \quad \cdots \quad A_1^{N_1-N_u}$
$B_1] = \begin{bmatrix} I_{N_u-n_A} \\ 0 \end{bmatrix}$. Denote $x_1 = [x_2^T \; x_3^T]^T$, where $\dim x_2 = N_u - n_A$ and $\dim x_3 = N_1 - N_u$. According to (4.5.11), $x_3(k+N_1|k) = 0$ is automatically satisfied. Therefore, considering deadbeat control of (4.3.2) is equivalent to considering deadbeat control of its partial states $[x_0^T \; x_2^T]$. Further, suppose $C_1 = [c_{11} \; c_{12} \; \cdots \; c_{1p}]$, consider $N_2 - N_1 = N_u - 1$ (i.e., $N_1 = n_B$, $N_2 - N_u = n_B - 1$), then (4.5.2) becomes

$$
\begin{bmatrix}
C_0 & c_{11} & c_{12} & \cdots & c_{1N_u-n_A} & \cdots & \cdots & c_{1p} \\
C_0 A_0 & 0 & c_{11} & \cdots & c_{1N_u-n_A-1} & \cdots & \cdots & c_{1p-1} \\
\vdots & \vdots & \vdots & \ddots & \vdots & \ddots & \ddots & \vdots \\
C_0 A_0^{N_u-n_A-1} & 0 & 0 & \cdots & c_{11} & \cdots & \cdots & c_{1p-N_u+n_A+1} \\
C_0 A_0^{N_u-n_A} & 0 & 0 & \cdots & 0 & c_{11} & \cdots & \cdots \\
\vdots & \vdots & \vdots & \ddots & \vdots & \vdots & \ddots & \vdots \\
C_0 A_0^{N_u-1} & 0 & 0 & \cdots & 0 & 0 & \cdots & *
\end{bmatrix}
$$
$$
\times \begin{bmatrix} x_0(k+N_1|k) \\ x_2(k+N_1|k) \\ 0 \end{bmatrix} = 0. \qquad (4.5.12)
$$

Since (A_0, C_0) is observable and $c_{11} \neq 0$, $\begin{bmatrix} C_0 & c_{11} & \cdots & c_{1N_u-n_A} \\ \vdots & \vdots & \ddots & \vdots \\ C_0 A_0^{N_u-n_A-1} & 0 & \cdots & c_{11} \\ \vdots & \vdots & \ddots & \vdots \\ C_0 A_0^{N_u-1} & 0 & \cdots & 0 \end{bmatrix}$

is nonsingular. Therefore, imposing (4.5.12) is equivalent to let $[x_0^T(k +$

$N_1|k)$ $\qquad\qquad\qquad\qquad\qquad\qquad\qquad\qquad\qquad\qquad x_2^T$

$(k+N_1|k)] = 0$. According to (4.5.11), $[\Delta u(k+n_A|k) \quad \cdots \quad \Delta u(k+N_u-1|k)]^T = x_2(k + N_1|k) = 0$. Therefore, (4.5.11) becomes

$$0 = A_0^{n_A} x_0(k) + W_{0,n_A} \Delta U_{n_A}(k). \tag{4.5.13}$$

According to (4.5.13), the optimal control law is given by:

$$\Delta u(k) = -\begin{bmatrix} 0 & \cdots & 0 & 1 \end{bmatrix}$$
$$\times \begin{bmatrix} B_0 & A_0 B_0 & \cdots & A_0^{n_A-1} B_0 \end{bmatrix}^{-1} A_0^{n_A} x_0(k) \tag{4.5.14}$$

which is Ackermann's formula for deadbeat control of $\{A_0, B_0, C_0\}$. Therefore, (4.3.2) will be deadbeat stable.

(b) $N_1 < N_u$. Firstly,

$$x(k + N_1|k) = A^{N_1} x(k) + W_{N_1} \Delta U_{N_1}(k). \tag{4.5.15}$$

Since $N_1 = n$ and $N_2 - N_u \geq n - 1$, $N_2 - N_1 \geq n + N_u - N_1 - 1$. Consider $N_2 - N_1 = n + N_u - N_1 - 1$, then (4.5.2) becomes

$$\begin{bmatrix} C & 0 & \cdots & 0 \\ CA & CB & \cdots & 0 \\ \vdots & \vdots & \ddots & \vdots \\ CA^{N_u-N_1} & CA^{N_u-N_1-1}B & \cdots & CB \\ \vdots & \vdots & \ddots & \vdots \\ CA^{n+N_u-N_1-1} & CA^{n+N_u-N_1-2}B & \cdots & CA^{n-1}B \end{bmatrix}$$

$$\times \begin{bmatrix} x(k + N_1|k) \\ \Delta u(k + N_1|k) \\ \Delta u(k + N_1 + 1|k) \\ \vdots \\ \Delta u(k + N_u - 1|k) \end{bmatrix} = 0. \tag{4.5.16}$$

Substituting (4.5.15) into (4.5.16) obtains

$$\begin{bmatrix} W_o \\ CA^{N_1} \\ \vdots \\ CA^{N_u-1} \end{bmatrix} A^{N_1} x(k) + \begin{bmatrix} \begin{bmatrix} W_0 W_{N_1} \\ CA^{N_1} \\ \vdots \\ CA^{N_u-1} \end{bmatrix} W_{N_1} & \begin{matrix} G_1 \\ \\ G_2 \end{matrix} \end{bmatrix} \times \Delta U_{N_u}(k) = 0,$$

$$\tag{4.5.17}$$

where G_1 and G_2 are matrices of the corresponding parts in (4.5.16).

Denote J as

$$J = \sum_{i=N_0}^{N_1-1} q_i y(k + i|k)^2 + \sum_{j=1}^{N_1} \lambda_j \Delta u^2(k + j - 1|k) + \sum_{j=N_1+1}^{N_u} \lambda_j \Delta u^2(k + j - 1|k)$$

$$= J_1 + \sum_{j=N_1+1}^{N_u} \lambda_j \Delta u^2(k + j - 1|k) = J_1 + J_2. \tag{4.5.18}$$

According to the optimality principle,

$$\min J \geq \min J_1 + \min J_2 \geq \min J_1.$$

Hence, $[\Delta u(k + N_1|k), \cdots, \Delta u(k + N_u - 1|k)] = 0$ is the best choice for minimizing J. By this choice, (4.5.17) is simplified as $W_0 A^{N_1} x(k) + W_0 W_{N_1} \Delta U_{N_1}(k) = 0$. Hence, the optimal control law is given by:

$$\Delta u(k) = -\begin{bmatrix} 0 & \cdots & 0 & 1 \end{bmatrix} \begin{bmatrix} B & AB & \cdots & A^{N_1-1}B \end{bmatrix} A^{N_1} x(k). \quad (4.5.19)$$

Consider system (4.3.2), since $N_1 = n_B = n$, (4.5.19) is Ackermann's formula for deadbeat control. $\qquad\square$

Lemma 4.5.4. *Under the following conditions the closed-loop system of GPC with terminal equality constraint is deadbeat stable:*

$$n_A > n_B, \ N_u \geq n_A, \ N_1 = n_B, \ N_2 - N_u \geq n_B - 1. \quad (4.5.20)$$

Proof. (a) $N_u = n_A$. Firstly, since $N_1 < N_u$,

$$x(k + N_1|k) = A^{N_1} x(k) + W_{N_1} \Delta U_{N_1}(k). \quad (4.5.21)$$

Since $N_1 = n_B$ and $N_2 - N_u \geq n_B - 1$, $N_2 - N_1 \geq N_u - 1 = n - 1$. Consider $N_2 - N_1 = N_u - 1$, then (4.5.2) becomes

$$\begin{bmatrix} C \\ CA \\ \vdots \\ CA^{N_u-N_1} \\ \vdots \\ CA^{N_u-1} \end{bmatrix} x(k + N_1|k) + \begin{bmatrix} 0 & \cdots & 0 \\ CB & \cdots & 0 \\ \vdots & \ddots & \vdots \\ CA^{N_u-N_1-1}B & \cdots & CB \\ \vdots & \ddots & \vdots \\ CA^{N_u-2}B & \cdots & CA^{N_1-1}B \end{bmatrix}$$

$$\times \begin{bmatrix} \Delta u(k + N_1|k) \\ \Delta u(k + N_1 + 1|k) \\ \vdots \\ \Delta u(k + N_u - 1|k) \end{bmatrix} = 0. \quad (4.5.22)$$

Denote $q = n_A - n_B$. Since the last q elements of C are zeros, by the special forms of A and B, it is easy to conclude that $CA^{-h}B = 0, \forall h \in \{1, 2, \ldots, q\}$. Therefore, (4.5.22) can be re-expressed as

$$\begin{bmatrix} C \\ CA \\ \vdots \\ CA^{N_u-1} \end{bmatrix} x(k + N_1|k) + \begin{bmatrix} CA^{-1}B & \cdots & CA^{-N_u+N_1}B \\ CB & \cdots & CA^{-N_u+N_1+1}B \\ \vdots & \ddots & \vdots \\ CA^{N_u-2}B & \cdots & CA^{N_1-1}B \end{bmatrix}$$

$$\times \begin{bmatrix} \Delta u(k + N_1|k) \\ \Delta u(k + N_1 + 1|k) \\ \vdots \\ \Delta u(k + N_u - 1|k) \end{bmatrix} = 0. \quad (4.5.23)$$

According to Cayley-Hamilton's Theorem, for any integer j, $[CA^{N_u-1+j} \quad CA^{N_u-2+j}B \quad \cdots \quad CA^{N_1-1+j}B]$ can be represented as a linear combination of the rows in

$$
\begin{bmatrix}
C & CA^{-1}B & \cdots & CA^{-N_u+N_1}B \\
CA & CB & \cdots & CA^{-N_u+N_1+1}B \\
\vdots & \vdots & \ddots & \vdots \\
CA^{N_u-1} & CA^{N_u-2}B & \cdots & CA^{N_1-1}B
\end{bmatrix}.
$$

Therefore, (4.5.23) is equivalent to

$$
\begin{bmatrix}
CA^{N_u-N_1} & CA^{N_u-N_1-1}B & \cdots & CAB & CB \\
CA^{N_u-N_1+1} & CA^{N_u-N_1}B & \cdots & CA^2B & CAB \\
\vdots & \vdots & \ddots & \vdots & \vdots \\
CA^{n+N_u-N_1-1} & CA^{n+N_u-N_1-2}B & \cdots & CA^nB & CA^{n-1}B
\end{bmatrix}
$$

$$
\times
\begin{bmatrix}
x(k+N_1|k) \\
\Delta u(k+N_1|k) \\
\Delta u(k+N_1+1|k) \\
\vdots \\
\Delta u(k+N_u-1|k)
\end{bmatrix}
= 0. \tag{4.5.24}
$$

Substituting (4.5.21) into (4.5.24) obtains

$$
W_0 A^{N_u} x(k) + W_0 W_{N_u} \Delta U_{N_u}(k) = 0. \tag{4.5.25}
$$

Hence, the optimal control law is given by

$$
\Delta u(k) = -\begin{bmatrix} 0 & \cdots & 0 & 1 \end{bmatrix} \begin{bmatrix} B & AB & \cdots & A^{N_u-1}B \end{bmatrix} A^{N_u} x(k). \tag{4.5.26}
$$

Consider system (4.3.2), since $N_u = n_A = n$, (4.5.26) is Ackermann's formula for deadbeat control.

(b) $N_u > n_A$. For the same reasons as in Lemma 4.5.3 (b), it is best that $[\Delta u(k+n_A|k), \cdots, \Delta u(k+N_u-1|k)] = 0$. Hence, the same conclusion can be obtained.   □

Lemma 4.5.5. *Under the following conditions the closed-loop system of GPC with terminal equality constraint is deadbeat stable:*

$$
n_A > n_B, \ N_u = n_A, \ N_1 \geq n_B, \ N_2 - N_1 \geq n_A - 1. \tag{4.5.27}
$$

Proof. (a) $N_1 \geq N_u$. Firstly,

$$
x(k+N_1|k) = A^{N_1} x(k) + A^{N_1-N_u} W_{N_u} \Delta U_{N_u}(k). \tag{4.5.28}
$$

Similarly to Lemma 4.5.2, since (A, C) is observable, choosing $N_2 - N_1 \geq n_A - 1 = n - 1$ is equivalent to letting $x(k+N_1|k) = 0$. Then, because A is nonsingular, the optimal control law is given by:

$$\Delta u(k) = - \begin{bmatrix} 0 & \cdots & 0 & 1 \end{bmatrix} \begin{bmatrix} B & AB & \cdots & A^{N_u-1}B \end{bmatrix} A^{N_u} x(k).$$
$$(4.5.29)$$

Consider system (4.3.2), since $N_u = n_A = n$, (4.5.29) is Ackermann's formula for deadbeat control.

(b) $N_1 < N_u$. For $N_1 = n_B$, the conclusion follows from Lemma 4.5.4 (a). For $N_1 > n_B$, by invoking the similar reason and deduction, (4.5.2) is equivalent to (4.5.24) and the conclusion holds. ☐

Moreover, compared with Lemmas 4.5.2-4.5.5, it is easier to prove that, under either of the following two conditions the closed-loop system of GPC with terminal equality constraints is deadbeat stable:

(i) $n_A = n_B$, $N_u = n_A$, $N_1 \geq n_B$, $N_2 - N_1 \geq n_A - 1$;

(ii) $n_A = n_B$, $N_u \geq n_A$, $N_1 = n_B$, $N_2 - N_u \geq n_B - 1$.

Combining the above results we obtain the following conclusion:

Theorem 4.5.1. *Under either of the following two conditions the closed-loop system of GPC with terminal equality constraint is deadbeat stable:*

$$\begin{aligned} &(i) \quad N_u = n_A, \ N_1 \geq n_B, \ N_2 - N_1 \geq n_A - 1; \\ &(ii) \quad N_u \geq n_A, \ N_1 = n_B, \ N_2 - N_u \geq n_B - 1. \end{aligned} \qquad (4.5.30)$$

Remark 4.5.1. Consider the objective function of routine GPC (the same as in section 4.3). The deadbeat condition of routine GPC with $\lambda = 0$ is the same as (4.5.30). With deadbeat control, the output of system (4.3.1) (where $\xi(k) = 0$) will reach the setpoint in n_B samples by changing input n_A times. This is the quickest response that system (4.3.1) can achieve. Also, at this speed it is the unique response (for any initial state). Therefore, GPC with terminal equality constraint and routine GPC are equivalent under $\lambda = 0$ and (4.5.30).

Remark 4.5.2. For GPC with terminal equality constraint, by choosing the parameters to satisfy

$$N_u \geq N_A, \ N_1 \geq n_B, \ N_2 - N_u \geq n_B - 1, \ N_2 - N_1 \geq n_A - 1 \quad (4.5.31)$$

and properly selecting other controller parameters, the optimization problem has a unique solution and the closed-loop system is asymptotically stable. However, if the parameters of the input/output model are on-line identified, or there is model-plant mismatch, then stability should be re-considered.

Remark 4.5.3. With input/output constraint considered, under the condition of Theorem 4.5.1, if the initial optimization is feasible, then the closed-loop system is deadbeat stable and the real control moves are the same as the unconstrained case. This is due to the uniqueness of deadbeat control. In this situation, the hard constraints are inactive. On the other hand, if the hard constraints are active, then the controller parameters cannot be selected according to Theorem 4.5.1.

Remark 4.5.4. Systems with hard constraints can be controlled by GPC with terminal equality constraint. Select the parameters to satisfy

$$N_u = N_A, \ N_1 = n_B, \ N_2 - N_u = n_B - 1, \ N_2 - N_1 = n_A - 1 \quad (4.5.32)$$

and find the solution. If the optimization problem is infeasible, then increase N_1, N_2, N_u by 1, until the optimization becomes feasible. If the optimization problem is feasible, then implement the current control move, and decrease N_1, N_2, N_u by 1, but stop decreasing when (4.5.32) is satisfied. The final closed-loop response will be deadbeat.

Remark 4.5.5. The deadbeat property of GPC can be directly obtained by that of the SIORHC (CRHPC), rather than applying Kleinman's controller. Since by parameterizing as in (4.5.30) SIORHC (CRHPC) is feasible, if GPC is parameterized as (4.5.30) and $\lambda = 0$ is chosen, then the minimum cost value of GPC is $J^*(k) = 0$. $J^*(k) = 0$ implies that the closed-loop system is deadbeat stable.

Remark 4.5.6. Remark 4.5.1 shows that, when deadbeat controller is applied, the output $y(k)$ of $\tilde{A}(z^{-1})y(k) = \tilde{B}(z^{-1})\Delta u(k)$ will reach the setpoint value in n_B sampling instants, while the input $u(k)$ only needs to change n_A times. This is the inherent property of $\tilde{A}(z^{-1})y(k) = \tilde{B}(z^{-1})\Delta u(k)$, which is not limited to MPC. Therefore, the deadbeat properties of SIORHC (CRHPC) can be directly obtained by invoking this inherent property, rather than by applying various forms of "Ackermann's formula for deadbeat control." Certainly, the deductions in section 4.5 simultaneously give the control law of SIORHC (CRHPC).

Remark 4.5.7. Deadbeat stability of GPC not relating with the concrete modeling coefficients, as well as the deadbeat property of SIORHC (CRHPC), is very limited. We have not generalized the results to multivariable and uncertain systems. Some results for applying Kleinman's controller in the multivariable GPC are referred to [15] where, however, the state-space model is directly applied.

Chapter 5

Two-step model predictive control

Here, two-step control applies to a class of special systems, i.e., systems with input nonlinearities. Input nonlinearities include input saturation, dead zone, etc. Moreover, a system represented by the Hammerstein model is an often seen input nonlinear system. The Hammerstein model consists of a static nonlinear part followed by a dynamic linear part; see [54]. Some nonlinear processes such as pH neutralization, high purity distillation, etc., can be represented by the Hammerstein model. The predictive control strategies for input nonlinear systems (mainly referred to input saturation and Hammerstein nonlinearity) can be classified into two categories.

One category takes the nonlinear model as a whole (overall category, e.g., [1]), incorporates the nonlinear part into the objective function and directly solves the control moves. Note that input saturation is usually taken as the constraint in the optimization. For this category, the control law calculation is rather complex and it is more difficult for real application.

The other category utilizes nonlinear separation technique (separation category, e.g., [30]), i.e., firstly calculates the intermediate variable utilizing the linear sub-model and predictive control, then computes the actual control move via nonlinear inversion. Note that, input saturation can be regarded as a kind of input nonlinearity, and also can be regarded as a constraint in the optimization. The nonlinear separation with respect to the Hammerstein model invokes the special structure of the Hammerstein model, and groups the controller designing problem into the linear control, which is much simpler than the overall category. In many control problems, the target is to make the system output track the setpoint as soon as possible (or, drive the systems state to the origin as soon as possible); the weighting on the control move is to restrict the magnitude or the variation of the control move. Hence, although the control moves are not directly incorporated into the optimization, the separation strategy is often more practical then the overall strategy.

Two-step model predictive control (TSMPC): For the Hammerstein model with input saturation, first utilize the linear sub-model and unconstrained MPC algorithm to compute the desired intermediate variable, then solve the nonlinear algebraic equation (group) (represented by Hammerstein nonlinearity) to obtain the control action, and utilize desaturation to satisfy the input saturation constraint. Since the computational time can be greatly reduced, TSMPC is very suitable for the fast control requirement, especially for the actual system where the model is on-line identified. When the linear part adopts GPC, we call it TSGPC. The reserved nonlinear item in the closed-loop of TSMPC is static.

In TSMPC, if the intermediate variable is exactly implemented by the actual control moves passing the nonlinear part of the system, then stability of the whole system is guaranteed by stability of the linear subsystem. However, in a real application, this ideal case is hard to ensure. The control move may saturate, and the solution of the nonlinear algebraic equation (group) unavoidably introduces solution error.

This chapter is mainly referred to in [8]. For sections 5.1 and 5.2, also refer to in [14], [27]. For section 5.3 also refer to in [19]. For sections 5.4, 5.5 and 5.6 also refer to in [23], [20]. For sections 5.7, 5.8 and 5.9 also refer to in [24], [26].

5.1   Two-step GPC

The Hammerstein model is composed of a static nonlinear model followed by a dynamic linear sub-model. The static nonlinearity is

$$v(k) = f\left(u(k)\right),\ f(0) = 0, \tag{5.1.1}$$

where u is the input, v the intermediate variable; in the literature, f is usually called invertible nonlinearity. Linear part adopts the controlled auto-regressive moving average (CARMA) model,

$$a(z^{-1})y(k) = b(z^{-1})v(k-1), \tag{5.1.2}$$

where y is the output, $a_{na} \neq 0$, $b_{nb} \neq 0$; $\{a, b\}$ is irreducible. Other details are referred to Chapter 4 (u is revised as v).

5.1.1   Case unconstrained systems

First, utilize (5.1.2) for designing the linear generalized predictive control (LGPC) such that the desired $v(k)$ is obtained. Adopt the following cost function:

$$J(k) = \sum_{i=N_1}^{N_2} [y(k+i|k) - y_s(k+i)]^2 + \sum_{j=1}^{N_u} \lambda \Delta v^2(k+j-1|k). \tag{5.1.3}$$

In this chapter, usually $y_s(k+i) = \omega$, $\forall i > 0$. Thus, the control law of LGPC is

$$\Delta v(k) = d^T(\vec{\omega} - \overleftarrow{f})$$ (5.1.4)

where $\vec{\omega} = [\omega, \omega, \cdots, \omega]^T$ and $\overleftarrow{f}$ is a vector composed of the past intermediate variable and output and the current output. Details are referred to Chapter 4 (with u revised as v).

Then, use

$$v^L(k) = v^L(k-1) + \Delta v(k)$$ (5.1.5)

to calculate $u(k)$ which is applied to the real plant, i.e., solve the following equation:

$$f(u(k)) - v^L(k) = 0,$$ (5.1.6)

with the solution denoted as

$$u(k) = g\left(v^L(k)\right).$$ (5.1.7)

When the above method was firstly proposed, it was called nonlinear GPC (NLGPC; see [65]).

5.1.2   Case with input saturation constraint

The input saturation constraint is usually inevitable in the real applications. Now, suppose the control move is restricted by the saturation constraint $|u| \le U$, where U is a positive scalar. After $\Delta v(k)$ is obtained by applying (5.1.4), solve the equation

$$f(\hat{u}(k)) - v^L(k) = 0$$ (5.1.8)

to decide $\hat{u}(k)$, with the solution denoted as

$$\hat{u}(k) = \hat{f}^{-1}(v^L(k)).$$ (5.1.9)

Then, the desaturation is invoked to obtain the actual control move $u(k) = \mathrm{sat}\{\hat{u}(k)\}$, where $\mathrm{sat}\{s\} = \mathrm{sign}\{s\}\min\{|s|, U\}$, denoted as (5.1.7).

The above control strategy is called type-I two-step GPC (TSGPC-I).

In order to handle the input saturation, one can also transform the input saturation constraint to the constraint on the intermediate variable. Then, another TSGPC strategy is obtained. Firstly, use the constraint on u, i.e., $|u| \le U$, to determine the constraint on v, i.e., $v_{\min} \le v \le v_{\max}$. After $\Delta v(k)$ is obtained by applying (5.1.4), let

$$\hat{v}(k) = \begin{cases} v_{\min}, & v^L(k) \le v_{\min} \\ v^L(k), & v_{\min} < v^L(k) < v_{\max} \\ v_{\max}, & v^L(k) \ge v_{\max} \end{cases}.$$ (5.1.10)

Then, solve the nonlinear algebraic equation

$$f(u(k)) - \hat{v}(k) = 0$$ (5.1.11)

and let the solution $u(k)$ satisfy saturation constraint, denoted as

$$u(k) = \hat{g}\left(\hat{v}(k)\right) \tag{5.1.12}$$

which can also be denoted as (5.1.7). This control strategy is called type-II TSGPC (TSGPC-II).

Remark 5.1.1. After the constraint on the intermediate variable is obtained, we can design another type of nonlinear separation GPC, called NSGPC. In NSGPC, solving $\Delta v(k)$ no longer adopts (5.1.4), but is through the following optimization problem:

$$\min_{\Delta v(k|k),\cdots,\Delta v(k+N_u-1|k)} J(k) = \sum_{i=N_1}^{N_2} [y(k+i|k) - y_s(k+i)]^2$$

$$+ \sum_{j=1}^{N_u} \lambda \Delta v^2(k+j-1|k), \tag{5.1.13}$$

s.t. $\Delta v(k+l|k) = 0,\ l \geq N_u,\ v_{\min} \leq v(k+j-1|k) \leq v_{\max},\ j \in \{1,\ldots,N_u\}$.
$$\tag{5.1.14}$$

Other computation and notations are the same as NLGPC. By this, we can easily find the difference between TSGPC and NSGPC.

As addressed above, NLGPC, TSGPC-I and TSGPC-II are all called as TSGPC. Now, suppose the real plant is "static nonlinearity + dynamic linear model" and the nonlinear sub-model is $v(k) = f_0\left(u(k)\right)$. We call the process determining $u(k)$ via $v^L(k)$ as nonlinear inversion. An ideal inversion will achieve $f_0 \circ g = 1$, i.e.,

$$v(k) = f_0(g(v^L(k))) = v^L(k). \tag{5.1.15}$$

If $f_0 \neq f$ or $f \neq g^{-1}$, then it is difficult to achieve $f_0 = g^{-1}$. In fact, it is usually impossible to achieve $f_0 = g^{-1}$. If there is no input saturation, theoretically, finding $u(k)$ via $v^L(k)$ is determined by the magnitude of $v^L(k)$ and the formulation of f. It is well-known that, even for the monotonic function $v = f\left(u\right)$, its inversion function $u = f^{-1}\left(v\right)$ does not necessarily exist for all the possible values of v. In the real applications, because of computational time and computational accuracy, the algebraic equation may be unable to be exactly solved. Hence, in general, the approximate solution to the algebraic equation is adopted. When there is input saturation, the effect of desaturation may incur $v(k) \neq v^L(k)$. In summary, due to the inaccuracy in equation solving, desaturation and modeling error, etc., the $v^L(k)$ obtained through the linear model may be unable to be implemented, and what is implemented is the $v(k)$.

The structure of TSGPC is shown in Figure 5.1.1. When $f_0 = g = 1$, Figure 5.1.1 is the block diagram of LGPC (see Chapter 4). The internal model

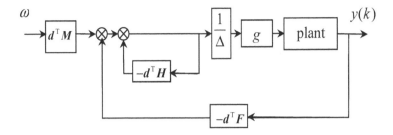

Figure 5.1.1: The original block diagram of TSGPC.

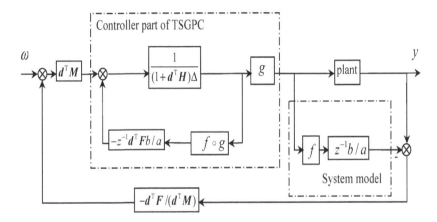

Figure 5.1.2: The internal model control structure of TSGPC.

control structure of Figure 5.1.1 is shown in Figure 5.1.2. When $f_0 = g = 1$, Figure 5.1.2 is the internal model control structure of LGPC (see Chapter 4). In the following section, we will analyze closed-loop stability of TSGPC when $f_0 \neq g^{-1}$.

5.2 Stability of two-step GPC

Since the reserved nonlinear item in the closed-loop system is $f_0 \circ g$, the inaccuracy in the nonlinear sub-model and the nonlinearity of the real actuator can also be incorporated into $f_0 \circ g$. Hence, stability results of TSGPC are the robustness results.

5.2.1 Results based on Popov's Theorem

Lemma 5.2.1. *(Popov's stability Theorem) Suppose $G(z)$ in the Figure 5.2.1 is stable and $0 \leq \varphi(\vartheta)\vartheta \leq K_\varphi \vartheta$. Then the closed-loop system is stable if $\frac{1}{K_\varphi} + Re\{G(z)\} > 0, \ \forall |z| = 1$.*

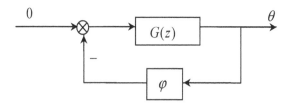

Figure 5.2.1: The static nonlinear feedback form, 1.

Re$\{\cdot\}$ refers to the real part of a complex and $|z|$ is mode of the complex z. Applying Lemma 5.2.1, we can obtain the following stability result of TSGPC.

Theorem 5.2.1. *(TSGPC's stability) Suppose the linear sub-model applied by TSGPC is accurate and there exist two constants $k_1, k_2 > 0$ such that*

(i) *the roots of $a(1 + d^T H)\Delta + (1 + k_1)z^{-1}d^T Fb = 0$ are all located in the unit circle;*

(ii)

$$\frac{1}{k_2 - k_1} + Re\left\{\frac{z^{-1}d^T Fb}{a(1 + d^T H)\Delta + (1 + k_1)z^{-1}d^T Fb}\right\} > 0, \ \forall |z| = 1. \tag{5.2.1}$$

Then the closed-loop system of TSGPC is stable if the following is satisfied:

$$k_1\vartheta^2 \leq (f_0 \circ g - 1)(\vartheta)\vartheta \leq k_2\vartheta^2. \tag{5.2.2}$$

Proof. Suppose, without loss of generality, $\omega = 0$. Transform Figure 5.1.1 into Figures 5.2.2, 5.2.3 and 5.2.4. If the system shown in Figure 5.2.4 is stable, then the original system is stable. Suppose the feedback item $f_0 \circ g - 1$ in Figure 5.2.4 satisfies (5.2.2). For utilizing Popov's stability Theorem, take

$$0 \leq \psi(\vartheta)\vartheta \leq (k_2 - k_1)\vartheta^2 = K_\psi\vartheta^2 \tag{5.2.3}$$

where

$$\psi(\vartheta) = (f_0 \circ g - 1 - k_1)(\vartheta). \tag{5.2.4}$$

The block diagram is transformed into Figure 5.2.5. Now, the characteristic equation of the linear part becomes

$$a(1 + d^T H)\Delta + (1 + k_1)z^{-1}d^T Fb = 0. \tag{5.2.5}$$

According to Lemma 5.2.1, Theorem 5.2.1 can be obtained. $\qquad\square$

Remark 5.2.1. For given λ, N_1, N_2 and N_u, we may find multiple sets of $\{k_0, k_3\}$, such that for $\forall k_1 \in \{k_0, k_3\}$, the roots of $a(1 + d^T H)\Delta + (1 +$

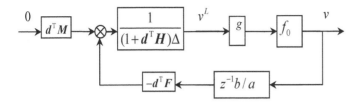

Figure 5.2.2: The block diagram with output v.

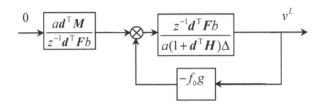

Figure 5.2.3: The static nonlinear feedback form, 2.

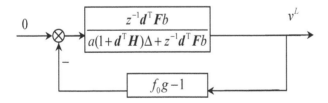

Figure 5.2.4: The static nonlinear feedback form, 3.

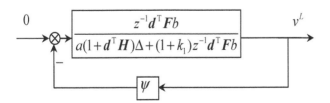

Figure 5.2.5: The static nonlinear feedback form, 4.

$k_1)z^{-1}d^T Fb = 0$ are all located in the unit circle. In this way, $[k_1, k_2] \subseteq [k_0, k_3]$ satisfying the conditions (i) and (ii) in Theorem 5.2.1 may be innumerable. Suppose that the nonlinear item in the real system satisfies

$$k_2^0 \vartheta^2 \leq (f_0 \circ g - 1)(\vartheta)\vartheta \leq k_2^0 \vartheta^2 \qquad (5.2.6)$$

where $k_1^0, k_2^0 > 0$ are constants, Then Theorem 5.2.1 means: if any set of $\{k_1, k_2\}$ satisfies

$$[k_1, k_2] \supseteq [k_1^0, k_2^0], \qquad (5.2.7)$$

the corresponding system is stable.

In fact, with (5.2.6) known, verifying stability can directly apply the following conclusion.

Corollary 5.2.1. *(TSGPC's stability) Suppose the linear sub-model applied by TSGPC is accurate and the nonlinear item satisfies (5.2.6). Then, under the following two conditions the closed-loop system of TSGPC will be stable:*

(i) *all the roots of $a(1 + d^T H)\Delta + (1 + k_1^0)z^{-1}d^T Fb = 0$ are located in the unit circle;*

(ii)

$$\frac{1}{k_2^0 - k_1^0} + Re\left\{\frac{z^{-1}d^T Fb}{a(1 + d^T H)\Delta + (1 + k_1^0)z^{-1}d^T Fb}\right\} > 0, \ \forall |z| = 1.$$
$$(5.2.8)$$

Remark 5.2.2. Theorem 5.2.1 and Corollary 5.2.1 do not require the corresponding LGPC to be stable, i.e., they do not require all the roots of $a(1 + d^T H)\Delta + z^{-1}d^T Fb = 0$ to be located in the unit circle. This is an advantage of the above stability result. Considering the relationship between TSGPC and LGPC shown in Figures 5.1.1 and 5.1.2, $0 \in [k_1^0, k_2^0]$ will have many advantages, but $0 \in [k_1^0, k_2^0]$ means that the corresponding LGPC is stable.

5.2.2   Two algorithms for finding controller parameters

Theorem 5.2.1 and Corollary 5.2.1 can also be applied to the design of the controller parameters λ, N_1, N_2 and N_u to stabilize the system. In the following we discuss two cases in the form of algorithms.

Algorithm 5.1 (Given $\{k_1^0, k_2^0\}$, design the controller parameters λ, N_1, N_2, N_u to stabilize the closed-loop system of TSGPC.)

Step 1. Search N_1, N_2, N_u and λ by variable alternation method, within their permissible (with respect to computational burden, etc.) ranges. If the search is finished, then terminate the whole algorithm, else choose one set of N_1, N_2, N_u, λ and determine $a(1 + d^T H)\Delta + z^{-1}d^T Fb$.

Step 2. Apply Jury's criterion to examine whether all roots of $a(1+d^T H)\Delta + (1+k_1^0)z^{-1}d^T Fb = 0$ are located in the unit circle. If not, then go to Step 1.

Step 3. Transform $-z^{-1}d^T Fb/[a(1+d^T H)\Delta + (1+k_1^0)z^{-1}d^T Fb]$ into irreducible form, denoted as $G(k_1^0, z)$.

Step 4. Substitute $z = \sigma + \sqrt{1-\sigma^2}i$ into $G(k_1^0, z)$ to obtain $\text{Re}\{G(k_1^0, z)\} = G_R(k_1^0, \sigma)$.

Step 5. Let $M = \max_{\sigma\in[-1,1]} G_R(k_1^0, \sigma)$. If $k_2^0 \le k_1^0 + \frac{1}{M}$, then terminate, else go to Step 1.

If the open-loop system has no eigenvalues outside of the unit circle, generally Algorithm 5.1 can obtain satisfactory λ, N_1, N_2, N_u. Otherwise, satisfactory λ, N_1, N_2, N_u may not be found for all given $\{k_1^0, k_2^0\}$ and in this case, one can restrict the degree of desaturation, i.e., try to increase k_1^0. The following algorithm can be used to determine a smallest k_1^0.

Algorithm 5.2 (Given desired $\{k_1^0, k_2^0\}$, determine the controller parameters λ, N_1, N_2, N_u such that $\{k_{10}^0, k_2^0\}$ satisfies stability requirements and $k_{10}^0 - k_1^0$ is minimized.)

Step 1. Let $k_{10}^{0,\text{old}} = k_2^0$.

Step 2. Same as Step 1 in Algorithm 5.1.

Step 3. Utilize root locus or Jury's criterion to decide $\{k_0, k_3\}$ such that $[k_0, k_3] \supset [k_{10}^{0,\text{old}}, k_2^0]$ and all roots of $a(1+d^T H)\Delta + (1+k_1)z^{-1}d^T Fb = 0$ are located in the unit circle, for $\forall k_1 \in [k_0, k_3]$. If such $\{k_0, k_3\}$ does not exist, then go to Step 2.

Step 4. Search k_{10}^0 in the range $k_{10}^0 \in \left[\max\{k_0, k_1^0\}, k_{10}^{0,\text{old}}\right]$, by increasing it gradually. If the search is finished, then go to Step 2, else transform $-z^{-1}d^T Fb/[a(1+d^T H)\Delta + (1+k_{10}^0)z^{-1}d^T Fb]$ into irreducible form, denoted as $G(k_{10}^0, z)$.

Step 5. Substitute $z = \sigma + \sqrt{1-\sigma^2}i$ into $G(k_{10}^0, z)$ to obtain $\text{Re}\{G(k_{10}^0, z)\} = G_R(k_{10}^0, \sigma)$.

Step 6. Let $M = \max_{\sigma\in[-1,1]} G_R(k_{10}^0, \sigma)$. If $k_2^0 \le k_{10}^0 + \frac{1}{M}$ and $k_{10}^0 \le k_{10}^{0,\text{old}}$, then take $k_{10}^{0,\text{old}} = k_{10}^0$ and denote $\{\lambda, N_1, N_2, N_u\}^* = \{\lambda, N_1, N_2, N_u\}$, go to Step 2. Else, go to Step 4.

Step 7. On finishing the search, let $k_{10}^0 = k_{10}^{0,\text{old}}$ and $\{\lambda, N_1, N_2, N_u\} = \{\lambda, N_1, N_2, N_u\}^*$.

5.2.3   Determination of bounds for the real nonlinearity

In the above, given $\{k_1^0, k_2^0\}$, we have described the algorithms for determining the controller parameters. In the following we briefly illustrate how to decide $\{k_1^0, k_2^0\}$ so as to bring Theorem 5.2.1 and Corollary 5.2.1 into play. We know that $f_0 \circ g \neq 1$ may be due to the following reasons:

(I) desaturation effect;

(II) solution error of nonlinear algebraic equation, including the case where an approximate solution is given since no accurate real-valued solution exists;

(III) inaccuracy in modeling of the nonlinearity;

(IV) execution error of the actuator in a real system.

Suppose TSGPC-II is adopted. Then $f_0 \circ g$ is shown in Figure 5.2.6. Further, suppose

(i) no error exists in solving nonlinear equation;

(ii) $k_{0,1} f(\vartheta)\vartheta \leq f_0(\vartheta)\vartheta \leq k_{0,2} f(\vartheta)\vartheta$ for all $v_{\min} \leq v \leq v_{\max}$;

(iii) the desaturation level satisfies $k_{s,1}\vartheta^2 \leq \text{sat}(\vartheta)\vartheta \leq \vartheta^2$,

then

(A) $f_0 \circ g = f_0 \circ \hat{g} \circ \text{sat}$;

(B) $k_{0,1}\text{sat}(\vartheta)\vartheta \leq f_0 \circ g(\vartheta)\vartheta \leq k_{0,2}\text{sat}(\vartheta)\vartheta$;

(C) $k_{0,1}k_{s,1}\vartheta^2 \leq f_0 \circ g(\vartheta)\vartheta \leq k_{0,2}\vartheta^2$,

and finally, $k_1^0 = k_{0,1}k_{s,1} - 1$ and $k_2^0 = k_{0,2} - 1$.

5.3   Region of attraction by using two-step GPC

For a fixed $k_{s,1}$, if all the conditions in Corollary 5.2.1 are satisfied, then the closed-loop system is stable. However, when TSGPS is applied for input saturated systems, $k_{s,1}$ will change along with the level of desaturation. In the last section, the issue that $\{k_1^0, k_2^0\}$ changes with $k_{s,1}$ is not handled. This issue is directly involved with the region of attraction for the closed-loop system, which needs to be discussed by the state space equation.

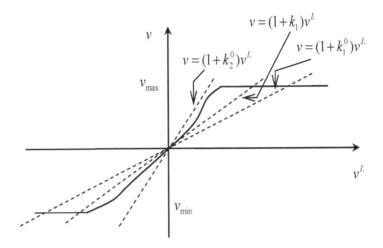

Figure 5.2.6: The sketch map of nonlinear item $f_0 \circ g$.

5.3.1 State space description of the controller

Transform (5.1.2) into the following state space model:

$$x(k+1) = Ax(k) + B\Delta v(k), \; y(k) = Cx(k) \qquad (5.3.1)$$

where $x \in \Re^n$. More details are referred to in section 4.3. For $0 < i \le N_2$ and $0 < j \le N_2$, take

$$q_i = \begin{cases} 1, & N_1 \le i \le N_2 \\ 0, & i < N_1 \end{cases} , \quad \lambda_j = \begin{cases} \lambda, & 1 \le j \le N_u \\ \infty, & j > N_u \end{cases} . \qquad (5.3.2)$$

Moreover, take a vector L such that $CL = 1$ (since $C \ne 0$, such an L exists but is not unique). Then the cost function (5.1.3) of LGPC can be equivalently transformed into the following cost function of LQ problem (refer to [40]):

$$J(k) = [x(k+N_2) - Ly_s(k+N_2)]^T C^T q_{N_2} C [x(k+N_2) - Ly_s(k+N_2)]$$
$$+ \sum_{i=0}^{N_2-1} \left\{ [x(k+i) - Ly_s(k+i)]^T C^T q_i C [x(k+i) - Ly_s(k+i)] \right.$$
$$\left. + \lambda_{i+1} \Delta v(k+i)^T \right\} . \qquad (5.3.3)$$

The LQ control law is

$$\Delta v(k) = - \left(\lambda + B^T P_1 B \right)^{-1} B^T [P_1 Ax(k) + r(k+1)], \qquad (5.3.4)$$

where P_1 can be obtained by the following Riccati iteration:

$$\begin{aligned} P_i &= q_i C^T C + A^T P_{i+1} A - A^T P_{i+1} B \left(\lambda_{i+1} + B^T P_{i+1} B \right)^{-1} B^T P_{i+1} A, \; P_{N_2} \\ &= C^T C, \end{aligned} \qquad (5.3.5)$$

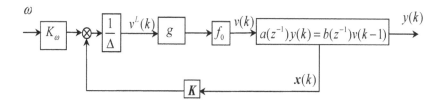

Figure 5.3.1: The equivalent block diagram of TSGPC.

and $r(k+1)$ can be calculated by

$$r(k+1) = - \sum_{i=N_1}^{N_2} \Psi^T(i,1) C^T y_s(k+i), \tag{5.3.6}$$

$$\Psi(1,1) = I, \Psi(j,1) = \prod_{i=1}^{j-1} \left[A - B \left(\lambda_{i+1} + B^T P_{i+1} B \right)^{-1} B^T P_{i+1} A \right], \ \forall j > 1. \tag{5.3.7}$$

Denote (5.3.4) as

$$\Delta v(k) = Kx(k) + K_r r(k+1) = [K \ \ K_r] \left[x(k)^T \ \ r(k+1)^T \right]^T. \tag{5.3.8}$$

Take $y_s(k+i) = \omega, \ \forall i > 0$. Then,

$$v^L(k) = v^L(k-1) + Kx(k) + K_\omega y_s(k+1), \tag{5.3.9}$$

where $K_\omega = -K_r \sum_{i=N_1}^{N_2} \Psi^T(i,1) C^T$. Figure 5.3.1 shows the equivalent block diagram of TSGPC.

5.3.2   Stability relating with the region of attraction

When (5.2.6) is satisfied, let $\delta \in Co\{\delta_1, \delta_2\} = Co\left[k_1^0 + 1, \ k_2^0 + 1\right]$, i.e., $\delta = \xi \delta_1 + (1-\xi)\delta_2$, where ξ is any value satisfying $0 \leq \xi \leq 1$. If we use δ to replace $f_0 \circ g$, then since δ is a scalar, it can move in the block diagram. Hence, Figure 5.3.1 is transformed into Figure 5.3.2. It is easy to know that, if the uncertain system in Figure 5.3.2 is robustly stable, then the closed-loop system of the original TSGPC is stable.

Now, we deduce the extended state space model of the system in Figure 5.3.2. Firstly,

$$\begin{aligned} v(k) &= v(k-1) + \delta Kx(k) + \delta K_\omega y_s(k+1) \\ &= [1 \ \ \delta K \ \ \delta K_\omega] \left[v(k-1) \ \ x(k)^T \ \ y_s(k+1) \right]^T. \end{aligned} \tag{5.3.10}$$

Since

$$y_s(k+2) = y_s(k+1), \ x(k+1) = (A + \delta BK)x(k) + \delta BK_\omega y_s(k+1), \tag{5.3.11}$$

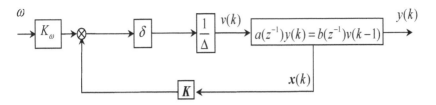

Figure 5.3.2: The uncertain system representation of TSGPC.

there is

$$
\begin{bmatrix} v(k) \\ x(k+1) \\ y_s(k+2) \end{bmatrix} = \begin{bmatrix} 1 & \delta K & \delta K_\omega \\ 0 & A+\delta BK & \delta BK_\omega \\ 0 & 0 & 1 \end{bmatrix} \begin{bmatrix} v(k-1) \\ x(k) \\ y_s(k+1) \end{bmatrix}. \tag{5.3.12}
$$

Denote (5.3.12) as

$$
x^{\mathrm{E}}(k+1) = \Phi(\delta)x^{\mathrm{E}}(k), \tag{5.3.13}
$$

and call $x^{\mathrm{E}} \in \Re^{n+2}$ as the extended state.

Both Theorem 5.2.1 and Corollary 5.2.1 are not related with the region of attraction. The region of attraction Ω of TSGPC with respect to the equilibrium point (u_e, y_e) is specially defined as the set of initial extended state $x^{\mathrm{E}}(0)$ that satisfies the following conditions:

$$
\forall x^{\mathrm{E}}(0) \in \Omega \subset \Re^{n+2}, \quad \lim_{k \to \infty} u(k) = u_e, \quad \lim_{k \to \infty} y(k) = y_e. \tag{5.3.14}
$$

For given $v(-1)$ and ω, the region of attraction Ω_x of TSGPC with respect to the equilibrium point (u_e, y_e) is specially defined as the set of initial extended state $x(0)$ that satisfies the following conditions:

$$
\forall x(0) \in \Omega_x \subset \Re^n, \quad \lim_{k \to \infty} u(k) = u_e, \quad \lim_{k \to \infty} y(k) = y_e. \tag{5.3.15}
$$

According to the above descriptions and Corollary 5.2.1, we can easily obtain the following result.

Theorem 5.3.1. *(TSGPC's stability) Suppose the linear part of the plant model for TSGPC is the same as the plant dynamics, and*

(i) *for $\forall x^{\mathrm{E}}(0) \in \Omega$, $\forall k > 0$, the level of desaturation $k_{s,1}$ is such that $f_0 \circ g$ satisfies (5.2.6);*

(ii) *the roots of $a(1 + d^T H)\Delta + (1 + k_1^0)z^{-1}d^T Fb = 0$ are all located inside of the unit circle;*

(iii) *(5.2.8) is satisfied.*

Then, the equilibrium point (u_e, y_e) of TSGPC is stable with a region of attraction Ω.

Remark 5.3.1. The main difference between Theorem 5.3.1 and Corollary 5.2.1 is that Theorem 5.3.1 introduces the region of attraction. The issues that can be handled by Theorem 5.3.1 include

    A. given $\left\{k_1^0, k_2^0, \lambda, N_1, N_2, N_u\right\}$ satisfying all the conditions in Corollary 5.2.1, determine the region of attraction Ω for the closed-loop system;

    B. given $\left\{k_{0,1}, k_{0,2}\right\}$ and the desired region of attraction Ω, search $\left\{\lambda, N_1, N_2, N_u\right\}$ satisfying all the conditions in Corollary 5.2.1.

    By adopting suitable controller parameters, the conditions (ii)-(iii) in Theorem 5.3.1 can be satisfied. Then, if the level of desaturation is sufficiently small, then the condition (i) can also be satisfied. Condition (i) can be satisfied only if $x^E(0)$ belongs to a certain set. This set is the region of attraction for the closed-loop system of TSGPC.

5.3.3   Computation of the region of attraction

Denote $\Phi_1 = \begin{bmatrix} 1 & K & K_\omega \end{bmatrix}$. In (5.3.12),

$$\Phi(\delta) \in Co\left\{\Phi^{(1)}, \Phi^{(2)}\right\}$$

$$= Co\left\{\begin{bmatrix} 1 & \delta_1 K & \delta_1 K_\omega \\ 0 & A+\delta_1 BK & \delta_1 BK_\omega \\ 0 & 0 & 1 \end{bmatrix}, \begin{bmatrix} 1 & \delta_2 K & \delta_2 K_\omega \\ 0 & A+\delta_2 BK & \delta_2 BK_\omega \\ 0 & 0 & 1 \end{bmatrix}\right\}.$$

$$(5.3.16)$$

Suppose all the conditions in Corollary 5.2.1 are satisfied, then we can adopt the following algorithm to calculate the region of attraction.

    Algorithm 5.3 (The theoretical method for calculating the region of attraction)

Step 1. Decide $k_{s,1}$ that satisfies all the conditions in Corollary 5.2.1 (if TSGPC-II is adopted, then $k_{s,1} = (k_1^0 + 1)/k_{0,1}$). Let

$$S_0 = \left\{\theta \in \mathfrak{R}^{n+2} \mid \Phi_1 \theta \leq v_{\max}/k_{s,1}, \ \Phi_1 \theta \geq v_{\min}/k_{s,1}\right\}$$
$$= \left\{\theta \in \mathfrak{R}^{n+2} \mid F^{(0)} \theta \leq g^{(0)}\right\}, \qquad (5.3.17)$$

where $g^{(0)} = \begin{bmatrix} v_{\max}/k_{s,1} \\ -v_{\min}/k_{s,1} \end{bmatrix}$, $F^{(0)} = \begin{bmatrix} \Phi_1 \\ -\Phi_1 \end{bmatrix}$. Let $j = 1$. In this step, if the extremums $v_{\min}^L$ and $v_{\max}^L$ of v^L are given, then we can let

$$S_0 = \left\{\theta \in \mathfrak{R}^{n+2} \mid \Phi_1 \theta \leq v_{\max}^L, \ \Phi_1 \theta \geq v_{\min}^L\right\} = \left\{\theta \in \mathfrak{R}^{n+2} \mid F^{(0)} \theta \leq g^{(0)}\right\}.$$
$$(5.3.18)$$

Step 2. Let

$$N_j = \left\{ \theta \in \Re^{n+2} | F^{(j-1)} \Phi^{(l)} \theta \le g^{(j-1)}, \; l = 1,2 \right\} \tag{5.3.19}$$

and

$$S_j = S_{j-1} \bigcap N_j = \left\{ \theta \in \Re^{n+2} | F^{(j)} \theta \le g^{(j)} \right\}. \tag{5.3.20}$$

Step 3. If $S_j = S_{j-1}$, then let $S = S_{j-1}$ and STOP. Else, let $j = j+1$ and turn to Step 2.

The region of attraction calculated by Algorithm 5.3 is also called the "maximal output admissible set" of the following system:

$$x^{\mathrm{E}}(k+1) = \Phi(\delta) x^{\mathrm{E}}(k), \; v^L(k) = \Phi_1 x^{\mathrm{E}}(k),$$
$$v_{\min}/k_{s,1} \le v^L(k) \le v_{\max}/k_{s,1} \; (\text{or } v^L_{\min} \le v^L(k) \le v^L_{\max}).$$

For maximal output admissible set, one can refer to, e.g., [32]; note that, here, the "output" refers to output $v^L(k)$ of the above system, rather than output y of the system (5.1.2); "admissible" refers to satisfaction of constraints. In Algorithm 5.3, the iterative method is adopted: define S_0 as the zero-step admissible set, then S_1 is 1-step admissible set, $\ldots$, S_j is j-step admissible set; the satisfaction of constraints means that the constraints are always satisfied irrespective of how many steps the sets have evolved.

In Algorithm 5.3, the following concept is involved.

Definition 5.3.1. *If there exists $d > 0$ such that $S_d = S_{d+1}$, then S is finite-determined. Then, $S = S_d$ and $d^* = \min \{d | S_d = S_{d+1}\}$ is the determinedness index (or, the output admissibility index).*

Since the judgment of $S_j = S_{j-1}$ can be transformed into the optimization problem, Algorithm 5.3 can be transformed into the following algorithm.

Algorithm 5.4 (The iterative algorithm for calculating region of attraction)

Step 1. Decide $k_{s,1}$ satisfying all the conditions in Corollary 5.2.1. Calculate S_0 according to (5.3.17) or (5.3.18). Take $j = 1$.

Step 2. Solve the following optimization problem:

$$\max_{\theta} J_{i,l}(\theta) = \left(F^{(j-1)} \Phi^{(l)} \theta - g^{(j-1)} \right)_i, \; i \in \{1, \ldots, n_j\}, \; l \in \{1,2\} \tag{5.3.21}$$

such that the following constraint is satisfied:

$$F^{(j-1)} \theta - g^{(j-1)} \le 0, \tag{5.3.22}$$

where n_j is the number of rows in $F^{(j-1)}$ and $(\cdot)_i$ denotes the i-th row. Let $J^*_{i,l}$ be the optimum of $J_{i,l}(\theta)$. If

$$J^*_{i,l} \le 0, \forall l \in \{1,2\}, \forall i \in \{1, \ldots, n_j\},$$

then STOP and take $d^* = j - 1$; else, continue.

Step 3. Calculate N_j via (5.3.19), and S_j via (5.3.20). Let $j = j + 1$ and turn to Step 2.

Remark 5.3.2. $J_{i,l}^* \le 0$ indicates that, when (5.3.22) is satisfied, $F^{(j-1)}\Phi^{(l)}\theta \le g^{(j-1)}$ is also satisfied. In S_j calculated by (5.3.19)-(5.3.20), there can be redundant inequalities; these redundant inequalities can be removed by the similar optimizations.

In real applications, it may not be possible to find a finite number of inequalities to precisely express the region of attraction S, i.e., d^* is not a finite value. It may also happen that d^* is finite but is very large, so that the convergence of the Algorithms 5.3 and 5.4 is very slow. In order to speed up the convergence, or, when the algorithms do not converge, approximate the region of attraction, one can introduce $\varepsilon > 0$. Denote $\tilde{1} = [1, 1, \cdots, 1]^T$ and, in (5.3.19), let

$$N_j = \left\{ \theta \in \Re^{n+2} | F^{(j-1)}\Phi^{(l)}\theta \le g^{(j-1)} - \varepsilon \tilde{1}, \ l = 1, 2 \right\}. \tag{5.3.23}$$

Algorithm 5.5 (The ε-iteration algorithm for calculating the region of attraction) All the details are the same as Algorithm 5.4 except that N_j is calculated by (5.3.23).

5.3.4   Numerical example

The linear part of the system is

$$y(k) - 2y(k - 1) = v(k - 1).$$

Take $N_1 = 1$, $N_2 = N_u = 2$, $\lambda = 10$. Then, $k_0 = 0.044$, $k_3 = 1.8449$ are obtained such that, when $k_1 \in [k_0, k_3]$, the condition (i) of Theorem 5.2.1 is satisfied. Take $k_1 = 0.287$, then the largest k_2 satisfying condition (ii) of Theorem 5.2.1 is $k_2 = 1.8314$; this is the set of $\{k_1, k_2\}$ satisfying $[k_1, k_2] \subseteq [k_0, k_3]$ such that $k_2 - k_1$ is maximized.

Take the Hammerstein nonlinearity as

$$f_0(\theta) = 2.3f(\theta) + 0.5\sin f(\theta), \ \ f(\theta) = \text{sign}\{\theta\}\theta \sin\left(\frac{\pi}{4}\theta\right).$$

The input constraint is $|u| \le 2$. Let the solution to the algebraic equation be utterly accurate. Then by utilizing the expression of f, it is known that $|\hat{v}| \le 2$. Let the level of desaturation satisfy $k_{s,1} = 3/4$, then according to the above description, it is known that $1.35\theta^2 \le f_0 \circ g(\theta)\theta \le 2.8\theta^2$, i.e, $k_1^0 = 0.35$ and $k_2^0 = 1.8$.

Under the above parameterizations, by applying Corollary 5.2.1 it is known that the system can be stabilized within a certain region of the initial extended state.

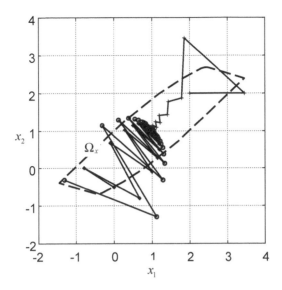

Figure 5.3.3: Region of attraction and closed-loop state trajectory of TSGPC.

Take the two system states as $x_1(k) = y(k)$ and $x_2(k) = y(k - 1)$. In Figure 5.3.3, the region in the dotted line is the region of attraction Ω_x for $v(-1) = 0$ and $\omega = 1$, which is calculated according to Algorithm 5.4. Take the three sets of initial values as

(A) $y(-1) = 2$, $y(0) = 2$, $v(-1) = 0$;

(B) $y(-1) = -1.3$, $y(0) = -0.3$, $v(-1) = 0$;

(C) $y(-1) = 0$, $y(0) = -0.5$, $v(-1) = 0$.

The setpoint value is $\omega = 1$. According to Theorem 5.3.1, the system should be stable. In Figure 5.3.3, the state trajectories shown with solid lines indicate closed-loop stability.

Notice that, Ω_x is the projection of a cross-section of Ω on the x_1-x_2 plane, which is not an invariant set. The overall projection of Ω on x_1-x_2 plane is much larger than Ω_x.

5.4 Two-step state feedback MPC (TSMPC)

Consider the following discrete-time system,

$$x(k + 1) = Ax(k) + Bv(k), \quad y(k) = Cx(k), \quad v(k) = \phi\left(u(k)\right), \qquad (5.4.1)$$

where $x \in \mathfrak{R}^n$, $v \in \mathfrak{R}^m$, $y \in \mathfrak{R}^p$, $u \in \mathfrak{R}^m$ are the state, intermediate variable, output and input, respectively; ϕ represents the relationship between the input and intermediate variable with $\phi(0) = 0$. Moreover, the following assumptions are given for TSMPC.

Assumption 5.4.1. *The state x is measurable.*

Assumption 5.4.2. *The pair (A, B) is stabilizable.*

Assumption 5.4.3. $\phi = f \circ \text{sat}$, *where f is the invertible static nonlinearity and* sat *represents the following input saturation (physical) constraint:*

$$-\underline{u} \leq u(k) \leq \bar{u} \tag{5.4.2}$$

where $\underline{u} := [\underline{u}_1, \underline{u}_2, \cdots, \underline{u}_m]^T$, $\bar{u} := [\bar{u}_1, \bar{u}_2, \cdots, \bar{u}_m]^T$, $\underline{u}_i > 0$, $\bar{u}_i > 0$, $i \in \{1, 2, \ldots, m\}$.

In two-step state feedback model predictive control (TSMPC), firstly consider the linear subsystem $x(k+1) = Ax(k) + Bv(k)$, $y(k) = Cx(k)$ and define the following cost function

$$J(N, x(k)) = \sum_{i=0}^{N-1} \left[\|x(k+i|k)\|_Q^2 + \|v(k+i|k)\|_R^2 \right] + \|x(k+N|k)\|_{Q_N}^2 , \tag{5.4.3}$$

where $Q \geq 0$, $R > 0$ are symmetric matrix; $Q_N > 0$ is the terminal state weighting matrix. At each time k, the following optimization problem should be solved:

$$\min_{\tilde{v}(k|k)} J(N, x(k)), \text{ s.t. } x(k+i+1|k) = Ax(k+i|k) + Bv(k+i|k),$$

$$i \geq 0, \ x(k|k) = x(k) \tag{5.4.4}$$

to get the optimal solution

$$\tilde{v}(k|k) = \left[v(k|k)^T, v(k+1|k)^T, \cdots, v(k+N-1|k)^T \right]^T . \tag{5.4.5}$$

This is a finite-horizon standard LQ problem, to which we can apply the following Riccati iteration:

$$P_j = Q + A^T P_{j+1} A - A^T P_{j+1} B \left(R + B^T P_{j+1} B \right)^{-1} B^T P_{j+1} A, \ 0$$
$$\leq j < N, \ P_N = Q_N. \tag{5.4.6}$$

The LQ control law is

$$v(k+i|k) = - \left(R + B^T P_{i+1} B \right)^{-1} B^T P_{i+1} Ax(k+i|k), \ i \in \{0, \ldots, N-1\}.$$

Since MPC utilizes the receding horizon strategy, only $v(k|k)$ in (5.4.5) is to be implemented. And the optimization (5.4.4) is repeated at time $k+1$ to obtain $\tilde{v}(k+1|k+1)$. Hence, the predictive control law is given by

$$v(k|k) = - \left(R + B^T P_1 B \right)^{-1} B^T P_1 Ax(k). \tag{5.4.7}$$

Note that the control law (5.4.7) may be unable to be implemented via $u(k)$, so is a "desired intermediate variable" and is denoted as:

$$v^L(k) = Kx(k) = -\left(R + B^T P_1 B\right)^{-1} B^T P_1 Ax(k). \qquad (5.4.8)$$

In the second step of TSMPC, $\hat{u}(k)$ is obtained by solving the algebraic equation $v^L(k) - f(\hat{u}(k)) = 0$, which is denoted as $\hat{u}(k) = \hat{f}^{-1}(v^L(k))$. For different f's, different methods for solving the equation can be utilized. In order to reduce the computational burden, often the equation needs not to be solved accurately. The control input $u(k)$ can be obtained by desaturating $\hat{u}(k)$ with $u(k) = \text{sat}\{\hat{u}(k)\}$ such that (5.4.2) is satisfied (applying desaturation avoids windup in real applications), and is denoted as $u(k) = g(v^L(k))$.

Thus,

$$v(k) = \phi\left(\text{sat}\{\hat{u}(k)\}\right) = (\phi \circ \text{sat} \circ \hat{f}^{-1})(v^L(k)) = (f \circ \text{sat} \circ g)(v^L(k)$$

and is denoted as

$$v(k) = h(v^L(k)).$$

The control law in terms of $v(k)$ will be

$$v(k) = h(v^L(k)) = h\left(-\left(R + B^T P_1 B\right)^{-1} B^T P_1 Ax(k)\right) \qquad (5.4.9)$$

and the closed-loop representation of the system will be

$$x(k+1) = Ax(k) + Bv(k) = \left[A - B\left(R + B^T P_1 B\right)^{-1} B^T P_1 A\right] x(k)$$
$$+ B[h\left(v^L(k)\right) - v^L(k)]. \qquad (5.4.10)$$

If the reserved nonlinear item $h = \tilde{1} = [1, 1, \cdots, 1]^T$, in (5.4.10) $[h(v^L(k)) - v^L(k)]$ will disappear and the system will become linear. But this generally cannot be guaranteed, since h may include

(i) the solution error of nonlinear equation;

(ii) the desaturation that makes $v^L(k) \neq v(k)$.

In real applications, it is usual that $h \neq \tilde{1}$, $v^L(k) \neq v(k)$. So we make the following assumptions on h.

Assumption 5.4.4. *The nonlinearity h satisfies*

$$\|h(s)\| \geq b_1 \|s\|, \quad \|h(s) - s\| \leq |b - 1| \cdot \|s\|, \quad \forall \|s\| \leq \Delta, \qquad (5.4.11)$$

where b and b_1 are scalars.

Assumption 5.4.5. *For decentralized f (i.e., one element of u has relation with one and only one of the elements in v, and the relationship is sequential), h satisfies*

$$b_{i,1}s_i^2 \leq h_i(s_i)s_i \leq b_{i,2}s_i^2, \quad i \in \{1, \ldots, m\}, \quad \forall |s_i| \leq \Delta, \qquad (5.4.12)$$

where $b_{i,2}$ and $b_{i,1}$ are positive scalars.

Since $h_i(s_i)$ and s_i has the same sign,

$$|h_i(s_i) - s_i| = ||h_i(s_i)| - |s_i|| \le \max\left\{|b_{i,1} - 1|, |b_{i,2} - 1|\right\} \cdot |s_i|.$$

Denote

$$b_1 = \min\left\{b_{1,1}, b_{2,1}, \cdots, b_{m,1}\right\},$$
$$|b - 1| = \max\left\{|b_{1,1} - 1|, \cdots, |b_{m,1} - 1|, |b_{1,2} - 1|, \cdots, |b_{m,2} - 1|\right\}. \quad (5.4.13)$$

Then (5.4.11) can be deduced from (5.4.12). The higher the degree of desaturation, the smaller will be $b_{i,1}$ and b_1. Therefore, with b_1 given, $\|h(s)\| \ge b_1 \|s\|$ in (5.4.11) will mainly represent a restriction on the degree of desaturation.

In the following, let us consider the usual input magnitude constraint, and take the SISO system as an example, to show the method for estimating b_1 and $|b - 1|$.

Note that

$$\text{sat}\{\hat{u}\} = \begin{cases} \underline{u}, & \hat{u} \le \underline{u} \\ \hat{u}, & \underline{u} \le \hat{u} \le \bar{u} \\ \bar{u}, & \hat{u} \ge \bar{u} \end{cases}. \quad (5.4.14)$$

Suppose

(A) the solution error of the nonlinear algebraic equation $v^L = f(\hat{u})$ is restricted, which can be represented as $\underline{b}\left(v^L\right)^2 \le f \circ \hat{f}^{-1}(v^L)v^L \le \bar{b}\left(v^L\right)^2$, where $\underline{b}$ and $\bar{b}$ are positive scalars;

(B) the real system design satisfies $\underline{v}^L \le v^L \le \bar{v}^L$ (where $\max\left\{-\underline{v}^L, \bar{v}^L\right\} = \Delta$) and, under this condition, there is always real solution to $v^L = f(\hat{u})$;

(C) denote $\underline{v} = f(\underline{u})$ and $\bar{v} = f(\bar{u})$, then $\underline{v}^L \le \underline{v} < 0$ and $0 < \bar{v} \le \bar{v}^L$;

(D) If $\hat{f}^{-1}(v^L) \le \underline{u}$, then $v^L \le \underline{v}$, and if $\hat{f}^{-1}(v^L) \ge \bar{u}$, then $v^L \ge \bar{v}$.

Denote $b_s = \min\left\{\underline{v}/\underline{v}^L, \bar{v}/\bar{v}^L\right\}$. For the studied system, it is easy to show that

$$h(v^L) = f \circ \text{sat} \circ g(v^L) = f \circ \text{sat} \circ \hat{f}^{-1}(v^L)$$
$$= \begin{cases} f(\underline{u}), & \hat{f}^{-1}(v^L) \le \underline{u} \\ f \circ \hat{f}^{-1}(v^L), & \underline{u} \le \hat{f}^{-1}(v^L) \le \bar{u} \\ f(\bar{u}), & \hat{f}^{-1}(v^L) \ge \bar{u} \end{cases}$$
$$= \begin{cases} \underline{v}, & \hat{f}^{-1}(v^L) \le \underline{u} \\ f \circ \hat{f}^{-1}(v^L), & \underline{u} \le \hat{f}^{-1}(v^L) \le \bar{u} \\ \bar{v}, & \hat{f}^{-1}(v^L) \ge \bar{u} \end{cases}. \quad (5.4.15)$$

By estimating the bounds for the three cases in (5.4.15), we obtain

$$\begin{cases} b_s(v^L)^2 \le \underline{v}v^L \le \left(v^L\right)^2, & \hat{f}^{-1}(v^L) \le \underline{u} \\ \underline{b}\left(v^L\right)^2 \le f \circ \hat{f}^{-1}(v^L)v^L \le \bar{b}\left(v^L\right)^2, & \underline{u} \le \hat{f}^{-1}(v^L) \le \bar{u} \\ b_s\left(v^L\right)^2 \le \bar{v}v^L \le \left(v^L\right)^2, & \hat{f}^{-1}(v^L) \ge \bar{u} \end{cases}. \quad (5.4.16)$$

Combining (5.4.15) and (5.4.16) yields

$$\min\{b_s, \underline{b}\}(v^L)^2 \leq h(v^L)v^L \leq \max\{1, \bar{b}\}(v^L)^2. \tag{5.4.17}$$

Hence,

$$b_1 = \min\{b_s, \underline{b}\}, \ b_2 = \max\{1, \bar{b}\}, \ |b - 1| = \max\{|b_1 - 1|, |b_2 - 1|\}. \tag{5.4.18}$$

Note that (A) and (B) are basic conditions for estimating h, and (C) and (D) are the special assumptions on the nonlinearity but without loss of generality. Under other cases, the methods for estimating the bounds of h can be analogously obtained. In real applications, one can take advantage of the concrete situation for estimating h.

Remark 5.4.1. The four assumptions (A)-(D) are only suitable for deducing (5.4.17)-(5.4.18). Assumptions 5.4.1-5.4.5 are suitable for all the contents of TSMPC.

It should be noted that, for the same h, with different Δ's, different b_1's and b's (or, $b_{i,1}$'s and $b_{i,2}$'s) can be obtained.

5.5 Stability of TSMPC

Definition 5.5.1. *A region Ω^N is the null controllable region (see [34]) of system (5.4.1), if*

(i) *$\forall x(0) \in \Omega^N$, there exists an admissible control sequence ($\{u(0), u(1), \cdots\}$, $-\underline{u} \leq u(i) \leq \bar{u}, \forall i \geq 0$) such that $\lim_{k \to \infty} x(k) = 0$;*

(ii) *$\forall x(0) \notin \Omega^N$, there does not exist an admissible control sequence such that $\lim_{k \to \infty} x(k) = 0$.*

According to Definition 5.5.1, for any setting of $\{\lambda, Q_N, N, Q\}$ and any equation solution error, the region of attraction of system (5.4.10) (denoted as Ω) satisfies $\Omega \subseteq \Omega^N$.

In the following, for simplicity we take $R = \lambda I$.

Theorem 5.5.1. *(Exponential stability of TSMPC) Consider system (5.4.1) with two-step predictive controller (5.4.8)-(5.4.9). Suppose*

(i) *the choosing of $\{\lambda, Q_N, N, Q\}$ makes $Q - P_0 + P_1 > 0$;*

(ii) *$\forall x(0) \in \Omega \subset \Re^n, \forall k \geq 0$,*

$$-\lambda h(v^L(k))^T h(v^L(k)) + [h(v^L(k)) - v^L(k)]^T (\lambda I + B^T P_1 B)$$
$$\times [h(v^L(k)) - v^L(k)] \leq 0. \tag{5.5.1}$$

Then the equilibrium $x = 0$ of the closed-loop system (5.4.10) is locally exponentially stable with a region of attraction Ω.

Proof. Define a quadratic function $V(k) = x(k)^T P_1 x(k)$. For $x(0) \in \Omega$, applying (5.4.6), (5.4.8) and (5.4.10) we have

$$V(k+1) - V(k)$$

$$=x(k)^T \left[A - B \left(\lambda I + B^T P_1 B \right)^{-1} B^T P_1 A \right]^T P_1$$

$$\times \left[A - B \left(\lambda I + B^T P_1 B \right)^{-1} B^T P_1 A \right] x(k)$$

$$- x(k)^T P_1 x(k) + 2\lambda x(k)^T A^T P_1 B \left(\lambda I + B^T P_1 B \right)^{-1} \left[h \left(v^L(k) \right) - v^L(k) \right]$$

$$+ \left[h \left(v^L(k) \right) - v^L(k) \right]^T B^T P_1 B \left[h \left(v^L(k) \right) - v^L(k) \right]$$

$$=x(k)^T \left[-Q + P_0 - P_1 - A^T P_1 B \left(\lambda I + B^T P_1 B \right)^{-1} \right.$$

$$\times \left. \lambda \left(\lambda I + B^T P_1 B \right)^{-1} B^T P_1 A \right] x(k)$$

$$+ 2\lambda x(k)^T A^T P_1 B \left(\lambda I + B^T P_1 B \right)^{-1} \left[h \left(v^L(k) \right) - v^L(k) \right]$$

$$+ \left[h \left(v^L(k) \right) - v^L(k) \right]^T B^T P_1 B \left[h \left(v^L(k) \right) - v^L(k) \right]$$

$$=x(k)^T \left(-Q + P_0 - P_1 \right) x(k) - \lambda \left(v^L(k) \right)^T v^L(k)$$

$$- 2\lambda \left(v^L(k) \right)^T \left[h \left(v^L(k) \right) - v^L(k) \right] + \left[h \left(v^L(k) \right) - v^L(k) \right]^T$$

$$\times B^T P_1 B \left[h \left(v^L(k) \right) - v^L(k) \right]$$

$$=x(k)^T \left(-Q + P_0 - P_1 \right) x(k) - \lambda h \left(v^L(k) \right)^T h \left(v^L(k) \right)$$

$$+ \left[h \left(v^L(k) \right) - v^L(k) \right]^T \left(\lambda I + B^T P_1 B \right) \left[h \left(v^L(k) \right) - v^L(k) \right].$$

Note that in the above we have utilized the following fact:

$$\left[A - B \left(\lambda I + B^T P_1 B \right)^{-1} B^T P_1 A \right]^T P_1 B$$

$$=A^T P_1 B \left[I - \left(\lambda I + B^T P_1 B \right)^{-1} B^T P_1 B \right] = \lambda A^T P_1 B \left(\lambda I + B^T P_1 B \right)^{-1}.$$

Under conditions (i) and (ii), it is clear that

$$V(k+1) - V(k) \le -\sigma_{\min} \left(Q - P_0 + P_1 \right) x(k)^T x(k) < 0, \ \forall x(k) \ne 0$$

(where $\sigma_{\min}(\cdot)$ denotes the minimum eigenvalue). Therefore, $V(k)$ is Lyapunov function for exponential stability. □

The conditions in Theorem 5.5.1 just reflect the essential idea of two-step design. Condition (i) is a requirement on the linear control law (5.4.8), while condition (ii) is an extra requirement on h. Generally, decreasing the equation solution error and Δ benefits satisfaction of (5.5.1). From the proof of Theorem 5.5.1, it is easy to know that (i) is a sufficient stability condition for unconstrained linear system because, with $h = \hat{1}$, (5.5.1) becomes $-\lambda v^L(k)^T v^L(k) \le 0$, i.e., condition (ii) is always satisfied.

Since (5.5.1) is not easy to check, the following two corollaries are given.

Corollary 5.5.1. *(Exponential stability of TSMPC) Consider system* (5.4.1) *with two-step predictive controller* (5.4.8)-(5.4.9). *Suppose*

(i) $Q - P_0 + P_1 > 0$;

(ii) $\forall x(0) \in \Omega \subset \Re^n$, $\left\| v^L(k) \right\| \leq \Delta$ *for all* $k \geq 0$;

(iii)

$$-\lambda \left[b_1^2 - (b-1)^2 \right] + (b-1)^2 \sigma_{\max} \left(B^T P_1 B \right) \leq 0. \tag{5.5.2}$$

Then the equilibrium $x = 0$ *of the closed-loop system* (5.4.10) *is locally exponentially stable with a region of attraction* Ω.

Proof. Applying (5.4.11), it follows that

$$
\begin{aligned}
&- \lambda h(s)^T h(s) + (h(s) - s)^T \left(\lambda I + B^T P_1 B \right) (h(s) - s) \\
\leq &- \lambda b_1^2 s^T s + (b-1)^2 \sigma_{\max} \left(\lambda I + B^T P_1 B \right) s^T s \\
= &- \lambda b_1^2 s^T s + \lambda (b-1)^2 s^T s + (b-1)^2 \sigma_{\max} \left(B^T P_1 B \right) s^T s \\
= &- \lambda \left[b_1^2 - (b-1)^2 \right] s^T s + (b-1)^2 \sigma_{\max} \left(B^T P_1 B \right) s^T s.
\end{aligned}
$$

Hence, if (5.5.2) is satisfied, (5.5.1) is also satisfied (where $s = v^L(k)$). □

Corollary 5.5.2. *(Exponential stability of TSMPC) Consider system* (5.4.1) *with two-step predictive controller* (5.4.8)-(5.4.9). *Suppose*

(i) $Q - P_0 + P_1 > 0$;

(ii) $\forall x(0) \in \Omega \subset \Re^n$, $\left| v_i^L(k) \right| \leq \Delta$ *for all* $k \geq 0$;

(iii) *the nonlinearity* f *is decentralized and*

$$-\lambda (2b_1 - 1) + (b-1)^2 \sigma_{\max} \left(B^T P_1 B \right) \leq 0. \tag{5.5.3}$$

Then the equilibrium $x = 0$ *of the closed-loop system* (5.4.10) *is locally exponentially stable with a region of attraction* Ω.

Proof. According to (5.4.12), $s_i \left[h_i(s_i) - s_i \right] \geq s_i \left[b_{i,1} s_i - s_i \right]$, $i \in \{1, \ldots, m\}$. Then,

$$
\begin{aligned}
&- \lambda s^T s - 2\lambda s^T (h(s) - s) + (h(s) - s)^T B^T P_1 B (h(s) - s) \\
\leq &- \lambda s^T s - 2\lambda \sum_{i=1}^{m} (b_{i,1} - 1) s_i^2 + (h(s) - s)^T B^T P_1 B (h(s) - s) \\
\leq &- \lambda s^T s - 2\lambda (b_1 - 1) s^T s + (b-1)^2 \sigma_{\max} \left(B^T P_1 B \right) s^T s \\
= &- \lambda (2b_1 - 1) s^T s + (b-1)^2 \sigma_{\max} \left(B^T P_1 B \right) s^T s.
\end{aligned}
$$

Hence, if (5.5.3) is satisfied, (5.5.1) is also satisfied (where $s = v^L(k)$). □

Proposition 5.5.1. *(Exponential stability of TSMPC) In Corollary 5.5.1 (Corollary 5.5.2), if the inequalities in conditions (i) and (iii) are substituted by:*

$$Q - P_0 + P_1 + \eta A^T P_1 B(\lambda I + B^T P_1 B)^{-2} B^T P_1 A > 0$$

where, $\eta = \lambda \left[b_1^2 - (b-1)^2 \right] - (b-1)^2 \sigma_{\max} \left(B^T P_1 B \right)$ *(Corollary 5.5.1) or* $\eta = \lambda (2b_1 - 1) - (b-1)^2 \sigma_{\max} \left(B^T P_1 B \right)$ *(Corollary 5.5.2), then the conclusion still holds.*

Proof. According to the proofs of Corollaries 5.5.1-5.5.2,

$$
\begin{aligned}
&- \lambda h \left(v^L(k) \right)^T h \left(v^L(k) \right) + \left[h \left(v^L(k) \right) - v^L(k) \right]^T \left(\lambda I + B^T P_1 B \right) \\
&\times \left[h \left(v^L(k) \right) - v^L(k) \right] \leq -\eta \left(v^L(k) \right)^T \left(v^L(k) \right).
\end{aligned}
$$

According to (5.4.8) and the proof of Theorem 5.5.1,

$$
\begin{aligned}
V(k+1) - V(k) \leq &-x(k)^T \left[Q - P_0 + P_1 + \eta A^T P_1 B(\lambda I + B^T P_1 B)^{-2} B^T P_1 A \right] \\
&\times x(k).
\end{aligned}
$$

Then, similar to Theorem 5.5.1, the conclusion holds.   □

Remark 5.5.1. Stability conclusion in Proposition 5.5.1 is less conservative than Corollaries 5.5.1-5.5.2, and is not necessarily more conservative than Theorem 5.5.1. However, for controller parameter tuning, applying Proposition 5.5.1 is not as straightforward as applying Corollaries 5.5.1-5.5.2.

Remark 5.5.2. If $f = \tilde{1}$, i.e., there is only input saturation, then $b_{i,2} = 1$, $(b-1)^2 = (b_1 - 1)^2$ and both (5.5.2) and (5.5.3) will become $-\lambda (2b_1 - 1) + (b_1 - 1)^2 \sigma_{\max} \left(B^T P_1 B \right) \leq 0$.

Denote (5.5.2) and (5.5.3) as:

$$-\lambda + \beta \sigma_{\max} \left(B^T P_1 B \right) \leq 0 \tag{5.5.4}$$

where, $\beta = (b-1)^2 / [b_1^2 - (b-1)^2]$ for (5.5.2) and $\beta = (b-1)^2 / (2b_1 - 1)$ for (5.5.3).

As for the region of attraction in Theorem 5.5.1, Corollaries 5.5.1 and 5.5.2, we give the following easily manipulable ellipsoidal one.

Corollary 5.5.3. *(Region of attraction of TSMPC) Consider system (5.4.1) with two-step predictive controller (5.4.8)-(5.4.9). If*

(i) $Q - P_0 + P_1 > 0$;

(ii) the choosing of $\{\Delta, b_1, b\}$ *satisfies (5.5.2),*

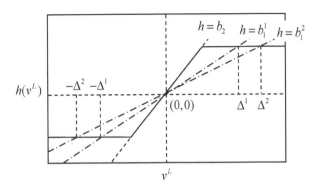

Figure 5.5.1: Parameter selection in the reserved nonlinear item.

then the region of attraction Ω for the equilibrium $x = 0$ of the closed-loop system (5.4.10) is not smaller than

$$S_c = \left\{ x \,\middle|\, x^T P_1 x \leq c \right\}, \ c = \frac{\Delta^2}{\left\| (\lambda I + B^T P_1 B)^{-1} B^T P_1 A P_1^{-1/2} \right\|^2}. \tag{5.5.5}$$

Proof. Transform the linear system (A, B, C) into $(\bar{A}, \bar{B}, \bar{C})$ with nonsingular transformation $\bar{x} = P_1^{1/2}$. Then, $\forall x(0) \in S_c$, $\|\bar{x}(0)\| \leq \sqrt{c}$ and

$$\left\| v^L(0) \right\| = \left\| (\lambda I + B^T P_1 B)^{-1} B^T P_1 A x(0) \right\|$$

$$= \left\| (\lambda I + B^T P_1 B)^{-1} B^T P_1 A P_1^{-1/2} \bar{x}(0) \right\|$$

$$\leq \left\| (\lambda I + B^T P_1 B)^{-1} B^T P_1 A P_1^{-1/2} \right\| \|\bar{x}(0)\| \leq \Delta.$$

Under (i) and (ii), all the conditions in Corollary 5.5.1 are satisfied at time $k = 0$ if $x(0) \in S_c$. Furthermore, according to the proof of Theorem 5.5.1, $x(1) \in S_c$ if $x(0) \in S_c$. Hence, for $\forall x(0) \in S_c$, $\left\| v^L(1) \right\| \leq \Delta$. This shows that all the conditions in Corollary 5.5.1 are satisfied at time $k = 1$, and by analogy they are also satisfied for any $k > 1$. $\qquad\square$

As mentioned in the last section, choosing different Δ may result in different b_1 and b (corresponding to (5.5.2)). So we can choose the largest possible Δ. Take a single input system with only symmetric saturation constraint (as in Figure 5.5.1) as an example. By Remark 5.5.2, we can choose the smallest b_1 satisfying $b_1 > 1/2$ and $(2b_1 - 1)/(b_1 - 1)^2 \geq B^T P_1 B/\lambda$, then Δ is determined according to Figure 5.5.1.

Apparently the region of attraction for given controller parameters may be too small if we have no desired region of attraction prior to the controller design. To design a controller with desired region of attraction, the concept of "semi-global stabilization" could be technically involved.

If A has no eigenvalues outside of the unit circle, semi-global stabiliza-tion (see [45], [46]) means the design of a feedback control law that results in a region of attraction that includes any a priori given compact set in n-dimensional space . If A has eigenvalues outside of the unit circle, semi-global stabilization (see [33]) means the design of a feedback control law that results in a region of attraction that includes any a priori given compact subset of the null controllable region.

The next section will derive the semi-global stabilization techniques for TSMPC. If A has no eigenvalues outside of the unit circle, TSMPC can be de-signed (tuning $\{\lambda, Q_N, N, Q\}$) to have an arbitrarily large region of attraction. Otherwise, a set of $\{\lambda, Q_N, N, Q\}$ can be chosen to obtain a set of domains of attraction with their union contained in the null controllable region.

5.6 Design of the region of attraction of TSMPC based on semi-global stability

5.6.1 Case system matrix has no eigenvalue outside of the unit circle

Theorem 5.6.1. *(Semi-global stability of TSMPC) Consider system (5.4.1) with two-step predictive controller (5.4.8)-(5.4.9). Suppose*

(i) A has no eigenvalues outside of the unit circle;

(ii) $b_1 > |b - 1| > 0$, i.e., $\beta > 0$.

Then, for any bounded set $\Omega \subset \mathfrak{R}^n$, there exist $\{\lambda, Q_N, N, Q\}$ such that the equilibrium $x = 0$ for the closed-loop system (5.4.10) is locally exponentially stable with a region of attraction Ω.

Proof. We show how $\{\lambda, Q_N, N, Q\}$ can be chosen to satisfy condition (i)-(iii) in Corollary 5.5.1. First of all, choose $Q > 0$ and an arbitrary N. In the following, we elaborate on how to choose λ and Q_N.

(s1) Choose Q_N to satisfy

$$Q_N = Q_1 + Q + A^T Q_N A - A^T Q_N B \left(\lambda I + B^T Q_N B\right)^{-1} B^T Q_N A \quad (5.6.1)$$

where $Q_1 \geq 0$ is an arbitrary symmetric matrix. Eq. (5.6.1) means that Riccati iteration (5.4.6) satisfies $P_{N-1} \leq P_N = Q_N$ and has monotonic decreasing property (see [56]), so $P_0 \leq P_1$ and $Q - P_0 + P_1 > 0$. Therefore, condition (i) in Corollary 5.5.1 can be satisfied for any λ. Changing λ, Q_N is also changed corresponding to (5.6.1).

(s2) The fake algebraic Riccati equation (see [56]) of (5.4.6) is

$$P_{j+1} = (Q + P_{j+1} - P_j) + A^T P_{j+1} A \quad (5.6.2)$$
$$- A^T P_{j+1} B \left(\lambda I + B^T P_{j+1} B\right)^{-1} B^T P_{j+1} A,$$
$$0 \leq j < N, \ P_N = Q_N, \ P_N - P_{N-1} = Q_1. \quad (5.6.3)$$

Multiply both sides of (5.6.3) by λ^{-1}, then

$$\bar{P}_{j+1} = (\lambda^{-1}Q + \bar{P}_{j+1} \tag{5.6.4}$$
$$- \bar{P}_j) + A^T \bar{P}_{j+1}A - A^T \bar{P}_{j+1}B \left(I + B^T \bar{P}_{j+1}B\right)^{-1} B^T \bar{P}_{j+1}A,$$
$$0 \le j < N, \ \bar{P}_N = \lambda^{-1}Q_N, \ \bar{P}_N - \bar{P}_{N-1} = \lambda^{-1}Q_1, \tag{5.6.5}$$

where $\bar{P}_{j+1} = \lambda^{-1}P_{j+1}$, $0 \le j < N$. As $\lambda \to \infty$, $\bar{P}_{j+1} \to 0$, $0 \le j < N$ (see [46]). Since $\beta > 0$, there exists a suitable λ_0^* such that whenever $\lambda \ge \lambda_0^*$, $\beta\sigma_{\max}\left(B^T\bar{P}_1B\right) \le 1$, i.e. condition (iii) in Corollary 5.5.1 can be satisfied.

(s3) Further, choose an arbitrary constant $\alpha > 1$, then there exists $\lambda_1^* \ge \lambda_0^*$ such that whenever $\lambda \ge \lambda_1^*$,

$$\bar{P}_1^{1/2}B\left(I + B^T\bar{P}_1B\right)^{-1}B^T\bar{P}_1^{1/2} \le (1 - 1/\alpha)\,I. \tag{5.6.6}$$

For $j = 0$, left and right multiplying both sides of (5.6.5) by $\bar{P}_1^{-1/2}$ and applying (5.6.6) obtains

$$\bar{P}_1^{-1/2}A^T\bar{P}_1A\bar{P}_1^{-1/2} \le \alpha I - \alpha\bar{P}_1^{-1/2}\left(\lambda^{-1}Q + \bar{P}_1 - \bar{P}_0\right)\bar{P}_1^{-1/2} \le \alpha I,$$

i.e., $\left\|\bar{P}_1^{1/2}A\bar{P}_1^{-1/2}\right\| \le \sqrt{\alpha}$.

For any bounded Ω, choose $\bar{c}$ such that

$$\bar{c} \ge \sup_{x \in \Omega, \ \lambda \in [\lambda_1^*, \infty)} x^T\lambda^{-1}P_1x.$$

So $\Omega \subseteq \bar{S}_{\bar{c}} = \left\{x | x^T\lambda^{-1}P_1x \le \bar{c}\right\}$.

Denote $(\bar{A}, \bar{B}, \bar{C})$ as the transformed system of (A, B, C) by nonsingular transformation $\bar{x} = \bar{P}_1^{1/2}x$, then there exists a sufficiently large $\lambda^* \ge \lambda_1^*$ such that $\forall \lambda \ge \lambda^*$ and $\forall x(0) \in \bar{S}_{\bar{c}}$,

$$\left\|\left(\lambda I + B^T P_1 B\right)^{-1}B^T P_1 Ax(0)\right\| = \left\|\left(I + \bar{B}^T\bar{B}\right)^{-1}\bar{B}^T\bar{A}\bar{x}(0)\right\|$$
$$\le \left\|\left(I + \bar{B}^T\bar{B}\right)^{-1}\bar{B}^T\right\|\sqrt{\alpha\bar{c}} \le \Delta$$

because $\left\|\left(I + \bar{B}^T\bar{B}\right)^{-1}\bar{B}^T\right\|$ tends to be smaller when λ is increased.

Hence, for $\forall x(0) \in \Omega$, condition (ii) in Corollary 5.5.1 can be satisfied at time $k = 0$, and according to the proof of Corollary 5.5.3 it can also be satisfied for all $k > 0$.

In a word, if we choose Q_N by (5.6.1), $Q > 0$, N arbitrary and $\lambda^* \le \lambda < \infty$, then the closed-loop system is locally exponentially stable with a region of attraction Ω. $\square$

Remark 5.6.1. In the proof of Theorem 5.6.1, both Ω and $\bar{S}_{\bar{c}}$ are regions of attraction with respect to $x = 0$ of system (5.4.10). For further explanation, we introduce the maximal region of attraction Ω^0 which contains any region of attraction with respect to $x = 0$ of system (5.4.10). Therefore, in Theorem 5.6.1, "with a region of attraction Ω" can be substituted by "with Ω contained in its maximal region of attraction."

Corollary 5.6.1. *(Semi-global stability of TSMPC) Suppose*

(i) *A has no eigenvalues outside of the unit circle;*

(ii) *the nonlinear equation is solved sufficiently accurate such that, in the absence of input saturation constraint, there exist suitable $\{\Delta = \Delta^0, b_1, b\}$ satisfying $b_1 > |b - 1| > 0$.*

Then the conclusion in Theorem 5.6.1 still holds.

Proof. In the absence of input saturation constraint, determining $\{b_1, b\}$ for given Δ (or given $\{b_1, b\}$, determining Δ) is independent of the controller parameter $\{\lambda, Q_N, N, Q\}$. When there is input saturation, still choose $\Delta = \Delta^0$. Then the following two cases may happen:

Case 1: $b_1 > |b - 1| > 0$ as $\lambda = \lambda_0$. Decide the parameters as in the proof of Theorem 5.6.1, except that $\lambda_0^* \geq \lambda_0$.

Case 2: $|b - 1| \geq b_1 > 0$ as $\lambda = \lambda_0$. Apparently the reason lies in that the control action is too much restricted by the saturation constraint. By the same reason as in the proof of Theorem 5.6.1 and by (5.4.8) we know that, for any bounded Ω, there exists $\lambda_2^* \geq \lambda_0$ such that

$$\forall \lambda \geq \lambda_2^* \text{ and } \forall x(k) \in \Omega, \ \hat{u}(k) \text{ does not violate the saturation constraint.}$$

This process is equivalent to decrease Δ and redetermine $\{b_1, b\}$ such that $b_1 > |b - 1| > 0$.

In a word, if the region of attraction has not been satisfactory with $\lambda = \lambda_0$, then it can be satisfied by choosing $\max\{\lambda^*, \lambda_2^*\} \leq \lambda < \infty$ and suitable $\{Q_N, N, Q\}$. $\square$

Although a method is implicitly presented in the proofs of Theorem 5.6.1 and Corollary 5.6.1 for tuning the controller parameters, a large λ tends to be achieved. Then, when the desired region of attraction Ω is large, the obtained controller will be very conservative. Actually, we do not have to choose λ as in Theorem 5.6.1 and Corollary 5.6.1. Moreover, a series of λ's can be chosen and the following controller algorithm can be applied.

Algorithm 5.6 (The λ-switching algorithm of TSMPC)
Off-line, complete the following steps 1-3:

Step 1. Choose adequate b_1, b and obtain the largest possible Δ, or, choose adequate Δ and obtain the smallest possible $|b - 1|$ and largest possible b_1 (refer to section 5.4 for estimating the bounds, and (5.5.4)). Calculate $\beta > 0$.

Step 2. Choose Q, N, Q_N as in the proof of Theorem 5.6.1.

Step 3. Gradually increase λ, until (5.5.4) is satisfied at $\lambda = \underline{\lambda}$. Increase λ to obtain $\lambda^M > \cdots > \lambda^2 > \lambda^1 \geq \underline{\lambda}$. The parameter λ^i corresponds to Con_i and the region of attraction S^i (S^i calculated by Corollary 5.5.3, $i \in \{1, 2, \ldots, M\}$). The inclusion condition $S^1 \subset S^2 \subset \cdots \subset S^M$ is satisfied and S^M should contain the desired region of attraction Ω.

On-line, at each time k,

A) if $x(k) \in S^1$, then choose Con_1;

B) if $x(k) \in S^i$, $x(k) \notin S^{i-1}$, then choose Con_i, $i \in \{2, 3, \ldots, M\}$.

5.6.2   Case system matrix has eigenvalues outside of the unit circle

In this case, semi-global stabilization cannot be implemented in a simple manner as in the above case. However, a set of ellipsoidal domains of attraction S^i, $i \in \{1, 2, \ldots, M\}$ can be achieved via a set of controllers with respect to different parameter sets $\{\lambda, Q_N, N, Q\}^i$ and the region of attraction S^i. In the following we give the algorithm.

Algorithm 5.7 (The method for parameter search in TSMPC)

Step 1. Refer to Step 1 of Algorithm 5.6. Set $S = \{0\}$, $i = 1$.

Step 2. Select $\{Q_N, N, Q\}$ (changing them alternatively).

Step 3. Determine $\{S_c, \lambda, Q_N, N, Q\}$ via the following Steps 3.1-3.3:

Step 3.1. Check if (5.5.4) is satisfied. If not, tune λ to satisfy (5.5.4).

Step 3.2. Check if $Q - P_0 + P_1 > 0$ is satisfied. If it is satisfied, go to Step 3.3; else tune $\{Q_N, N, Q\}$ to satisfy it and go to Step 3.1.

Step 3.3. Determine P_1 and determine c by $\left\| (\lambda I + B^T P_1 B)^{-1} B^T P_1 A P_1^{-1/2} \right\|$ $\sqrt{c} = \Delta$. Then the region of attraction for the real system will include the level set $S_c = \{x | x^T P_1 x \leq c\}$.

Step 4. Set $\{\lambda, Q_N, N, Q\}^i = \{\lambda, Q_N, N, Q\}$, $S^i = S_c$ and $S = S \bigcup S^i$.

Step 5. Check if S contains the desired region of attraction Ω. If it is, go to Step 6; else set $i = i + 1$ and go to Step 2.

Step 6. Set $M = i$ and STOP.

Three cases may happen in Algorithm 5.7:

(A) the trivial case that a single S^i is found to satisfy $S^i \supseteq \Omega$;

(B) a set of S^i ($i \in \{1, 2, \ldots, M\}$ and $M > 1$) are found satisfying $\bigcup_{i=1}^{M} S^i \supseteq \Omega$;

(C) S^i satisfying $\bigcup_{i=1}^{M} S^i \supseteq \Omega$ cannot be found with a finite number M (in real application, M is prescribed to be not larger than an M_0).

For case (B), the following controller switching algorithm can be applied (which is somewhat different from Algorithm 5.6):

Algorithm 5.8 (Switching algorithm of TSMPC)

Off-line, apply Algorithm 5.7 to choose a set of ellipsoidal domains $S^1, S^2, \cdots$, S^M satisfying $\bigcup_{i=1}^{M} S^i \supseteq \Omega$. Arrange S^i in a proper way that results in $S^{(1)}, S^{(2)}, \cdots, S^{(M)}$ with corresponding controllers $Con_{(i)}$, $i \in \{1, 2, \ldots, M\}$. It is not necessary that $S^{(j)} \subseteq S^{(j+1)}$ for any $j \in \{1, 2, \ldots, M-1\}$.

On-line, at each time k,

A) if $x(k) \in S^{(1)}$, then choose $Con_{(1)}$;

B) if $x(k) \in S^{(i)}$, $x(k) \notin S^{(l)}$, $\forall l < i$, then choose $Con_{(i)}$, $i \in \{2, 3, \ldots, M\}$.

For case (C), we can take one of the following strategies:

(i) Decrease the equation solution error and re-determine $\{\Delta, b_1, b\}$.

(ii) When the state lies outside of $\bigcup_{i=1}^{M_0} S^i$, adopt the nonlinear separation method as in Remark 5.1.1 (optimizing $\tilde{v}(k|k)$ considering the constraint on the intermediate variable). For this reason, we should first transform the saturation constraint on u into the constraint on v. For complex nonlinearity, obtaining the constraint on v could be very difficult, and it is even possible that nonlinear constraint is encountered. If f is de-centralized, as that in Assumption 5.4.5, then it is easy to obtain the constraint on the intermediate variable.

(iii) When the state lies outside of $\bigcup_{i=1}^{M_0} S^i$, substitute $v(k+i|k)$ in (5.4.3) by $u(k+i|k)$ and apply the pure nonlinear MPC to calculate the control action (i.e., adopt MPC based on the nonlinear prediction model and nonlinear optimization).

Remark 5.6.2. The techniques in Algorithms 5.3 and 5.4 can be utilized to calculate the region of attraction of TSMPC. Even, the case when the linear sub-model has uncertainties can be considered; see [52].

5.6.3   Numerical example

First consider the case that A has no eigenvalue outside of the unit circle. The linear subsystem is $A = \begin{bmatrix} 1 & 0 \\ 1 & 1 \end{bmatrix}$, $B = \begin{bmatrix} 1 \\ 0 \end{bmatrix}$. The invertible static nonlinearity is

$$f(\vartheta) = 4/3\vartheta + 4/9\vartheta \text{sign}\,\{\vartheta\} \sin(40\vartheta),$$

and the input constraint is $|u| \leq 1$. A simple solution $\hat{u} = 3/4v^L$ is applied to the algebraic equation. The resultant h is shown in Figure 5.6.1. Choose

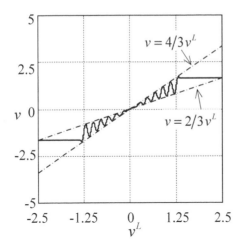

Figure 5.6.1: Curve of h.

$b_1 = 2/3$ and $b_2 = 4/3$ as in Figure 5.6.1, then $\beta = 1/3$. Choose $\Delta = f(1)/b_1 = 2.4968$.

The initial state is $x(0) = [10, -33]^T$. Choose $N = 4$, $Q = 0.1I$, $Q_N = 0.11I + A^T Q_N A - A^T Q_N B \left(\lambda + B^T Q_N B\right)^{-1} B^T Q_N A$.

Choose $\lambda = 0.225,\ 0.75,\ 2,\ 10,\ 50$, then the domains of attraction determined by Corollary 5.5.3 are, respectively,

$$S_c^1 = \left\{ x \middle| x^T \begin{bmatrix} 0.6419 & 0.2967 \\ 0.2967 & 0.3187 \end{bmatrix} x \leq 1.1456 \right\},$$

$$S_c^2 = \left\{ x \middle| x^T \begin{bmatrix} 1.1826 & 0.4461 \\ 0.4461 & 0.3760 \end{bmatrix} x \leq 3.5625 \right\},$$

$$S_c^3 = \left\{ x \middle| x^T \begin{bmatrix} 2.1079 & 0.6547 \\ 0.6547 & 0.4319 \end{bmatrix} x \leq 9.9877 \right\},$$

$$S_c^4 = \left\{ x \middle| x^T \begin{bmatrix} 5.9794 & 1.3043 \\ 1.3043 & 0.5806 \end{bmatrix} x \leq 62.817 \right\},$$

$$S_c^5 = \left\{ x \middle| x^T \begin{bmatrix} 18.145 & 2.7133 \\ 2.7133 & 0.8117 \end{bmatrix} x \leq 429.51 \right\}.$$

Figure 5.6.2 depicts S_c^1, S_c^2, S_c^3, S_c^4 and S_c^5 from inside to outside. $x(0)$ lies in S_c^5.

The rule of Algorithm 5.6 in the simulation is:

- if $x(k) \in S_c^1$, then $\lambda = 0.225$;

- else if $x(k) \in S_c^2$, then $\lambda = 0.75$;

- else if $x(k) \in S_c^3$, then $\lambda = 2$;

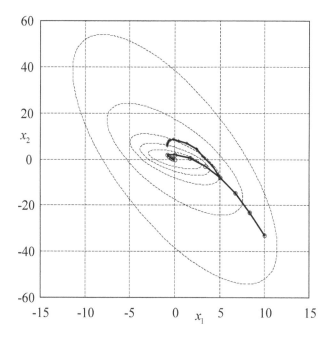

Figure 5.6.2: The closed-loop state trajectories when A has no eigenvalue outside of the unit circle.

- else if $x(k) \in S_c^4$, then $\lambda = 10$;

- else $\lambda = 50$.

The simulation result is shown in Figure 5.6.2. The line with "o" is the state trajectory with Algorithm 5.6, while the line with "*" is the state trajectory with $\lambda = 50$. With Algorithm 5.6, the trajectory is very close to the origin after 15 simulation samples, but when $\lambda = 50$ is always adopted, the trajectory has just reached to the boundary of S_c^2 after 15 simulation samples.

Further, consider the case that A has eigenvalues outside of the unit circle. The linear subsystem is $A = \begin{bmatrix} 1.2 & 0 \\ 1 & 1.2 \end{bmatrix}$ and $B = \begin{bmatrix} 1 \\ 0 \end{bmatrix}$. The nonlinearities, the solution of equation and the corresponding $\{b_1, b_2, \Delta\}$ are the same as above. We obtain three ellipsoidal regions of attraction S^1, S^2, S^3, the corresponding parameter sets are:

$$\{\lambda, Q_N, Q, N\}^1 = \left\{ 8.0, \begin{bmatrix} 1 & 0 \\ 0 & 1 \end{bmatrix}, \begin{bmatrix} 0.01 & 0 \\ 0 & 1.01 \end{bmatrix}, 12 \right\},$$

$$\{\lambda, Q_N, Q, N\}^2 = \left\{ 2.5, \begin{bmatrix} 1 & 0 \\ 0 & 1 \end{bmatrix}, \begin{bmatrix} 0.9 & 0 \\ 0 & 0.1 \end{bmatrix}, 4 \right\},$$

$$\{\lambda, Q_N, Q, N\}^3 = \left\{ 1.3, \begin{bmatrix} 3.8011 & 1.2256 \\ 1.2256 & 0.9410 \end{bmatrix}, \begin{bmatrix} 1.01 & 0 \\ 0 & 0.01 \end{bmatrix}, 4 \right\}.$$

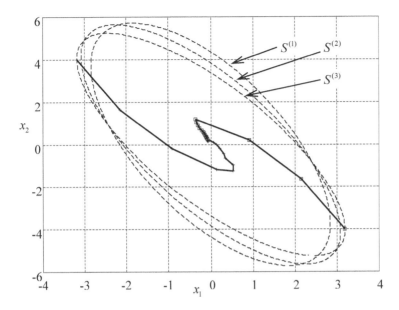

Figure 5.6.3: The closed-loop state trajectories when A has eigenvalue outside of the unit circle.

Arrange S^1, S^2, S^3 and the parameter sets in the following way:

$$S^{(1)} = S^3, \ S^{(2)} = S^2, \ S^{(3)} = S^1,$$
$$\{\lambda, Q_N, Q, N\}^{(1)} = \{\lambda, Q_N, Q, N\}^3, \ \{\lambda, Q_N, Q, N\}^{(2)} = \{\lambda, Q_N, Q, N\}^2,$$
$$\{\lambda, Q_N, Q, N\}^{(3)} = \{\lambda, Q_N, Q, N\}^1.$$

Choose two sets of initial state as $x(0) = [-3.18, \ 4]^T$ and $x(0) = [3.18, \ -4]^T$. $x(0) \in S^{(3)}$. With Algorithm 5.8 applied, the resultant state trajectories are shown in Figure 5.6.3.

5.7 Two-step output feedback model predictive control (TSOFMPC)

Consider the system model (5.4.1), with notations the same as above. In TSOFMPC, suppose (A, B, C) is completely controllable and observable. Moreover, suppose f is not an exact model of the nonlinearity, the true nonlinearity is f_0 and it is possible that $f \neq f_0$. The first step in TSOFMPC only considers the linear subsystem. The state estimation is $\tilde{x}$ and the prediction model is

$$\tilde{x}(k+1|k) = A\tilde{x}(k|k) + Bv(k|k), \ \tilde{x}(k|k) = \tilde{x}(k). \tag{5.7.1}$$

Define the objective function as

$$J(N, \tilde{x}(k)) = \|\tilde{x}(k+N|k)\|_{P_N}^2 + \sum_{j=0}^{N-1} \left[\|\tilde{x}(k+j|k)\|_Q^2 + \|v(k+j|k)\|_R^2 \right]$$

(5.7.2)

where Q, R are the same as in TSMPC; $P_N \geq 0$ is the weighting matrix of the terminal state. The Riccati iteration

$$P_j = Q + A^T P_{j+1} A - A^T P_{j+1} B \left(R + B^T P_{j+1} B \right)^{-1} B^T P_{j+1} A, \; j < N \quad (5.7.3)$$

is adopted to obtain the predictive control law

$$v^*(k|k) = - \left(R + B^T P_1 B \right)^{-1} B^T P_1 A \tilde{x}(k). \quad (5.7.4)$$

Noting that $v^*(k|k)$ in (5.7.4) may be impossible to implement via a real control input, we formalize it as

$$v^L(k) \triangleq K\tilde{x}(k) = - \left(R + B^T P_1 B \right)^{-1} B^T P_1 A \tilde{x}(k). \quad (5.7.5)$$

The second step in TSOFMPC is the same as in TSMPC. Hence, the actual intermediate variable is

$$v(k) = h(v^L(k)) = f_0 \left(\text{sat} \{u(k)\} \right) = f_0 \circ \text{sat} \circ g(v^L(k)). \quad (5.7.6)$$

Equation (5.7.6) is the control law of TSOFMPC in terms of the intermediate variable.

Now, we turn our attention to designing the state observer. First, we suppose v is not measurable, and that $f \neq f_0$. Then, v is not available (not exactly known), and the observer can be designed based on v^L as follows

$$\tilde{x}(k+1) = (A - LC) \tilde{x}(k) + Bv^L(k) + Ly(k). \quad (5.7.7)$$

When v is measurable or $f = f_0$, v is available (exactly known) and an estimator simpler than (5.7.7) can be adopted. The case in which v is available will be discussed later in section 5.9.

L in (5.7.7) is the observer gain matrix, defined as

$$L = A\tilde{P}_1 C^T \left(R_o + C\tilde{P}_1 C^T \right)^{-1} \quad (5.7.8)$$

where $\tilde{P}_1$ can be iterated from

$$\tilde{P}_j = Q_o + A\tilde{P}_{j+1} A^T - A\tilde{P}_{j+1} C^T \left(R_o + C\tilde{P}_{j+1} C^T \right)^{-1} C\tilde{P}_{j+1} A^T, \; j < N_o \quad (5.7.9)$$

where R_o, Q_o, $\tilde{P}_{N_o}$ and N_o are taken as tunable parameters.

By (5.4.1) and (5.7.7), denoting $e = x - \tilde{x}$ we can obtain the following closed-loop system:

$$\begin{cases} x(k+1) = (A+BK)\,x(k) - BKe(k) + B\left[h(v^L(k)) - v^L(k)\right] \\ e(k+1) = (A-LC)\,e(k) + B\left[h(v^L(k)) - v^L(k)\right] \end{cases} . \quad (5.7.10)$$

When $h = \tilde{1}$, the nonlinear item in (5.7.10) will disappear, and the studied problem will become linear. However, because of desaturation, the error encountered in solving equation, the modeling error of nonlinearity, etc., $h = \tilde{1}$ cannot hold.

5.8 Stability of TSOFMPC

Lemma 5.8.1. *Suppose X and Y are matrices, while s and t are vectors, all with appropriate dimensions, then*

$$2s^T XYt \le \gamma s^T XX^T s + 1/\gamma \cdot t^T Y^T Yt, \ \forall \gamma > 0. \quad (5.8.1)$$

Define $v^x(k) \triangleq Kx(k)$ and $v^e(k) \triangleq Ke(k)$, so that $v^x(k) = v^L(k) + v^e(k)$. In the following, we take $R = \lambda I$.

Theorem 5.8.1. *(Stability of TSOFMPC) For system represented by (5.4.1), TSOFMPC (5.7.5)-(5.7.7) is adopted. Suppose there exist positive scalars γ_1 and γ_2 such that the system design satisfies*

(i) $Q > P_0 - P_1$;

(ii) $(1 + 1/\gamma_2)\left(-Q_o + \tilde{P}_0 - LR_oL^T\right) - \tilde{P}_1$
$\quad < -\lambda^{-1}\,(1 + 1/\gamma_1)\,A^T P_1 B\left(\lambda I + B^T P_1 B\right)^{-1} B^T P_1 A;$

(iii) $\forall\left[x(0)^T, e(0)^T\right] \in \Omega \subset \Re^{2n},\ \forall k \ge 0,$
$\quad -\lambda h(v^L(k))^T h(v^L(k)) + \left[h(v^L(k)) - v^L(k)\right]^T$
$\quad \times \left[(1 + \gamma_1)\,(\lambda I + B^T P_1 B) + (1 + \gamma_2)\,\lambda B^T \tilde{P}_1 B\right]\left[h(v^L(k)) - v^L(k)\right] \le$
$\quad 0.$

Then, the equilibrium $\{x = 0, e = 0\}$ of the closed-loop system is exponentially stable and the region of attraction is Ω.

Proof. Choose a quadratic function as $V(k) = x(k)^T P_1 x(k) + \lambda e(k)^T \tilde{P}_1 e(k)$. Applying (5.7.10), we obtain the following:

$$V(k+1) - V(k)$$
$$= \left\{(A+BK)\,x(k) + B\left[h(v^L(k)) - v^L(k)\right]\right\}^T P_1\left\{(A+BK)\,x(k) \right.$$
$$\left. + B\left[h(v^L(k)) - v^L(k)\right]\right\}$$

$$- x(k)^T P_1 x(k) - 2e(k)^T K^T B^T P_1 \left\{ (A + BK) x(k) + B \left[h(v^L(k)) - v^L(k) \right] \right\}$$

$$+ e(k)^T K^T B^T P_1 BK e(k) + \lambda \left\{ (A - LC) e(k) + B \left[h(v^L(k)) - v^L(k) \right] \right\}^T \tilde{P}_1$$

$$\times \left\{ (A - LC) e(k) + B \left[h(v^L(k)) - v^L(k) \right] \right\} - \lambda e(k)^T \tilde{P}_1 e(k)$$

$$= x(k)^T \left(-Q + P_0 - P_1 - \lambda K^T K \right) x(k) - 2\lambda x(k)^T K^T \left[h(v^L(k)) - v^L(k) \right]$$

$$+ \left[h(v^L(k)) - v^L(k) \right]^T$$

$$\times B^T P_1 B \left[h(v^L(k)) - v^L(k) \right] - 2e(k)^T K^T B^T P_1 \left\{ (A + BK) x(k) \right.$$

$$+ B \left[h(v^L(k)) - v^L(k) \right] \right\}$$

$$+ e(k)^T K^T B^T P_1 BK e(k) + \lambda \left\{ (A - LC) e(k) + B \left[h(v^L(k)) - v^L(k) \right] \right\}^T \tilde{P}_1$$

$$\times \left\{ (A - LC) e(k) + B \left[h(v^L(k)) - v^L(k) \right] \right\} - \lambda e(k)^T \tilde{P}_1 e(k)$$

$$= x(k)^T \left(-Q + P_0 - P_1 \right) x(k) - \lambda v^x(k)^T v^x(k) - 2\lambda v^x(k)^T \left[h(v^L(k)) - v^L(k) \right]$$

$$+ \left[h(v^L(k)) - v^L(k) \right]^T$$

$$\times B^T P_1 B \left[h(v^L(k)) - v^L(k) \right] - 2e(k)^T K^T B^T P_1 (A + BK) x(k)$$

$$- 2e(k)^T K^T B^T P_1 B \left[h(v^L(k)) - v^L(k) \right]$$

$$+ e(k)^T K^T B^T P_1 BK e(k) + \lambda e(k)^T (A - LC)^T \tilde{P}_1 (A - LC) e(k)$$

$$+ 2\lambda e(k)^T (A - LC)^T \tilde{P}_1 B \left[h(v^L(k)) - v^L(k) \right]$$

$$+ \lambda \left[h(v^L(k)) - v^L(k) \right]^T B^T \tilde{P}_1 B \left[h(v^L(k)) - v^L(k) \right] - \lambda e(k)^T \tilde{P}_1 e(k)$$

$$= x(k)^T \left(-Q + P_0 - P_1 \right) x(k) - \lambda v^x(k)^T v^x(k) - 2\lambda v^x(k)^T \left[h(v^L(k)) - v^L(k) \right]$$

$$+ \left[h(v^L(k)) - v^L(k) \right]^T$$

$$\times B^T \left(P_1 + \lambda \tilde{P}_1 \right) B \left[h(v^L(k)) - v^L(k) \right] + 2\lambda v^e(k)^T v^x(k) - 2v^e(k)^T B^T P_1 B$$

$$\times \left[h(v^L(k)) - v^L(k) \right] + v^e(k)^T B^T P_1 B v^e(k)$$

$$+ \lambda e(k)^T \left[(A - LC)^T \tilde{P}_1 (A - LC) - \tilde{P}_1 \right] e(k)$$

$$+ 2\lambda e(k)^T (A - LC)^T \tilde{P}_1 B \left[h(v^L(k)) - v^L(k) \right]$$

$$= x(k)^T \left(-Q + P_0 - P_1 \right) x(k) - \lambda v^e(k)^T v^e(k) - \lambda v^L(k)^T v^L(k) - 2\lambda v^e(k)^T v^L(k)$$

$$- 2\lambda v^e(k)^T \left[h(v^L(k)) - v^L(k) \right] - 2\lambda v^L(k)^T \left[h(v^L(k)) - v^L(k) \right]$$

$$+ \left[h(v^L(k)) - v^L(k) \right]^T B^T \left(P_1 + \lambda \tilde{P}_1 \right) B \left[h(v^L(k)) - v^L(k) \right]$$

$$+ 2\lambda v^e(k)^T v^e(k) + 2\lambda v^e(k)^T v^L(k) - 2v^e(k)^T B^T P_1 B \left[h(v^L(k)) - v^L(k) \right]$$

$$+ v^e(k)^T B^T P_1 B v^e(k) + \lambda e(k)^T \left[(A - LC)^T \tilde{P}_1 (A - LC) - \tilde{P}_1 \right] e(k)$$

$$+ 2\lambda e(k)^T (A - LC)^T \tilde{P}_1 B \left[h(v^L(k)) - v^L(k) \right]$$

$$=x(k)^T\left(-Q+P_0-P_1\right)x(k)-\lambda h(v^L(k))^T h(v^L(k))+\left[h(v^L(k))-v^L(k)\right]^T$$
$$\times\left(\lambda I+B^T P_1 B+\lambda B^T\tilde{P}_1 B\right)\left[h(v^L(k))-v^L(k)\right]$$
$$-2v^e(k)^T\left(\lambda I+B^T P_1 B\right)\left[h(v^L(k))-v^L(k)\right]+v^e(k)^T\left(\lambda I+B^T P_1 B\right)v^e(k)$$
$$+\lambda e(k)^T\left[(A-LC)^T\tilde{P}_1(A-LC)-\tilde{P}_1\right]$$
$$e(k)+2\lambda e(k)^T(A-LC)^T\tilde{P}_1 B\left[h(v^L(k))-v^L(k)\right].$$

By applying Lemma 5.8.1 twice, we obtain

$$V(k+1)-V(k)$$
$$\leq x(k)^T\left(-Q+P_0-P_1\right)x(k)-\lambda h(v^L(k))^T h(v^L(k))$$
$$+\left[h(v^L(k))-v^L(k)\right]^T\left[(1+\gamma_1)\left(\lambda I+B^T P_1 B\right)+(1+\gamma_2)\lambda B^T\tilde{P}_1 B\right]$$
$$\times\left[h(v^L(k))-v^L(k)\right]+(1+1/\gamma_1)v^e(k)^T\left(\lambda I+B^T P_1 B\right)v^e(k)$$
$$+\lambda e(k)^T\left[(1+1/\gamma_2)(A-LC)^T\tilde{P}_1(A-LC)-\tilde{P}_1\right]e(k)$$
$$=x(k)^T\left(-Q+P_0-P_1\right)x(k)-\lambda h(v^L(k))^T h(v^L(k))$$
$$+\left[h(v^L(k))-v^L(k)\right]^T\left[(1+\gamma_1)\left(\lambda I+B^T P_1 B\right)+(1+\gamma_2)\lambda B^T\tilde{P}_1 B\right]$$
$$\times\left[h(v^L(k))-v^L(k)\right]+(1+1/\gamma_1)v^e(k)^T\left(\lambda I+B^T P_1 B\right)v^e(k)$$
$$+(1+1/\gamma_2)\lambda e(k)^T\left(-Q_o+\tilde{P}_0-LR_o L^T\right)e(k)-\lambda e(k)^T\tilde{P}_1 e(k).$$

With conditions (i)-(iii) satisfied, $V(k+1)-V(k)<0$, $\forall\left[x(k)^T,e(k)^T\right]\neq 0$. Hence, $V(k)$ is Lyapunov function that proves exponential stability. □

Conditions (i)-(ii) in Theorem 5.8.1 are the requirements imposed on R, Q, P_N, N, R_o, Q_o, $\tilde{P}_{N_o}$ and N_o, while (iii) is the requirement imposed on h. Generally, decreasing the equation solution error, γ_1 and γ_2 will improve satisfaction of (iii). When $h=\tilde{1}$, (i)-(ii) are stability conditions of the linear system.

If (i)-(ii) are satisfied but $h\neq\tilde{1}$, then we can obtain more sensible conditions under (iii). For this reason we adopt Assumptions 5.4.4-5.4.5.

Corollary 5.8.1. *(Stability of TSOFMPC) For systems represented by* (5.4.1), *TSOFMPC* (5.7.5)-(5.7.7) *is adopted, where h satisfies* (5.4.11). *Suppose*

(A1) $\forall\left[x(0),e(0)\right]\in\Omega\subset\mathfrak{R}^{2n}$, $\left\|v^L(k)\right\|\leq\Delta$ *for all* $k\geq 0$;

(A2) there exist positive scalars γ_1 and γ_2 such that the system design satisfies (i)-(ii) in Theorem 5.8.1 and

$$(iii)\quad -\lambda\left[b_1^2-(1+\gamma_1)(b-1)^2\right]$$
$$+(b-1)^2\sigma_{\max}\left((1+\gamma_1)B^T P_1 B+(1+\gamma_2)\lambda B^T\tilde{P}_1 B\right)\leq 0.$$

Then, the equilibrium $\{x = 0,\ e = 0\}$ of the closed-loop system is exponentially stable with a region of attraction Ω.

Proof. Applying condition (iii) in Theorem 5.8.1 and (5.4.11) to obtain the following result:

$$
\begin{aligned}
&-\lambda h(s)^T h(s) + [h(s) - s]^T \left[(1 + \gamma_1)\left(\lambda I + B^T P_1 B\right) + (1 + \gamma_2)\lambda B^T \tilde{P}_1 B \right] \\
&[h(s) - s] \\
&\leq -\lambda b_1^2 s^T s + (b - 1)^2\, \sigma_{\max}\left((1 + \gamma_1)\left(\lambda I + B^T P_1 B\right) + (1 + \gamma_2)\lambda B^T \tilde{P}_1 B \right) s^T s \\
&= -\lambda b_1^2 s^T s + (1 + \gamma_1)\lambda\,(b - 1)^2\, s^T s + (b - 1)^2\, \sigma_{\max} \\
&\quad \times \left((1 + \gamma_1) B^T P_1 B + (1 + \gamma_2)\lambda B^T \tilde{P}_1 B \right) s^T s \\
&= -\lambda\left[b_1^2 - (1 + \gamma_1)(b - 1)^2 \right] s^T s + (b - 1)^2\, \sigma_{\max} \\
&\quad \times \left((1 + \gamma_1) B^T P_1 B + (1 + \gamma_2)\lambda B^T \tilde{P}_1 B \right) s^T s.
\end{aligned}
$$

Hence, if condition (iii) is satisfied, then condition (iii) in Theorem 5.8.1 will also be satisfied. This proves stability. $\square$

Remark 5.8.1. We can substitute (iii) in Corollary 5.8.1 by the following more conservative condition:
$$
-\lambda\left[b_1^2 - (1 + \gamma_1)(b - 1)^2 - (1 + \gamma_2)(b - 1)^2\, \sigma_{\max}\left(B^T \tilde{P}_1 B \right) \right]
$$
$$
+ (1 + \gamma_1)(b - 1)^2\, \sigma_{\max}\left(B^T P_1 B \right) \leq 0.
$$
The above condition will be useful in the following.

About the region of attraction in Theorem 5.8.1 and Corollary 5.8.1, we have the following result.

Theorem 5.8.2. *(Region of attraction of TSOFMPC) For systems represented by (5.4.1), TSOFMPC (5.7.5)-(5.7.7) is adopted where h satisfies (5.4.11). Suppose there exist positive scalars Δ, γ_1 and γ_2 such that conditions (i)-(iii) in Corollary 5.8.1 are satisfied. Then, the region of attraction for the closed-loop system is not smaller than*

$$
S_c = \left\{ (x, e) \in \Re^{2n} | x^T P_1 x + \lambda e^T \tilde{P}_1 e \leq c \right\}, \tag{5.8.2}
$$

where

$$
c = (\Delta/d)^2, \quad d = \left\| (\lambda I + B^T P_1 B)^{-1} B^T P_1 A \left[P_1^{-1/2}, \quad -\lambda^{-1/2} \tilde{P}_1^{-1/2} \right] \right\|. \tag{5.8.3}
$$

Proof. Having satisfied conditions (i)-(iii) in Corollary 5.8.1, we need only to verify that $\forall\, [x(0),\ e(0)] \in S_c$, $\left\| v^L(k) \right\| \leq \Delta$ for all $k \geq 0$.

We adopt two nonsingular transformations:

$$\bar{x} = P_1^{1/2}x, \ \bar{e} = \lambda^{1/2}\tilde{P}_1^{1/2}e.$$

Then, $\forall [x(0), e(0)] \in S_c$, $\left\| [\bar{x}(0)^T \ \bar{e}(0)^T] \right\| \leq \sqrt{c}$ and

$$\begin{aligned}
\left\| v^L(0) \right\| &= \left\| (\lambda I + B^T P_1 B)^{-1} B^T P_1 A \left[x(0) - e(0) \right] \right\| \\
&\leq \left\| \left[(\lambda I + B^T P_1 B)^{-1} B^T P_1 A P_1^{-1/2} \right. \right. \\
&\qquad \left. \left. - (\lambda I + B^T P_1 B)^{-1} B^T P_1 A \lambda^{-1/2} \tilde{P}_1^{-1/2} \right] \right\| \\
&\quad \times \left\| [\bar{x}(0)^T \ \bar{e}(0)^T] \right\| \leq \Delta.
\end{aligned} \qquad (5.8.4)$$

Thus, all the conditions in Corollary 5.8.1 are satisfied at time $k = 0$ if $[x(0), e(0)] \in S_c$. According to the proof of Theorem 5.8.1, $[x(1), e(1)] \in S_c$ if $[x(0), e(0)] \in S_c$. Therefore, $\left\| v^L(1) \right\| \leq \Delta$ for all $[x(0), e(0)] \in S_c$, which shows that all the conditions in Corollary 5.8.1 are satisfied at time $k = 1$. By analogy, we can conclude that $\left\| v^L(k) \right\| \leq \Delta$ for all $[x(0), e(0)] \in S_c$. Thus, S_c is a region of attraction. $\qquad \square$

By applying Theorem 5.8.2, we can tune the control parameters so as to satisfy conditions (i)-(iii) in Corollary 5.8.1 and obtain the desired region of attraction. The following algorithm may serve as a guideline.

Algorithm 5.9 (Parameter tuning guideline for achieving the desired region of attraction Ω)

Step 1. Define the accuracy of the equation solution. Choose the initial Δ. Determine b_1 and b.

Step 2. Choose $\left\{ R_o, Q_o, \tilde{P}_{N_o}, N_o \right\}$ rendering a convergent observer.

Step 3. Choose $\{\lambda, Q, P_N, N\}$ (mainly Q, P_N, N) satisfying (i).

Step 4. Choose $\{\gamma_1, \gamma_2, \lambda, Q, P_N, N\}$ (mainly $\gamma_1, \gamma_2, \lambda$) satisfying (ii)-(iii). If they cannot be both satisfied, then go to one of Steps 1-3 (according to the actual situation).

Step 5. Check if (i)-(iii) are all satisfied. If they are not, go to Step 3. Otherwise, decrease γ_1 and γ_2 (maintaining satisfaction of (ii)) and increase Δ (b_1 is decreased accordingly, maintaining satisfaction of (iii)).

Step 6. Calculate c using (5.8.3). If $S_c \supseteq \Omega$, then STOP, else turn to Step 1.

Of course, this does not mean that any desired region of attraction can be obtained for any system. But if A has all its eigenvalues inside or on the circle, we have the following conclusion.

Theorem 5.8.3. *(Semi-global stability of TSOFMPC) For systems repre-sented by (5.4.1), TSOFMPC (5.7.5)-(5.7.7) is adopted where h satisfies (5.4.11). Suppose A has all its eigenvalues inside or on the circle, and there exist Δ and γ_1 such that $b_1^2 - (1 + \gamma_1)(b-1)^2 > 0$ in the absence of input sat-uration constraint. Then, for any bounded set $\Omega \subset \Re^{2n}$, the controller and the observer parameters can be adjusted to make the closed-loop system possess a region of attraction not smaller than Ω.*

Proof. In the absence of saturation, b_1 and b are determined independently of the controller parameters. Denote the parameters $\{\gamma_1, \Delta\}$ that make $b_1^2 - (1 + \gamma_1)(b-1)^2 > 0$ as $\{\gamma_1^0, \Delta^0\}$. When there is saturation still choose $\gamma_1 = \gamma_1^0$ and $\Delta = \Delta^0$. Then, the following two cases may occur:

Case 1: $b_1^2 - (1 + \gamma_1)(b-1)^2 > 0$ as $\lambda = \lambda_0$. Decide the parameters in the following way:

(A) Choose

$$P_N = Q + A^T P_N A - A^T P_N B \left(\lambda I + B^T P_N B\right)^{-1} B^T P_N A.$$

Then, $P_0 - P_1 = 0$. Furthermore, choose $Q > 0$, then $Q > P_0 - P_1$ and condition (i) in Corollary 5.8.1 will be satisfied for all λ and N. Choose an arbitrary N.

(B) Choose $R_o = \varepsilon I$, $Q_o = \varepsilon I > 0$, where ε is a scalar. Choose

$$\tilde{P}_{N_o} = Q_o + A\tilde{P}_{N_o}A^T - A\tilde{P}_{N_o}C^T \left(R_o + C\tilde{P}_{N_o}C^T\right)^{-1} C\tilde{P}_{N_o}A^T$$

and an arbitrary N_o, then, $\tilde{P}_0 - \tilde{P}_1 = 0$. On the other hand, if we denote

$$\tilde{P}_{N_o}^I = I + A\tilde{P}_{N_o}^I A^T - A\tilde{P}_{N_o}^I C^T \left(I + C\tilde{P}_{N_o}^I C^T\right)^{-1} C\tilde{P}_{N_o}^I A^T,$$

then there exist $\gamma_2 > 0$ and $\xi > 0$ such that

$$(1 + 1/\gamma_2) I - 1/\gamma_2 \tilde{P}_{N_o}^I \geq \xi I.$$

Since $\tilde{P}_{N_o} = \varepsilon \tilde{P}_{N_o}^I$ and $\tilde{P}_1 = \tilde{P}_{N_o}$,

$$(1 + 1/\gamma_2) Q_o - 1/\gamma_2 \tilde{P}_1 \geq \varepsilon \xi I$$

holds for all ε. Choose a sufficiently small ε such that

$$b_1^2 - (1 + \gamma_1)(b-1)^2 - (1 + \gamma_2)(b-1)^2 \sigma_{\max}\left(B^T \tilde{P}_1 B\right) > 0.$$

At this point,

$$(1 + 1/\gamma_2)(-Q_o + \tilde{P}_0 - LR_oL^T) - \tilde{P}_1$$
$$= (1 + 1/\gamma_2)(-Q_o - LR_oL^T) + 1/\gamma_2\tilde{P}_1 \leq -\varepsilon\xi I.$$

(C) Multiplying both sides of

$$P_1 = Q + A^T P_1 A - A^T P_1 B \left(\lambda I + B^T P_1 B\right)^{-1} B^T P_1 A$$

by λ^{-1}, then

$$\bar{P}_1 = \lambda^{-1} Q + A^T \bar{P}_1 A - A^T \bar{P}_1 B \left(I + B^T \bar{P}_1 B\right)^{-1} B^T \bar{P}_1 A.$$

Since A has all its eigenvalues inside or on the circle, we know that $\bar{P}_1 \to 0$ as $\lambda \to \infty$ (refer to [46]). Hence, there exists $\lambda_1 \geq \lambda_0$ such that $\forall \lambda \geq \lambda_1$,

(a) $\lambda^{-1} \left(1 + 1/\gamma_1\right) A^T P_1 B \left(\lambda I + B^T P_1 B\right)^{-1} B^T P_1 A$
$= \left(1 + 1/\gamma_1\right) A^T \bar{P}_1 B \left(I + B^T \bar{P}_1 B\right)^{-1} B^T \bar{P}_1 A < \varepsilon \xi I,$
i.e., condition (ii) in Corollary 5.8.1 is satisfied,

(b) $- \left[b_1^2 - (1 + \gamma_1)(b-1)^2 - (1 + \gamma_2)(b-1)^2 \sigma_{\max}\left(B^T \tilde{P}_1 B\right)\right]$
$+ (b-1)^2 (1 + \gamma_1) \sigma_{\max}\left(B^T \bar{P}_1 B\right) \leq 0,$
i.e., the inequality in Remark 5.8.1 is satisfied and in turn, condition (iii) in Corollary 5.8.1 is satisfied.

(D) Further, there exists $\lambda_2 \geq \lambda_1$ such that $\forall \lambda \geq \lambda_2$, $\left\|\bar{P}_1^{1/2} A \bar{P}_1^{-1/2}\right\| \leq \sqrt{2}$ (refer to [46]). Now, let

$$\bar{c} = \sup_{\lambda \in [\lambda_2, \infty), (x,e) \in \Omega} \left(x^T \bar{P}_1 x + e^T \tilde{P}_1 e\right),$$

then $\Omega \subseteq \bar{S}_{\bar{c}} = \left\{(x,e) \in \Re^{2n} | x^T \bar{P}_1 x + e^T \tilde{P}_1 e \leq \bar{c}\right\}$. Define two transformations: $\bar{x} = \bar{P}_1^{1/2} x$ and $\bar{e} = \tilde{P}_1^{1/2} e$. Then, there exists $\lambda_3 \geq \lambda_2$ such that $\forall \lambda \geq \lambda_3$ and $\forall \left[x(0)^T, e(0)^T\right] \in \bar{S}_{\bar{c}}$,

$$\left\|v^L(0)\right\| = \left\|(\lambda I + B^T P_1 B)^{-1} B^T P_1 A \left[x(0) - e(0)\right]\right\|$$
$$= \left\|\left[(\lambda I + B^T P_1 B)^{-1} B^T P_1 A,\right.\right.$$
$$\left.- (\lambda I + B^T P_1 B)^{-1} B^T P_1 A\right] \left[x(0)^T, e(0)^T\right]^T \right\|$$
$$= \left\|\left[(\lambda I + B^T P_1 B)^{-1} B^T P_1 A \bar{P}_1^{-1/2},\right.\right.$$
$$\left.- (\lambda I + B^T P_1 B)^{-1} B^T P_1 A \tilde{P}_1^{-1/2}\right] \left[\bar{x}(0)^T, \bar{e}(0)^T\right]^T \right\|$$
$$\leq \left\|\left[(\lambda I + B^T P_1 B)^{-1} B^T P_1 A \bar{P}_1^{-1/2},\right.\right.$$
$$\left.- (\lambda I + B^T P_1 B)^{-1} B^T P_1 A \tilde{P}_1^{-1/2}\right]\right\| \left\|\left[\bar{x}(0)^T, \bar{e}(0)^T\right]\right\|$$
$$\leq \left\|\sqrt{2}(I + B^T \bar{P}_1 B)^{-1} B^T \bar{P}_1^{1/2}, \; - (I + B^T \bar{P}_1 B)^{-1} B^T \bar{P}_1 A \tilde{P}_1^{-1/2}\right\|$$
$$\sqrt{\bar{c}} \leq \Delta.$$

Hence, for $\forall \left[x(0)^T, \ e(0)^T \right] \in \Omega$, the conditions in Corollary 5.8.1 can be satisfied at time $k = 0$, and by the proof of Theorem 5.8.2, they are also satisfied for $\forall k > 0$.

Through the above decision procedure, the designed controller will have the desired region of attraction.

Case 2: $b_1^2 - (1 + \gamma_1)(b - 1)^2 \leq 0$ as $\lambda = \lambda_0$, which apparently is due to the fact that the control action is restricted too much by the saturation constraint. For the same reason as in Case 1 (C), by (5.7.4), we know that for any bounded Ω, there exists sufficiently large $\lambda_4 \geq \lambda_0$ such that for $\forall \lambda \geq \lambda_4$ and $\forall \left[x(k)^T, e(k)^T \right] \in \Omega$, $\hat{u}(k)$ does not violate the saturation constraint. This process is equivalent to decreasing Δ and re-determining b_1 and b such that $b_1^2 - (1 + \gamma_1)(b - 1)^2 > 0$.

In a word, if the region of attraction is not satisfactory with $\lambda = \lambda_0$, then it can be satisfied by choosing $\lambda \geq \max\{\lambda_3, \lambda_4\}$ and suitable $\left\{ Q, P_N, N, R_o, Q_o, \tilde{P}_{N_o}, N_o \right\}$. □

In the proof of Theorem 5.8.3, we have emphasized the effect of tuning λ. If A has all its eigenvalues inside or on the circle, then by properly fixing other parameters, we can tune λ to obtain the arbitrarily large bounded region of attraction. This is very important because many industrial processes can be represented by stable models plus integrals. When A has no eigenvalue outside of the unit circle, but has eigenvalues on the unit circle, then the corresponding system can be critical stable or unstable; however, by slight controls this system can be stabilized.

5.9 TSOFMPC: case where the intermediate variable is available

When v is available (exactly known), the following state observer can be applied to obtain the estimation:

$$\tilde{x}(k + 1) = (A - LC)\,\tilde{x}(k) + Bv(k) + Ly(k). \qquad (5.9.1)$$

As stated in section 5.7, this case occurs when v is measurable or $f = f_0$. Note that if $f = f_0$, then $v(k) = f(u(k))$ can be obtained, i.e., $v(k)$ can be easily calculated by applying $u(k)$.

All other design details are the same as the case where v is "unavailable." In the following, we show that, when we substitute (5.7.7) with (5.9.1), the following more relaxed conclusion can be obtained.

Similarly to (5.7.10), the closed-loop system with observer (5.9.1) is

$$\begin{cases} x(k + 1) = (A + BK)\,x(k) - BKe(k) + B\left[h(v^L(k)) - v^L(k) \right] \\ e(k + 1) = (A - LC)\,e(k) \end{cases} . \qquad (5.9.2)$$

We choose Lyapunov function as $\hat{V}(k) = x(k)^T P_1 x(k) + e(k)^T \tilde{P}_1 e(k)$. Similarly as in section 5.8 we can obtain some stability results (for brevity we omit the proofs).

Theorem 5.9.1. *(Stability of TSOFMPC) For systems represented by (5.4.1), TSOFMPC (5.7.5), (5.7.6) and (5.9.1) is adopted. Suppose there exists a positive scalar γ such that the design of the controller satisfies the following conditions:*

(i) $Q > P_0 - P_1$;

(ii) $-Q_o + \tilde{P}_0 - \tilde{P}_1 - LR_oL^T < -(1 + 1/\gamma)\, A^T P_1 B\, (R + B^T P_1 B)^{-1}\, B^T P_1 A$;

(iii) $\forall\, [x(0)^T, e(0)^T] \in \Omega \subset \mathfrak{R}^{2n},\ \forall k \geq 0$,
$-h(v^L(k))^T R h(v^L(k))$
$+ (1 + \gamma) \left[h(v^L(k)) - v^L(k) \right]^T (R + B^T P_1 B) \left[h(v^L(k)) - v^L(k) \right] \leq 0.$

Then, the equilibrium $\{x = 0,\ e = 0\}$ of the closed-loop system is exponentially stable with a region of attraction Ω.

Corollary 5.9.1. *(Stability of TSOFMPC) For systems represented by (5.4.1), TSOFMPC (5.7.5), (5.7.6) and (5.9.1) is adopted, where h satisfies (5.4.11). Suppose*

(A1) $\forall\, [x(0), e(0)] \in \Omega \subset \mathfrak{R}^{2n},\ \|v^L(k)\| \leq \Delta$ *for all $k \geq 0$*;

(A2) *there exists a positive scalar γ such that the system design satisfies conditions (i)-(ii) in Theorem 5.9.1 and*

(iii) $-\lambda \left[b_1^2 - (1 + \gamma)(b - 1)^2 \right] + (1 + \gamma)(b - 1)^2\, \sigma_{\max}\left(B^T P_1 B \right) \leq 0.$

Then, the equilibrium $\{x = 0, e = 0\}$ of the closed-loop system is exponentially stable with a region of attraction Ω.

Theorem 5.9.2. *(Region of attraction of TSOFMPC) For systems represented by (5.4.1), TSOFMPC (5.7.5), (5.7.6) and (5.9.1) is adopted, where h satisfies (5.4.11). Suppose there exists a positive scalar Δ and γ such that the designed system satisfies conditions (i)-(iii) in Corollary 5.9.1. Then, the region of attraction for the closed-loop system is not smaller than*

$$\hat{S}_{\hat{c}} = \left\{ (x, e) \in \mathfrak{R}^{2n} | x^T P_1 x + e^T \tilde{P}_1 e \leq \hat{c} \right\}, \qquad (5.9.3)$$

where

$$\hat{c} = \left(\Delta / \hat{d} \right)^2, \quad \hat{d} = \left\| (\lambda I + B^T P_1 B)^{-1} B^T P_1 A \left[P_1^{-1/2}, \ -\tilde{P}_1^{-1/2} \right] \right\|. \qquad (5.9.4)$$

Theorem 5.9.3. *(Semi-global stability of TSOFMPC) For systems repre-sented by (5.4.1), TSOFMPC (5.7.5), (5.7.6) and (5.9.1) is adopted, where h satisfies (5.4.11). Suppose A has all its eigenvalues inside or on the circle, and there exist Δ and γ such that $b_1^2 - (1+\gamma)(b-1)^2 > 0$ in the absence of in-put saturation constraints. Then, for any bounded set $\Omega \subset \Re^{2n}$, the controller and the observer parameters can be adjusted to make the closed-loop system possess a region of attraction not smaller than Ω.*

For two-step MPC one can further refer to [4], [13]. For solving the non-linear algebraic equation (group) one can refer to [53].

Remark 5.9.1. The ellipsoidal regions of attraction, given for TSMPC and TSOFMPC, are relatively conservative with respect to its corresponding con-trol laws. In general, with a set of controller parameters given, the maximum region of attraction for TSMPC is non-ellipsoidal. If the closed-loop system is linear, then the region of attraction can be computed using Algorithms 5.3 and 5.4. However, if there is any uncertainty in the linear system, then the region of attraction calculated by Algorithms 5.3 and 5.4 is not the actual maximum region of attraction (i.e., when the state lies outside of the region of attraction calculated by Algorithms 5.3 and 5.4, the real system can still be stable).

Remark 5.9.2. For the nonlinear separation MPC for the Hammerstein model, if the nonlinear reversion is utterly accurate (the nonlinear algebraic equation is exactly solved), then the overall region of attraction is not affected by the nonlinear separation method. That is to say, the nonlinear separation itself cannot decrease the volume of the region of attraction.

Chapter 6

Sketch of synthesis approaches of MPC

In Chapter 1, we explained that, since industrial MPC lacks guarantee of stability, the synthesis approach of MPC was widely studied in the 1990s and some control algorithms which emerged in the early 1970s are also recognized as MPC. In the late 1990s, the basic frame of synthesis approach (especially for the state feedback case) reached maturity. People began to have largely different ideas about MPC, which is not restricted to industrial MPC.

Synthesis approach of MPC is inherited from, and has developed the traditional optimal control. Synthesis approach of MPC considers model uncertainty, input/output constraints, etc., based on the traditional optimal control. Therefore, although it is rare to see that synthesis approach is applied in process industries, we should not be surprised that synthesis approach will have taken effect in some engineering areas. As a kind of theory, the importance of synthesis approach should not be judged based on whether or not it can be soon applied. Rather, it should be judged based on whether or not it can solve some important issues in the whole control theory. To some extent, just because some important problems (stability, robustness, convergence, computational complexity) are yet to be solved, we cannot apply some advanced control algorithms.

For sections 6.1-6.5, one is referred to in [49], for the continuous-time system also to in [48]. For section 6.6 one is referred to in [5].

6.1 General idea: case discrete-time systems

6.1.1 Modified optimization problem

The nonlinear system is described by

$$x(k+1) = f(x(k), u(k)), \; y(k) = g(x(k)) \tag{6.1.1}$$

141

satisfying $f(0,0) = 0$. The state is measurable. Consider the following constraints

$$x(k) \in \mathcal{X}, \ u(k) \in \mathcal{U}, \ k \geq 0. \tag{6.1.2}$$

$\mathcal{X} \subseteq \mathfrak{R}^n$ is convex and closed, $\mathcal{U} \subseteq \mathfrak{R}^m$ convex and compact, satisfying $\mathcal{X} \supset \{0\}$, $\mathcal{U} \supset \{0\}$. At each time k, the cost is defined by

$$J_N(k) = \sum_{i=0}^{N-1} \ell(x(k+i|k), u(k+i|k)) + F(x(k+N|k)). \tag{6.1.3}$$

Suppose the stage cost $\ell(x,u) \geq c \left\| [x^T \ u^T] \right\|^2$, $\ell(0,0) = 0$. The following constraint is called terminal constraint (which is artificial)

$$x(k+N|k) \in \mathcal{X}_f \subseteq \mathcal{X}. \tag{6.1.4}$$

At each time k the following optimization problem is to be solved:

$$\min_{\tilde{u}_N(k)} J_N(k), \tag{6.1.5}$$

$$\text{s.t. } x(k+i+1|k) = f\left(x(k+i|k), u(k+i|k)\right), \ i \geq 0, \ x(k|k) = x(k), \tag{6.1.6}$$

$$x(k+i|k) \in \mathcal{X}, \ u(k+i|k) \in \mathcal{U}, \ i \in \{0,1,\dots,N-1\}, \ x(k+N|k) \in \mathcal{X}_f, \tag{6.1.7}$$

where, $\tilde{u}_N(k) = \{u(k|k), u(k+1|k), \cdots, u(k+N-1|k)\}$ is the decision variable. The solution to (6.1.5)-(6.1.7) is denoted as $\tilde{u}_N^*(k) = \{u^*(k|k), u^*(k+1|k), \cdots, u^*(k+N-1|k)\}$, and the corresponding optimal cost value as $J_N^*(k)$, the state as $x^*(k+i|k)$, $\forall i > 0$. Notice that more concrete expression of $J_N(k)$ should be $J_N(x(k))$ or $J_N(x(k), \tilde{u}_N(k))$, representing that $J_N(k)$ is function of $x(k)$ and $\tilde{u}_N(k)$.

In $\tilde{u}_N^*(k)$, only $u(k) = u^*(k|k)$ is implemented. Hence, the following implicit control law is formed

$$K_N(x(k)) = u^*(k|k).$$

Because N is finite, the minimum of (6.1.5)-(6.1.7) exists if f, ℓ, F are continuous, $\mathcal{U}$ compact, $\mathcal{X}_f$ and $\mathcal{X}$ closed.

Remark 6.1.1. Dynamic programming could, in principle, be used on (6.1.5)-(6.1.7) to determine a sequence $\{J_i(\cdot)\}$ of value functions and a sequence of control laws $\{K_i(\cdot)\}$. If that is the case, the closed-form of $K_N(\cdot)$ is obtained and there is no necessity for receding-horizon optimization. In real applications, this is generally impossible (except for linear time-invariant unconstrained systems). In MPC, usually receding-horizon solution of $u^*(k|k)$, rather than the off-line solution of $K_N(\cdot)$, is applied. Therefore, as has been continuously emphasized in this book, the difference between MPC and the traditional optimal control lies in the implementation.

Remark 6.1.2. Receding-horizon implementation is the unique feature of MPC. Hence, gain-scheduling controller, whichever branch of control theories it appears in, can always be regarded as MPC.

6.1.2 "Three ingredients" and the uniform ideas for stability proof

Usually, the optimum of the cost function (notice that the optimum is the result of the optimization; the optimum of the cost function is called value function) is served as Lyapunov function. Suppose the value function is continuous. Suppose in the terminal constraint set $\mathcal{X}_f$, there is local controller $K_f(\cdot)$. $F(\cdot)$, $\mathcal{X}_f$ and $K_f(\cdot)$ are "three ingredients" of synthesis approaches. Denote $\Delta J_N^*(k+1) = J_N^*(k+1) - J_N^*(k)$.

Method 1 (Direct method) Employ the value function as Lyapunov function, and obtain conditions on $F(\cdot)$, $\mathcal{X}_f$ and $K_f(\cdot)$ that ensure

$$\Delta J_N^*(k+1) + \ell(x(k), K_N(x(k))) \leq 0.$$

Usually, in Method 1, $J_N^*(k+1)$ is not calculated. Instead, the feasible solution of $\tilde{u}_N(k+1)$ is invoked to calculate $J_N(k+1)$, the upper bound of $J_N^*(k+1)$.

Method 2 (Monotonicity method) Employ the value function as Lyapunov function, obtain conditions on $F(\cdot)$, $\mathcal{X}_f$ and $K_f(\cdot)$ such that

$$\Delta J_N^*(k+1) + \ell(x(k), K_N(x(k))) \leq J_N^*(k+1) - J_{N-1}^*(k+1),$$

and show that the right sight of the above is negative-definite.

In most situations, Methods 1 and 2 are equivalent.

6.1.3 Direct method for stability proof

Definition 6.1.1. $S_i(\mathcal{X}, \mathcal{X}_f)$ *is called an i-step stabilizable set contained in* $\mathcal{X}$ *for the system (6.1.1), if $\mathcal{X}_f$ is a control invariant set of $\mathcal{X}$ and $S_i(\mathcal{X}, \mathcal{X}_f)$ contains all state in $\mathcal{X}$ for which there exists an admissible control sequence of length i which will drive the states of the system to $\mathcal{X}_f$ in i steps or less, while keeping the evolution of the state inside $\mathcal{X}$, i.e.,*

$$S_i(\mathcal{X}, \mathcal{X}_f) \triangleq \{x(0) \in \mathcal{X} : \exists u(0), u(1), \cdots, u(i-1) \in \mathcal{U},$$
$$\exists M \leq i \text{ such that } x(1), x(2), \cdots, x(M-1) \in \mathcal{X} \text{ and}$$
$$x(M), x(M+1), \cdots, x(i) \in \mathcal{X}_f, \mathcal{X}_f \text{ is invariant}\}.$$

For the above definition, the readers are referred to Definitions 1.6.1, 1.6.2 and 1.8.1.

According to Definition 6.1.1, $S_0(\mathcal{X}, \mathcal{X}_f) = \mathcal{X}_f$.

Suppose the following conditions are satisfied:

$$\mathcal{X}_f \subseteq \mathcal{X}, \ K_f(x) \in \mathcal{U}, \ f(x, K_f(x)) \in \mathcal{X}_f, \ \forall x \in \mathcal{X}_f.$$

If, at time k, solving (6.1.5)-(6.1.7) gives solution $\tilde{u}_N^*(k)$, then, at time $k+1$, the following is feasible for the optimization problem:

$$\tilde{u}_N(k+1) = \{u^*(k+1|k), \cdots, u^*(k+N-1|k), K_f(x^*(k+N|k))\}.$$

The state trajectory resulting from $\tilde{u}_N(k+1)$ is

$$\{x^*(k+1|k), \cdots, x^*(k+N|k), f(x^*(k+N|k), K_f(x^*(k+N|k)))\},$$

where $x^*(k+1|k) = x(k+1)$.

The cost value associated with $\tilde{u}_N(k+1)$ is

$$\begin{aligned}
J_N(k+1) = {} & J_N^*(k) - \ell(x(k), K_N(x(k))) - F(x^*(k+N|k)) \\
& + \ell(x^*(k+N|k), K_f(x^*(k+N|k))) \\
& + F(f(x^*(k+N|k), K_f(x^*(k+N|k)))).
\end{aligned} \qquad (6.1.8)$$

At time $k+1$, by re-optimization of $J_N(k+1)$, $J_N^*(k+1)$ is obtained which, according to the principle of optimality, $J_N^*(k+1){\leq}J_N(k+1)$. Suppose $\Delta F(f(x, K_f(x))) = F(f(x, K_f(x))) - F(x)$ satisfies

$$\Delta F(f(x, K_f(x))) + \ell(x, K_f(x)){\leq}0, \ \forall x \in \mathcal{X}_f. \qquad (6.1.9)$$

Combining (6.1.8) and (6.1.9) and considering $J_N^*(k+1) \leq J_N(k+1)$ yields

$$J_N^*(k+1){\leq}J_N^*(k) - \ell(x(k), K_N(x(k))). \qquad (6.1.10)$$

By considering the above deductions, we can give the following conditions for synthesis approaches of MPC:

(A1) $\mathcal{X}_f \subseteq \mathcal{X}$, $\mathcal{X}_f$ closed, $\mathcal{X}_f \supset \{0\}$ (state constraint is satisfied in $\mathcal{X}_f$);

(A2) $K_f(x) \in \mathcal{U}$, $\forall x \in \mathcal{X}_f$ (control constraint is satisfied in $\mathcal{X}_f$);

(A3) $f(x, K_f(x)) \in \mathcal{X}_f$, $\forall x \in \mathcal{X}_f$ ($\mathcal{X}_f$ is positively invariant under $K_f(\cdot)$);

(A4) $\Delta F(f(x, K_f(x))) + \ell(x, K_f(x)){\leq}0$, $\forall x \in \mathcal{X}_f$ ($F(\cdot)$ is a local Lyapunov function in $\mathcal{X}_f$).

The common situation is to select $\mathcal{X}_f$ as a level set of $F(\cdot)$, i.e., $\mathcal{X}_f = \{x|F(x) \leq \eta\}$ where η is a constant. If $\mathcal{X}_f$ is a level set of $F(\cdot)$, then (A4) implies (A3).

With (A1)-(A4) satisfied, asymptotic (exponential) stability of the closed-loop system can be proved by invoking some other "more fundamental" conditions (e.g., the controllability for some systems, the continuity for some systems, etc., which are inevitable even beyond MPC literature). Certainly, in general only sufficient (not necessary) stability conditions are obtained.

6.1.4   Monotonicity method for stability proof

By the principle of optimality, the following holds:

$$J_N^*(k) = \ell(x(k), K_N(x(k))) + J_{N-1}^*(k+1), \ \forall x(k) \in S_N(\mathcal{X}, \mathcal{X}_f).$$

Introduce $J_N^*(k+1)$. Then the above equation is equivalent to

$$\Delta J_N^*(k+1) + \ell(x(k), K_N(x(k))) = J_N^*(k+1) - J_{N-1}^*(k+1). \qquad (6.1.11)$$

Hence, if the following is satisfied:

$$J_N^*(k) \leq J_{N-1}^*(k), \ \forall x(k) \in S_{N-1}(\mathcal{X}, \mathcal{X}_f),$$

then Method 2 becomes Method 1.

Denote the optimization problem for optimizing $J_N(k)$ as $\mathbb{P}_N(k)$. If $\mathbb{P}_{N-1}(k+1)$ has the following solution:

$$\tilde{u}_{N-1}^*(k+1) = \{u^*(k+1|k+1), u^*(k+2|k+1), \cdots, u^*(k+N-1|k+1)\},$$

then a feasible solution to $\mathbb{P}_N(k+1)$ is

$$\tilde{u}_N(k+1) = \{u^*(k+1|k+1), \cdots, u^*(k+N-1|k+1), K_f(x^*(k+N|k+1))\},$$

where, by invoking the optimality principle,

$$u^*(k+i|k+1) = u^*(k+i|k), \ \forall i > 0, \ x^*(k+N|k+1) = x^*(k+N|k).$$

Applying the feasible solution of $\mathbb{P}_N(k+1)$ yields

$$\begin{aligned} J_N(k+1) =& J_{N-1}^*(k+1) + \ell(x^*(k+N|k), K_f(x^*(k+N|k))) \\ & - F(x^*(k+N|k)) + F(f(x^*(k+N|k), K_f(x^*(k+N|k)))) \end{aligned}$$
$$(6.1.12)$$

and $J_N(k+1)$ is the upper bound of $J_N^*(k+1)$. If condition (A4) is satisfied, then (6.1.12) implies $J_N^*(k+1) \leq J_{N-1}^*(k+1)$. Further, applying (6.1.11) yields (6.1.10).

For the monotonicity of the value function we have the following two conclusions.

Proposition 6.1.1. *Suppose $J_1^*(k) \leq J_0^*(k)$ for all $x(k) \in S_0(\mathcal{X}, \mathcal{X}_f)$. Then $J_{i+1}^*(k) \leq J_i^*(k)$ for all $x(k) \in S_i(\mathcal{X}, \mathcal{X}_f)$, $i \geq 0$.*

Proof. (By induction) Suppose $J_i^*(k) \leq J_{i-1}^*(k)$ for all $x(k) \in S_{i-1}(\mathcal{X}, \mathcal{X}_f)$. Consider two control laws $K_i(\cdot)$ and $K_{i+1}(\cdot)$. When $J_{i+1}^*(k)$ is optimized, $K_{i+1}(\cdot)$ is optimal and $K_i(\cdot)$ is not optimal. Hence, by the optimality principle the following holds:

$$\begin{aligned} J_{i+1}^*(x(k), \tilde{u}_{i+1}^*(k)) =& \ell(x(k), K_{i+1}(x(k))) + J_i^*(f(x(k), K_{i+1}(x(k))), \tilde{u}_i^*(k+1)) \\ \leq& \ell(x(k), K_i(x(k))) + J_i^*(f(x(k), K_i(x(k))), \tilde{u}_i^*(k+1)), \\ & \forall x(k) \in S_i(\mathcal{X}, \mathcal{X}_f). \end{aligned}$$

Hence

$$
\begin{aligned}
&J_{i+1}^*(k) - J_i^*(k) \\
&= \ell(x(k), K_{i+1}(x(k))) + J_i^*(k+1) - \ell(x(k), K_i(x(k))) - J_{i-1}^*(k+1) \\
&\leq \ell(x(k), K_i(x(k))) + J_i^*(f(x(k), K_i(x(k))), \tilde{u}_i^*(k+1)) - \ell(x(k), K_i(x(k))) \\
&\quad - J_{i-1}^*(k+1) \\
&= J_i^*(k+1) - J_{i-1}^*(k+1) \leq 0, \ \forall x(k) \in S_i(\mathcal{X}, \mathcal{X}_f).
\end{aligned}
$$

Note that $x(k) \in S_i(\mathcal{X}, \mathcal{X}_f)$ implies $x(k+1) \in S_{i-1}(\mathcal{X}, \mathcal{X}_f)$. $J_i(f(x(k), K_i(x(k))), \tilde{u}_i^*(k+1))$ is not necessarily optimal for $x(k) \in S_i(\mathcal{X}, \mathcal{X}_f)$, but is optimal for $x(k+1) \in S_{i-1}(\mathcal{X}, \mathcal{X}_f)$. $\qquad\square$

Proposition 6.1.2. *Suppose $F(\cdot)$, $\mathcal{X}_f$ and $K_f(\cdot)$ satisfy (A1)-(A4). Then $J_1^*(k) \leq J_0^*(k)$ for all $x(k) \in S_0(\mathcal{X}, \mathcal{X}_f)$.*

Proof. According to the optimality principle,

$$
\begin{aligned}
J_1^*(k) &= \ell(x(k), K_1(x(k))) + J_0^*(f(x(k), K_1(x(k)))) \\
&\leq \ell(x(k), K_f(x(k))) + J_0^*(f(x(k), K_f(x(k)))) \ \text{(by optimality of } K_1(\cdot)) \\
&= \ell(x(k), K_f(x(k))) + F(f(x(k), K_f(x(k)))) \ \text{(by definition of } J_0(\cdot)) \\
&\leq F(x(k)) = J_0^*(k) \ \text{(by (A4))}.
\end{aligned}
$$

Therefore, the conclusion holds. $\qquad\square$

Remark 6.1.3. With (A1)-(A4) satisfied, Method 2 indirectly utilizes Method 1. For this reason Method 1 is called "direct method." In synthesis approaches of MPC (not restricted to the model studied in this section), usually Method 1 is applied.

6.1.5 Inverse optimality

As a matter of fact, the value function of synthesis approach is also the infinite-horizon value function of a modified problem. The advantage of the infinite-horizon value function is that, if this value exists and finite, then closed-loop stability is naturally guaranteed (this is easily known by considering the positive definiteness of the cost function).

Equation (6.1.11) can be written in the form

$$
J_N^*(k) = \bar{\ell}(x(k), K_N(x(k))) + J_N^*(f(x(k), K_N(x(k)))) \tag{6.1.13}
$$

where

$$
\bar{\ell}(x(k), u(k)) \triangleq \ell(x(k), u(k)) + J_{N-1}^*(k+1) - J_N^*(k+1).
$$

If (A1)-(A4) are satisfied, then $\bar{\ell}(x, u) \geq \ell(x, u) \geq c \left\| [x^T \ u^T] \right\|^2$. Consider

$$
\bar{J}_\infty(k) = \sum_{i=0}^{\infty} \bar{\ell}(x(k+i|k), u(k+i|k)).
$$

Then, (6.1.13) is the Hamilton-Jacobi-Bellman algebraic equation corresponding to the optimal control with cost function $\bar{J}_\infty(k)$.

Remark 6.1.4. Corresponding to the original cost function $J_N(k)$, (6.1.13)
is called the fake Hamilton-Jacobi-Bellman algebraic equation. For a linear unconstrained system, the Hamilton-Jacobi-Bellman algebraic equation
is reduced to the algebraic Riccati equation, and the fake Hamilton-Jacobi-
Bellman algebraic equation is reduced to the fake algebraic Riccati equation
(see Chapter 5).

For MPC with cost function $\bar{J}_\infty(k)$, $K_N(\cdot)$ is the optimal solution. Hence,
the finite-horizon optimization problem (6.1.5)-(6.1.7) is equivalent to another
infinite-horizon optimization problem. This phenomenon in MPC is called
"inverse optimality." Here "optimality" refers to result by optimizing $J_\infty(k) = \sum_{i=0}^\infty \ell(x(k+i|k), u(k+i|k))$. When the optimality is for $J_\infty(k)$, rather than
for $\bar{J}_\infty(k)$, then "inverse optimality" is resulted ("inverse" indicates that, if
it is optimal for $\bar{J}_\infty(k)$, then it is not optimal for $J_\infty(k)$ and, by optimizing
$\bar{J}_\infty(k)$, the optimum can deviate from that of $J_\infty(k)$).

Since synthesis approach is equivalent to another infinite-horizon optimal controller, stability margin of synthesis approach is determined by the
corresponding infinite-horizon optimal controller. Satisfaction of "inverse optimality," although not advantageous for "optimality," is capable of showing
stability margin.

6.2 General idea: case continuous-time systems

More details are referred to discrete-time model. Suppose the system is described by

$$\dot{x}(t) = f(x(t), u(t)), \tag{6.2.1}$$

simplified by $\dot{x} = f(x, u)$. Consider the following cost function

$$J_T(t) = \int_0^T \ell(x(t+s|t), u(t+s|t))ds + F(x(t+T|t)). \tag{6.2.2}$$

The following optimization problem is to be solved:

$$\min_{\tilde{u}_T(t)} J_T(t), \tag{6.2.3}$$

$$\text{s.t. } \dot{x}(t+s|t) = f(x(t+s|t), u(t+s|t)), \ s\geq0, \ x(t|t) = x(t), \tag{6.2.4}$$

$$x(t+s|t) \in \mathcal{X}, \ u(t+s|t) \in \mathcal{U}, \ s \in [0,T], \ x(t+T|t) \in \mathcal{X}_f, \tag{6.2.5}$$

where $\tilde{u}_T(t)$ is the decision variable defined on the time interval $[0, T]$. The
implicit MPC law is denoted as

$$K_T(x(t)) = u^*(t) = u^*(t|t).$$

Given a function $\phi(x)$, let $\dot\phi(f(x,u))$ denote its directional derivative in the direction $f(x,u)$. Then $\dot\phi(f(x,u)) = \phi_x(x)f(x,u)$, where ϕ_x is the partial derivative of ϕ with respect to x.

The ingredients $F(\cdot)$, $\mathcal{X}_f$ and $K_f(\cdot)$ for the continuous-time case are required to satisfy

(B1) $\mathcal{X}_f \subseteq \mathcal{X}$, $\mathcal{X}_f$ closed, $\mathcal{X}_f \supset \{0\}$;

(B2) $K_f(x) \in \mathcal{U}, \forall x \in \mathcal{X}_f$;

(B3) $\mathcal{X}_f$ is positively invariant for $\dot x = f(x, K_f(x))$;

(B4) $\dot F(f(x, K_f(x))) + \ell(x, K_f(x)) \leq 0, \forall x \in \mathcal{X}_f$.

Condition (B4) implies (B3) if $\mathcal{X}_f$ is a level set of $F(\cdot)$.

The definition of the positively invariant set in continuous-time system is the same as that in the discrete-time case (refer to Chapter 1, which is the limit of that in the discrete-time case when the sampling interval tends to zero). As in Definition 6.1.1, we can define

$$S_\tau(\mathcal{X}, \mathcal{X}_f) \triangleq \{x(0) \in \mathcal{X} : \exists u(t) \in \mathcal{U}, \ t \in [0, \tau), \exists \tau_1 \leq \tau \text{ such that}$$
$$x(t) \in \mathcal{X}, \ t \in [0, \tau_1) \text{ and } x(t) \in \mathcal{X}_f, \ t \in [\tau_1, \tau], \ \mathcal{X}_f \text{ is invariant}\}.$$

Applying conditions (B1)-(B4) we can prove

$$\dot J_T^*(t) + \ell(x(t), K_T(x(t))) \leq 0, \ \forall x \in S_T(\mathcal{X}, \mathcal{X}_f).$$

Hence, by satisfying some other "more fundamental" conditions, asymptotic (exponential) stability of the closed-loop system can be proved. Certainly, the obtained stability results are usually sufficient, not necessary.

For monotonicity of the value function we have the following two conclusions.

Proposition 6.2.1. *Suppose* $(\partial/\partial\tau)J_{\tau=0}^*(t) \leq 0$ *for all* $x(t) \in S_0(\mathcal{X}, \mathcal{X}_f) = \mathcal{X}_f$. *Then,* $(\partial/\partial\tau)J_\tau^*(t) \leq 0$ *for all* $x(t) \in S_\tau(\mathcal{X}, \mathcal{X}_f)$, $\tau \in [0, T]$.

Proof. If $J_\tau^*(t)$ is continuously differentiable, then applying the principle of optimality yields

$$(\partial/\partial\tau)J_\tau^*(t) = \ell(x(t), K_\tau(x(t))) + (\partial/\partial x)J_\tau^*(t)f(x(t), K_\tau(x(t))).$$

Since $(\partial/\partial\tau)J_{\tau=0}^*(t) \leq 0$,

$$\lim_{\Delta\tau \to 0} \frac{1}{\Delta\tau} \left\{ \int_0^{\Delta\tau} \ell(x^*(t+s|t), u_0^*(t+s|t))ds + F(x^*(t+\Delta\tau|t))|_{\tilde u_0^*(t)} - F(x(t)) \right\}$$
$$\leq 0$$

where $u_0^*(t+s|t) = K_f(x^*(t+s|t))$, the subscript $\tilde u_0^*(t)$ represents "based on the control move $\tilde u_0^*(t)$."

When the optimization horizon is $\Delta\tau$, $\tilde{u}^*_{\Delta\tau}(t)$, rather than $\tilde{u}^*_0(t)$, is the optimal control sequence. $(\partial/\partial\tau)J^*_{\tau=0}(t) = 0$ only happens at $J^*_{\tau=0}(t) = 0$. For sufficiently small $\Delta\tau$, applying (6.2.2) yields

$$J^*_{\Delta\tau}(t) - J^*_0(t) = \int_0^{\Delta\tau} \ell(x^*(t+s|t), u^*_{\Delta\tau}(t+s|t))ds$$

$$+ F(x^*(t+\Delta\tau|t))|_{\tilde{u}^*_{\Delta\tau}(t)} - F(x(t))$$

$$\leq \int_0^{\Delta\tau} \ell(x^*(t+s|t), u^*_0(t+s|t))ds$$

$$+ F(x^*(t+\Delta\tau|t))|_{\tilde{u}^*_0(t)} - F(x(t)) \leq 0,$$

which shows the monotonicity of $J^*_\tau(t)$, i.e., with the increase of τ, $J^*_\tau(t)$ does not increase.

In general, if $(\partial/\partial\tau)J^*_\tau(t) \leq 0$ then

$$0 \geq \lim_{\Delta\tau \to 0} \frac{1}{\Delta\tau} \left\{ \int_\tau^{\tau+\Delta\tau} \ell(x^*(t+s|t), u^*_\tau(t+s|t))ds + F(x^*(t+\tau+\Delta\tau|t))|_{\tilde{u}^*_\tau(t)} \right.$$

$$\left. - F(x^*(t+\tau|t))|_{\tilde{u}^*_\tau(t)} \right\}.$$

When the optimization horizon is $\tau + \Delta\tau$, $\tilde{u}^*_{\tau+\Delta\tau}(t)$, rather than $\tilde{u}^*_\tau(t)$, is the optimal control sequence. Applying (6.2.2) yields

$$J^*_{\tau+\Delta\tau}(t) - J^*_\tau(t) = \int_0^{\tau+\Delta\tau} \ell(x^*(t+s|t), u^*_{\tau+\Delta\tau}(t+s|t))ds$$

$$+ F(x^*(t+\tau+\Delta\tau|t))|_{\tilde{u}^*_{\tau+\Delta\tau}(t)}$$

$$- \int_0^\tau \ell(x^*(t+s|t), u^*_\tau(t+s|t))ds - F(x^*(t+\tau|t))|_{\tilde{u}^*_\tau(t)}$$

$$\leq \int_\tau^{\tau+\Delta\tau} \ell(x^*(t+s|t), u^*_\tau(t+s|t))ds$$

$$+ F(x^*(t+\tau+\Delta\tau|t))|_{\tilde{u}^*_\tau(t)} - F(x^*(t+\tau|t))|_{\tilde{u}^*_\tau(t)}$$

$$\leq 0$$

which shows that for any $\Delta\tau > 0$, $J^*_{\tau+\Delta\tau}(t) \leq J^*_\tau(t)$, i.e., $(\partial/\partial\tau)J^*_\tau(t) \leq 0$. $\square$

If $J^*_\tau(t)$ is continuous, then the fake Hamilton-Jacobi equation is

$$(\partial/\partial x)J^*_\tau(t)f(x(t), K_\tau(x(t))) + \bar{\ell}(x(t), K_\tau(x(t))) = 0,$$

where
$$\bar{\ell}(x(t), u(t)) = \ell(x(t), u(t)) - (\partial/\partial\tau)J^*_\tau(t).$$

When $(\partial/\partial\tau)J^*_\tau(t) \leq 0$, $\bar{\ell}(x(t), u(t)) \geq \ell(x(t), u(t))$.

Proposition 6.2.2. *Suppose (B1)-(B4) are true. Then $(\partial/\partial\tau)J^*_{\tau=0}(t) \leq 0$ for all $x(t) \in S_0(\mathcal{X}, \mathcal{X}_f) = \mathcal{X}_f$.*

Proof. Invoking the principle of optimality we have

$$
\begin{aligned}
(\partial/\partial\tau)J^*_{\tau=0}(t) =& \ell(x(t), K_0(x(t))) + (\partial/\partial x)J^*_0(t)f(x(t), K_0(x(t))) \\
\leq& \ell(x(t), K_f(x(t))) + (\partial/\partial x)J^*_0(t)f(x(t), K_f(x(t))) \\
& \text{(by optimality of } K_0(\cdot)) \\
=& \ell(x(t), K_f(x(t))) + (\partial/\partial x)F(x(t))f(x(t), K_f(x(t))) \\
& \text{(by definition of } J_0(\cdot)) \\
\leq& 0 \text{ (by (B4)).}
\end{aligned}
$$

Therefore, the conclusion holds.   □

6.3   Realizations

6.3.1   Using terminal equality constraint

In Chapter 4, Kleinman's controller and Ackermann's formula for deadbeat control are both MPC adopting the terminal equality constraint. For example, for system

$$x(k+1) = Ax(k) + Bu(k), \tag{6.3.1}$$

consider the Kleinman's controller

$$u(k) = -R^{-1}B^T\left(A^T\right)^N\left[\sum_{h=0}^{N}A^hBR^{-1}B^T\left(A^T\right)^h\right]^{-1}A^{N+1}x(k) \tag{6.3.2}$$

where $R > 0$. Minimizing the cost function

$$J_N(k) = \sum_{i=0}^{N-1}\|u(k+i|k)\|_R^2, \text{ s.t. } x(k+N|k) = 0$$

which, if feasible, yields the stabilizing Kleinman's controller in the form of (6.3.2).

Substitute $x(k+N|k) \in \mathcal{X}_f$ in problem (6.1.5)-(6.1.7) with $x(k+N|k) = 0$, then MPC with terminal equality constraint is obtained. Applying terminal equality constraint amounts to taking

$$F(\cdot) = 0, \ \mathcal{X}_f = \{0\}, \ K_f(\cdot) = 0. \tag{6.3.3}$$

It is easy to verify that (6.3.3) satisfies (A1)-(A4).

6.3.2 Using terminal cost function

The terminal constraint is not imposed.

1. *linear, unconstrained systems*

Here $f(x(k), u(k)) = Ax(k) + Bu(k)$, and $\ell(x, u) = \|x\|_Q^2 + \|x\|_R^2$ where $Q > 0$, $R > 0$, (A, B) stabilizable. Since the system is unconstrained ($\mathcal{X} = \mathfrak{R}^n$, $\mathcal{U} = \mathfrak{R}^m$), conditions (A1)-(A3) are trivially satisfied. Let $K_f(x) = Kx$ stabilize the system, and let $P > 0$ satisfy Lyapunov equation

$$(A + BK)^T P (A + BK) - P + Q + K^T RK = 0.$$

Then, $F(x) \triangleq x^T Px$ satisfies condition (A4) (with equality). The three ingredients are

$$F(x) \triangleq x^T Px, \ \ \mathcal{X}_f = \mathfrak{R}^n, \ \ K_f(x) = Kx.$$

The closed-loop system is asymptotically (exponentially) stable with a region of attraction $\mathfrak{R}^n$.

2. *Linear, constrained, open-loop stable systems*

The system is the same as in 1 except that, in addition, the system is open-loop stable and constrained. $\mathcal{X}_f = \mathcal{X} = \mathfrak{R}^n$. It follows from (A2) that $K_f(\cdot)$, if linear, must satisfy $K_f(x) \equiv 0$. Thus, conditions (A1)-(A3) are trivially satisfied. Let $P > 0$ satisfy Lyapunov equation

$$A^T PA - P + Q = 0.$$

Then $F(x) \triangleq x^T Px$ satisfy (A4) with equality. The three ingredients are

$$F(x) \triangleq x^T Px, \ \ \mathcal{X}_f = \mathfrak{R}^n, \ \ K_f(x) \equiv 0.$$

The closed-loop system is asymptotically (exponentially) stable with a region of attraction $\mathfrak{R}^n$.

Remark 6.3.1. From semi-global stability in Chapter 5, by adopting one off-line calculated fixed linear feedback law, an input constrained open-loop stable linear system cannot be globally stabilized. However, by applying on-line solution of MPC, nonlinear (or time-varying) control laws are obtained. This point represents an important difference between MPC and the traditional state feedback control.

6.3.3 Using terminal constraint set

The terminal cost is not adopted. Usually, dual-mode control is invoked. In finite steps, the state is driven into $\mathcal{X}_f$. Inside of $\mathcal{X}_f$, the local controller stabilizes the system.

If, with the evolution of time, the control horizon N decreases by 1 in each sampling interval, then conditions (A1)-(A4) are trivially satisfied. If N is also a decision variable, then stability proof is easier than that of the fixed horizon case, which will be talked about later.

If fixed horizon approach is adopted, then the selection of the local controller should satisfy (A1)-(A3). In order to satisfy (A4), $\ell(x, K_f(x)) = 0$ is required inside of $\mathcal{X}_f$. A suitable choice is to substitute the original $\ell(x, u)$ with $\tilde{\ell}(x, u) \triangleq \alpha(x)\ell(x, u)$, where

$$\alpha(x) = \begin{cases} 0 & x \in \mathcal{X}_f \\ 1 & x \notin \mathcal{X}_f \end{cases}.$$

Thus, condition (A4) with equality is satisfied.

6.3.4 Using terminal cost function and terminal constraint set

This version attracts most attention in current literature. Ideally, $F(\cdot)$ should be chosen to be $J_\infty^*(k)$ since, in this case, the virtues of infinite-horizon optimal control are obtained. In general, this is only possible for linear systems.

1. *Linear, constrained systems*

One can take $F(x) \triangleq x^T P x$ as the value function of the infinite-horizon unconstrained LQR (refer to 1 in section 6.3.2), and take $K_f(x) = Kx$ as the solution to this infinite-horizon unconstrained LQR, and take $\mathcal{X}_f$ as the output admissible set for $x(k + 1) = (A + BK)x(k)$ (the output admissible set is the set inside of which constraints are satisfied; the region of attraction of TSGPC in Chapter 5 is an output admissible set). By this selection of the three ingredients, conditions (A1)-(A4) are satisfied ((A4) is satisfied with equality).

2. *Nonlinear, unconstrained systems*

For the linearized system $x(k + 1) = Ax(k) + Bu(k)$, select the local controller as $K_f(x) = Kx$, such that $x(k + 1) = (A + BK)x(k)$ is stabilized, and $x^T P x$ as the corresponding Lyapunov function. Select $\mathcal{X}_f$ as the level set of $x^T P x$. When $\mathcal{X}_f$ is sufficiently small, conditions (A1)-(A2) are satisfied and $x^T P x$ is Lyapunov function of $x(k + 1) = f(x(k), K_f(x(k)))$, with a region of attraction $\mathcal{X}_f$. Then, select $F(x) \triangleq \alpha x^T P x$ such that conditions (A3)-(A4) are satisfied. Although the unconstrained system is studied, the region of attraction is usually a subset of $\Re^n$. Notice that $x^T P x$ is Lyapunov function implies that $F(x)$ is the Lyapunov function.

The essential requirement for stability is that $F(x)$ should be a Lyapunov function of $x(k + 1) = f(x(k), K_f(x(k)))$ in the neighborhood of the origin; if this is the case, there exists a constraint set $\mathcal{X}_f \triangleq \{x | F(x) \leq r\}$ and a local control law $K_f(\cdot)$ such that (A1)-(A4) are satisfied.

3. *Nonlinear, constrained systems*

Choose $K_f(x) = Kx$ to stabilize the linearized system $x(k + 1) = (A + BK)x(k)$ and choose $\mathcal{X}_f$ to satisfy the set constraint $\mathcal{X}_f \subseteq \mathcal{X}$ and $K_f(\mathcal{X}_f) \subseteq \mathcal{U}$. These choices satisfy conditions (A1)-(A2). Choose $\mathcal{X}_f \triangleq \{x | F(x) \leq r\}$ where $F(x) \triangleq x^T P x$ is a control Lyapunov function for the system $x(k + 1) =$

$f(x(k), K_f(x(k)))$ satisfying Lyapunov equation

$$F(Ax + BK_f(x)) - F(x) + \tilde{\ell}(x, K_f(x)) = 0$$

where $\tilde{\ell}(x, u) \triangleq \beta \ell(x, u)$, $\beta \in (1, \infty)$. When r is sufficiently small, since the above Lyapunov equation utilizes β, a sufficient margin is provided to ensure the satisfaction of (A3)-(A4).

Certainly, it would be better to do as follows: instead of adopting linearized model, directly take $F(x)$ as the infinite-horizon value function of the nonlinear system $x(k + 1) = f(x(k), K_f(x(k)))$, and take $\mathcal{X}_f$ as the region of attraction of $x(k + 1) = f(x(k), K_f(x(k)))$, with $\mathcal{X}_f$ an invariant set. In such a way, conditions (A3)-(A4) are satisfied.

Remark 6.3.2. It should be noted that, rather than the exact rules, the implementation methods in this section are only some guidelines. In synthesizing MPC, various changes can happen. Readers may consider the feedback linearization method, which can transform a nonlinear system into a closed-loop linear system. Unfortunately, linear input/state constraints can become nonlinear by the feedback linearization.

Remark 6.3.3. In proving stability, $\Delta J_N^*(k + 1) + \ell(x(k), K_N(x(k))) \leq 0$ is adopted, where $\Delta J_N^*(k+1) = J_N^*(k+1) - J_N^*(k)$. In fact, for nonlinear systems it is impractical to find the exact value of $J_N^*(\cdot)$. $J_N^*(\cdot)$ can be an approximate value. $\Delta J_N^*(k+1) + \ell(x(k), K_N(x(k))) \leq 0$, as compared with $\Delta J_N^*(k+1) < 0$, is conservative and, thus, brings stability margin, which allows a difference between the approximated value and the theoretical optimum.

Remark 6.3.4. In general, in the optimization problem of a synthesis approach, $F(\cdot)$ and $\mathcal{X}_f$ are explicitly added. However, $K_f(\cdot)$ is not explicitly added, and only for stability proof. That is to say, in the so-called "three ingredients," usually only two ingredients appear in a synthesis approach. In some special synthesis approaches, when the state enters into the terminal constraint set, $K_f(\cdot)$ is explicitly adopted. In the receding-horizon optimization, $K_f(\cdot)$ can be utilized to calculate the initial value of $\tilde{u}_N(k)$. The key points of a synthesis approach can be concluded as "234" (2: $F(\cdot)$ and $\mathcal{X}_f$; 3 ingredients; 4 conditions (A1)-(A4)).

6.4 General idea: case uncertain systems (robust MPC)

For uncertain systems, there are three approaches to study the robustness:

(i) (Inherent robustness) Design MPC using nominal model, and analyze the controller when there is modeling uncertainty.

(ii) Consider all the possible realizations of the uncertainty, and adopt open-loop "min-max" MPC, to ensure closed-loop robust stability. MPC based on the open-loop optimization is the usual form of MPC, where a sequence of control move $\tilde{u}_N(k)$ is optimized.

(iii) Introduce feedback in the min-max optimal control problem.

Suppose the nonlinear system is described by

$$x(k+1) = f(x(k), u(k), w(k)), \; z(k) = g(x(k)). \tag{6.4.1}$$

The details are as former sections. The disturbance satisfies $w(k) \in \mathcal{W}(x(k), u(k))$ where $\mathcal{W}(x, u)$ is closed and $\mathcal{W}(x, u) \supset \{0\}$. Notice that, in this situation, the predictions of state and cost function will incorporate the disturbance.

If the estimated state is adopted (output feedback MPC), then the state estimation error can be modeled by $w(k) \in \mathcal{W}_k$.

6.4.1 Uniform idea for stability proof

The following cost function is to be minimized at each time k:

$$J_N(k) = \sum_{i=0}^{N-1} \ell(x(k+i|k), u(k+i|k), w(k+i)) + F(x(k+N|k)).$$

In a synthesis approach, all the possible realizations of w should be considered. Hence, stability conditions need to be strengthened. The three ingredients $F(\cdot)$, $\mathcal{X}_f$ and $K_f(\cdot)$ need to satisfy the following conditions:

(A1) $\mathcal{X}_f \subseteq \mathcal{X}$, $\mathcal{X}_f$ closed, $\mathcal{X}_f \supset \{0\}$;

(A2) $K_f(x) \in \mathcal{U}$, $\forall x \in \mathcal{X}_f$;

(A3a) $f(x, K_f(x), w) \in \mathcal{X}_f$, $\forall x \in \mathcal{X}_f$, $\forall w \in \mathcal{W}(x, K_f(x))$;

(A4a) $\Delta F(f(x, K_f(x), w)) + \ell(x, K_f(x), w) \leq 0$, $\forall x \in \mathcal{X}_f$, $\forall w \in \mathcal{W}(x, K_f(x))$.

By appropriately choosing the three ingredients, if $F(\cdot)$ is Lyapunov function of $x(k+1) = f(x(k), K_f(x(k)), w(k))$ in the neighborhood of the origin, then (A1)-(A2) and (A3a)-(A4a) can ensure that

$$\Delta J_N^*(k+1) + \ell(x(k), K_N(x(k)), w(k)) \leq 0,$$
$$x \text{ in an appropriate set, } \forall w \in \mathcal{W}(x, K_N(x)),$$

(or, when x lies in an appropriate set, for all $w \in \mathcal{W}(x, K_N(x))$ the value $J_N^*(k)$ is non-increasing with the increase of N).

Remark 6.4.1. For systems with disturbance or noise, the state may be unable to converge to the origin, and only convergence to a neighborhood of the origin $\Omega \subseteq \mathcal{X}_f$ is ensured, Ω being an invariant set. Certainly, whether or not the state will converge to the origin depends on the property of the disturbance or noise. For example, for $f(x(k), u(k), w(k)) = Ax(k) + Bu(k) + Dw(k)$, when $Dw(k)$ does not tend to zero with the evolution of k, $x(k)$ cannot converge to the origin; for $f(x(k), u(k), w(k)) = Ax(k) + Bu(k) + Dx(k)w(k)$, even when $w(k)$ does not tend to zero with the evolution of k, $x(k)$ can converge to the origin.

6.4.2   Open-loop min-max MPC

For nominal systems, $S_N(\mathcal{X}, \mathcal{X}_f)$ is defined. $S_N(\mathcal{X}, \mathcal{X}_f)$ is the positively invariant set for the system $x(k+1) = f(x(k), K_N(x(k)))$. For nominal systems, if $x(k) \in S_N(\mathcal{X}, \mathcal{X}_f)$ then $x(k+1) \in S_{N-1}(\mathcal{X}, \mathcal{X}_f) \subseteq S_N(\mathcal{X}, \mathcal{X}_f)$; for uncertain systems this property does not hold in general.

At each time k the following cost function is to be minimized:

$$J_N(k) = \max_{\tilde{w}_N(k) \in \tilde{W}_N(x(k), \tilde{u}_N(k))} V_N(x(k), \tilde{u}_N(k), \tilde{w}_N(k))$$

where $\tilde{w}_N(k) \triangleq \{w(k), w(k+1), \cdots, w(k+N-1)\}$ and $\tilde{W}_N(x(k), \tilde{u}_N(k))$ is the set of all possible realizations of the sequence of disturbance within the switching horizon,

$$V_N(x(k), \tilde{u}_N(k), \tilde{w}_N(k)) = \sum_{i=0}^{N-1} \ell(x(k+i|k), u(k+i|k), w(k+i))$$

$$+ F(x(k+N|k)),$$

$$x(k+i+1|k) = f(x(k+i|k), u(k+i|k), w(k+i)).$$

Other constraints in optimization are the same as the nominal case. Suppose the set in which the optimization problem is feasible is $S_N^{ol}(\mathcal{X}, \mathcal{X}_f) \subseteq S_N(\mathcal{X}, \mathcal{X}_f)$. The superscript "ol" represents open-loop.

Suppose the three ingredients $F(\cdot)$, $\mathcal{X}_f$ and $K_f(\cdot)$ satisfy conditions (A1)-(A2) and (A3a)-(A4a). Then there is a difficulty in stability proof, which is described as follows:

i. Suppose $x(k) \in S_N^{ol}(\mathcal{X}, \mathcal{X}_f)$ and the optimization problem has an optimal solution $\tilde{u}_N^{ol}(k) \triangleq \tilde{u}_N(k)$. For all $\tilde{w}_N(k) \in \tilde{W}_N(x(k), \tilde{u}_N(k))$, this optimal control sequence can drive all the possible states into $\mathcal{X}_f$ in not more than N steps.

ii. At time $k+1$, the control sequence $\{u^*(k+1|k), u^*(k+2|k), \cdots, u^*(k+N-1|k)\}$ can drive all the possible states into $\mathcal{X}_f$ in not more than $N-1$ steps. Hence, $x(k+1|k) \in S_{N-1}^{ol}(\mathcal{X}, \mathcal{X}_f)$.

iii. The problem is that, at time $k+1$, it may be unable to find the following control sequence:

$$\{u^*(k+1|k), u^*(k+2|k), \cdots, u^*(k+N-1|k), v\}$$

to serve as the feasible solution of the optimization problem. This is because $v \in \mathcal{U}$ has to satisfy

$$f(x^*(k+N|k), v, w(k+N)) \in \mathcal{X}_f, \ \forall \tilde{w}_N(k) \in \tilde{\mathcal{W}}_N(x(k), \tilde{u}_N^*(k)). \quad (6.4.2)$$

Condition (A3a) does not ensure that (6.4.2) holds with $v = K_f(x^*(k+N|k))$ (except for $N = 1$).

iv. At time $k+1$ a feasible control sequence does not exist and, hence, the upper bound of $J_N^*(k+1)$ cannot be obtained.

An alternative to overcome the above difficulty is to adopt the varying horizon strategy. In the varying horizon approach, besides $\tilde{u}_N(k)$, N is also a decision variable. Suppose the optimal solution $\{\tilde{u}_{N^*(k)}^*(k), N^*(k)\}$ is obtained at time k; then at time $k+1$, $\{\tilde{u}_{N^*(k)-1}^*(k), N^*(k)-1\}$ is a feasible solution. Thus, with conditions (A1)-(A2) and (A3a)-(A4a) satisfied, closed-loop stability can be proved, i.e., the following can be proved:

$$\Delta J_{N^*(k+1)}^*(k+1) + \ell(x(k), K_{N^*(k)}^*(x(k)), w(k)) \leq 0,$$
$$\forall x \in S_{N^*(k)}^{\text{ol}}(\mathcal{X}, \mathcal{X}_f) \backslash \mathcal{X}_f, \forall w \in \mathcal{W}(x, K_{N^*(k)}^*(x))$$

where $\Delta J_{N^*(k+1)}^*(k+1) = J_{N^*(k+1)}^*(k+1) - J_{N^*(k)}^*(k)$. Inside of $\mathcal{X}_f$, adopt $K_f(\cdot)$. Conditions (A1)-(A2) and (A3a)-(A4a) guarantee the existence of suitable $\mathcal{X}_f$ and $K_f(\cdot)$.

Notice that here the varying horizon does not represent $N(k+1) \leq N(k)$, $N(k+1) < N(k)$ or $N(k+1) = N(k)-1$. Rather, $N(k)$ is served as a decision variable.

6.5   Robust MPC based on closed-loop optimization

Although open-loop min-max MPC has a number of advantages, a deficiency of this kind of MPC is that it adopts open-loop prediction, i.e., $\tilde{u}_N(k)$ is utilized in the prediction to handle all possible realizations of the disturbance. This is unrealistic since, every sampling time a control move is implemented, the uncertainties of the state evolutions are shrinking.

In feedback MPC, the decision variable is not $\tilde{u}_N(k)$, but

$$\pi_N(k) \triangleq \{u(k), F_1(x(k+1|k)), \cdots, F_{N-1}(x(k+N-1|k))\}$$

where $F_i(\cdot)$ is the state feedback law, rather than the control move (of course, $u(k)$ needs not to be substituted with $F_0(\cdot)$ since $x(k)$ is known).

At each time k the following cost function is to be minimized:

$$J_N(k) = \max_{\tilde{w}_N(k) \in \tilde{\mathcal{W}}_N(x(k), \pi_N(k))} V_N(x(k), \pi_N(k), \tilde{w}_N(k))$$

where $\tilde{\mathcal{W}}_N(x(k), \pi_N(k))$ is the set of all possible realizations of the disturbance within the switching horizon N,

$$V_N(x(k), \pi_N(k), \tilde{w}_N(k)) = \sum_{i=0}^{N-1} \ell(x(k+i|k), F_i(x(k+i|k)), w(k+i))$$
$$+ F(x(k+N|k)),$$
$$x(k+i+1|k) = f(x(k+i|k), F_i(x(k+i|k)), w(k+i)),$$
$$x(k+1|k) = f(x(k), u(k), w(k)).$$

Other constraints for the optimization problem are the same as the nominal case. Suppose the set in which the optimization problem is feasible is $S_N^{\text{fb}}(\mathcal{X}, \mathcal{X}_f) \subseteq S_N(\mathcal{X}, \mathcal{X}_f)$. The superscript "fb" represents feedback.

Suppose, at time k, the optimization problem yields optimal solution

$$\pi_N^*(k) = \{u^*(k), F_1^*(x^*(k+1|k)), \cdots, F_{N-1}^*(x^*(k+N-1|k))\}.$$

Suppose conditions (A1)-(A2) and (A3a)-(A4a) are satisfied. Then, at $k+1$ the following is feasible:

$$\pi_N(k+1) = \{F_1^*(x^*(k+1|k)), \cdots, F_{N-1}^*(x^*(k+N-1|k)), K_f(x^*(k+N|k))\}$$

and it can be proved that

$$\Delta J_N^*(k+1) + \ell(x(k), K_N^*(x(k)), w(k)) \le 0,$$
$$\forall x \in S_N^{\text{fb}}(\mathcal{X}, \mathcal{X}_f) \backslash \mathcal{X}_f, \ \forall w \in \mathcal{W}(x, K_N^*(x)).$$

Thus, with certain "more fundamental" conditions satisfied, closed-loop stability can be proved.

Compared with open-loop min-max MPC, the advantages of feedback MPC includes $S_N^{\text{ol}}(\mathcal{X}, \mathcal{X}_f) \subseteq S_N^{\text{fb}}(\mathcal{X}, \mathcal{X}_f)$ and $S_N^{\text{fb}}(\mathcal{X}, \mathcal{X}_f) \subseteq S_{N+1}^{\text{fb}}(\mathcal{X}, \mathcal{X}_f)$. For open-loop min-max MPC, when N is increased, $S_N^{\text{ol}}(\mathcal{X}, \mathcal{X}_f)$ does not necessarily become enlarged. However, feedback MPC has its severe deficiency, i.e., the optimization problem involved is too complex and in general it is not solvable (except for some special cases).

6.6 A concrete realization: case continuous-time nominal systems

The above discussions are mainly concerned with discrete-time systems. In the following, we adopt a more concrete continuous-time system as an example.

This example complies with the "key points 234" of synthesis approach and, hence, more details are omitted here. Through this example, we show which are the "more fundamental" conditions besides the conditions (B1)-(B4).

Consider the system (6.2.1) and only consider the input constraint. Suppose

(C1) f is twice continuously differentiable and $f(0,0) = 0$. Thus, $(x, u) = (0,0)$ is an equilibrium point of the system;

(C2) $\mathcal{U}$ is compact, convex and $\mathcal{U} \supset \{0\}$;

(C3) system (6.2.1) has a unique solution for any initial state $x(0) = x_0$ and any piecewise continuous and right-continuous $\bar{u}(\cdot) : [0, \infty) \to \mathcal{U}$.

At time t consider the following optimization problem:

$$\min_{\tilde{u}_T(t)} J_T(t) = \int_0^T \left(\|x(t + s|t)\|_Q^2 + \|u(t + s|t)\|_R^2 \right) ds + \|x(t + T|t)\|_P^2 ,$$
$$(6.6.1)$$

$$\text{s.t. } (6.2.4), u(t + s|t) \in \mathcal{U}, \ s \in [0, T], \ x(t + T|t) \in \mathcal{X}_f, \tag{6.6.2}$$

where $Q > 0, \ R > 0$,

$$\|x(t + T|t)\|_P^2 \geq \int_T^\infty \left(\|x(t + s|t)\|_Q^2 + \|u(t + s|t)\|_R^2 \right) ds,$$
$$u(t + s|t) = K_f x(t + s|t), \ \forall x(t + T|t) \in \mathcal{X}_f. \tag{6.6.3}$$

The optimal solution of problem (6.6.1)-(6.6.2) is $\tilde{u}_T^*(t) : [t, t + T] \to \mathcal{U}$ (i.e., $u^*(t + s|t), \ s \in [0, T]$), with the corresponding cost value $J_T^*(t)$.

In the real applications, the optimization needs to be re-done after a certain time period. Suppose the optimization cycle is δ satisfying $\delta < T$. The actually implemented control move is

$$u^*(\tau) = u^*(\tau|t), \ \tau \in [t, t + \delta). \tag{6.6.4}$$

At time $t + \delta$, based on the newly measured $x(t + \delta)$, the optimization problem (6.6.1)-(6.6.2), with t replaced by $t + \delta$, is re-solved.

6.6.1 Determination of the three ingredients

Consider the Jacobian linearization of the system (6.2.1) at $(x, u) = (0, 0)$: $\dot{x} = Ax + Bu$. When (A, B) is stabilizable, there exists $u = K_f x$ such that $A_f = A + BK_f$ is asymptotically stable.

Lemma 6.6.1. *Suppose (A, B) is stabilizable. Then,*

(1) for $\beta \in (0, \infty)$ satisfying $\beta < -\lambda_{\max}(A_f)$ ($\lambda_{\max}(\cdot)$ is the maximum real part of the eigenvalue), the following Lyapunov equation:

$$(A_f + \beta I)^T P + P(A_f + \beta I) = -(Q + K_f^T R K_f) \qquad (6.6.5)$$

admits a unique positive-definite symmetric solution P;

(2) there exists a constant $\alpha \in (0, \infty)$ specifying a level set $\Omega_\alpha \triangleq \{x | x^T P x \leq \alpha\}$ such that

(i) $K_f x \in \mathcal{U}$, $\forall x \in \Omega_\alpha$;

(ii) Ω_α is invariant for the system $\dot{x} = f(x, K_f x)$;

(iii) for any $\bar{x} \in \Omega_a$, the infinite-horizon cost

$$J_\infty(t_1) = \int_{t_1}^{\infty} \left(\|x(s)\|_Q^2 + \|K_f x(s)\|_R^2 \right) ds,$$

$$\dot{x}(t) = f(x(t), K_f x(t)), \ \forall t \geq t_1, \ x(t_1) = \bar{x}$$

is bounded from above by $J_\infty(t_1) \leq \bar{x}^T P \bar{x}$.

Proof. When $\beta < -\lambda_{\max}(A_f)$, $A_f + \beta I$ is asymptotically stable and $Q + K_f^T R K_f$ is positive-definite. Hence, (6.6.5) has a unique positive-definite and symmetric solution P.

Since $\mathcal{U} \supset \{0\}$, for any $P > 0$, one can always find $\alpha_1 \in (0, \infty)$ such that $K_f x \in \mathcal{U}$, $\forall x \in \Omega_{\alpha_1}$. Let $\alpha \in (0, \alpha_1]$. Then $K_f x \in \mathcal{U}$, $\forall x \in \Omega_\alpha$. Hence, (i) holds.

For $\dot{x} = f(x, K_f x)$, applying $x^T P x$ yields

$$\frac{d}{dt} x(t)^T P x(t) = x(t)^T (A_f^T P + P A_f) x(t) + 2x(t)^T P \phi(x(t)), \qquad (6.6.6)$$

where $\phi(x) = f(x, K_f x) - A_f x$. Moreover,

$$x^T P \phi(x) \leq \|x^T P\| \cdot \|\phi(x)\| \leq \|P\| \cdot L_\phi \cdot \|x\|^2 \leq \frac{\|P\| \cdot L_\phi}{\lambda_{\min}(P)} \|x\|_P^2 \qquad (6.6.7)$$

where $\lambda_{\min}(\cdot)$ is the minimum real part of the eigenvalue,

$$L_\phi \triangleq \sup \left\{ \frac{\|\phi(x)\|}{\|x\|} \ \middle| \ x \in \Omega_\alpha, x \neq 0 \right\}.$$

Now choose $\alpha \in (0, \alpha_1]$ such that, in Ω_α, $L_\phi \leq \frac{\beta \cdot \lambda_{\min}(P)}{\|P\|}$. Then, (6.6.7) leads to

$$x^T P \phi(x) \leq \beta x^T P x. \qquad (6.6.8)$$

Substituting (6.6.8) into (6.6.6) yields

$$\frac{d}{dt} x(t)^T P x(t) \leq x(t)^T ((A_f + \beta I)^T P + P(A_f + \beta I)) x(t).$$

Further applying (6.6.5) yields

$$\frac{d}{dt}x(t)^T Px(t) \leq -x(t)^T (Q + K_f^T RK_f)x(t). \tag{6.6.9}$$

Equation (6.6.9) implies that (ii) holds.

Finally, for any $\bar{x} \in \Omega_a$, integrate (6.6.9) from $t = t_1$ to $t = \infty$, one obtains $J_\infty(t_1) \leq \bar{x}^T P\bar{x}$.   □

The three ingredients for synthesis are: $F(x(t + T|t)) = \|x(t + T|t)\|_P^2$, $\mathcal{X}_f = \Omega_a$, $K_f(x) = K_f x$.

Lemma 6.6.2. *For sufficiently small sampling time $\delta > 0$, if the optimization problem (6.6.1)-(6.6.2) has a feasible solution at $t = 0$, then this problem is feasible for any $t > 0$.*

Proof. Suppose at time t, a feasible solution $\tilde{u}_T^*(t)$ of (6.6.1)-(6.6.2) exists. At time $t + \delta$, the following is a feasible solution to (6.6.1)-(6.6.2):

$$u(s|t + \delta) = \begin{cases} u^*(s|t), & s \in [t + \delta, t + T] \\ K_f x(s|t + \delta), & s \in [t + T, t + \delta + T] \end{cases} \tag{6.6.10}$$

which is simplified as $\tilde{u}_T(t + \delta)$.

In real application, usually take the actual control move as

$$u^{\text{act}}(\tau) = u^*(t) = u^*(t|t), \quad \tau \in [t, t + \delta). \tag{6.6.11}$$

Thus, (6.6.10) is not necessarily feasible. However, since the state is continuous, it is easy to show that, for sufficiently small δ, (6.6.10) is still feasible.   □

6.6.2   Asymptotic stability

Lemma 6.6.3. *Suppose the optimization problem (6.6.1)-(6.6.2) is feasible at time $t = 0$. Then, for any $t > 0$ and $\tau \in (t, t + \delta]$ the optimal value function satisfies*

$$J_T^*(\tau) \leq J(x(\tau), \tilde{u}_T(\tau)) \leq J_T^*(t) - \int_t^\tau \left(\|x(s)\|_Q^2 + \|u^*(s)\|_R^2 \right) ds. \tag{6.6.12}$$

Proof. Suppose, at time t, (6.6.1)-(6.6.2) has a feasible solution $\tilde{u}_T^*(t)$ and the control is implemented according to (6.6.4). For any time $\tau \in (t, t + \delta]$, consider the following feasible control input:

$$u(s|\tau) = \begin{cases} u^*(s|t), & s \in [\tau, t + T] \\ K_f x(s|t + \delta), & s \in [t + T, \tau + T] \end{cases} \tag{6.6.13}$$

which is simplified as $\tilde{u}_T(\tau)$. We will calculate the cost value corresponding to (6.6.13), which is denoted as $J(x(\tau), \tilde{u}_T(\tau))$.

Lemma 6.6.1 can be applied. Applying (6.6.9), integrating from $t + T$ to $\tau + T$ yields

$$\left\| x(\tau + T | \tau) |_{\tilde{u}_T(\tau)} \right\|_P^2 \leq \left\| x^*(t + T | t) |_{\tilde{u}_T^*(t)} \right\|_P^2$$
$$- \int_{t+T}^{\tau+T} \left\| x(s | \tau) |_{\tilde{u}_T(\tau)} \right\|_{Q + K_f^T R K_f}^2 ds.$$

Thus,

$$J(x(\tau), \tilde{u}_T(\tau)) = \int_{\tau}^{\tau+T} \left(\left\| x(s | \tau) |_{\tilde{u}_T(\tau)} \right\|_Q^2 + \left\| u(s | \tau) \right\|_R^2 \right) ds + \left\| x(\tau + T | \tau) |_{\tilde{u}_T(\tau)} \right\|_P^2$$
$$= \int_{\tau}^{t+T} \left(\left\| x^*(s | t) \right\|_Q^2 + \left\| u^*(s | t) \right\|_R^2 \right) ds$$
$$+ \int_{t+T}^{\tau+T} \left\| x(s | \tau) |_{\tilde{u}_T(\tau)} \right\|_{Q + K_f^T R K_f}^2 ds + \left\| x(\tau + T | \tau) |_{\tilde{u}_T(\tau)} \right\|_P^2$$
$$\leq \int_{\tau}^{t+T} \left(\left\| x^*(s | t) \right\|_Q^2 + \left\| u^*(s | t) \right\|_R^2 \right) ds + \left\| x^*(t + T | t) |_{\tilde{u}_T^*(t)} \right\|_P^2.$$

Comparing with the optimum $J_T^*(t)$ yields

$$J(x(\tau), \tilde{u}_T(\tau)) \leq J_T^*(t) - \int_t^{\tau} \left(\left\| x^*(s | t) \right\|_Q^2 + \left\| u^*(s | t) \right\|_R^2 \right) ds.$$

Consider implementing the control moves according to (6.6.4). Then the above equation becomes

$$J(x(\tau), \tilde{u}_T(\tau)) \leq J_T^*(t) - \int_t^{\tau} \left(\left\| x(s) \right\|_Q^2 + \left\| u^*(s) \right\|_R^2 \right) ds.$$

Finally, considering the optimality of $J_T^*(\cdot)$ yields (6.6.12).   □

Remark 6.6.1. If, in real applications, (6.6.11) is applied, then readers may suspect the conclusion in Lemma 6.6.3. An alternative is to take, in the optimization problem, $T = (n_1 + 1)\delta, n_1 \geq 1$, and replace the decision variables with $\{u(t | t), u(t + \delta | t), \cdots, u(t + n_1 \delta | t)\}$ satisfying the constraint $u(\tau | t) = u(t + i\delta | t)$, $\tau \in [t + i\delta, t + (i + 1)\delta)$.

Theorem 6.6.1. *(Stability) Suppose*

(i) assumptions (C1)-(C3) are satisfied;

(ii) (A, B) is stabilizable;

(iii) for any $x_0 \in \Omega$, problem (6.6.1)-(6.6.2) has a feasible solution at $t = 0$.

Then, for a sufficiently small δ, the closed-loop system with (6.6.4) is asymptotically stable, with a region of attraction Ω.

Proof. $J_T^*(t) = J_T^*(x(t), \tilde{u}_T^*(t))$ has properties : (i) $J_T^*(0, \tilde{u}_T^*(t)) = 0$ and $J_T^*(t) > 0$, $\forall x(t) \neq 0$; (ii) there exists a constant $\gamma \in (0, \infty)$ such that $J_T^*(t) \leq \gamma$; (iii) for $0 \leq t_1 < t_2 \leq \infty$, $J_T^*(t_2) - J_T^*(t_1) \leq - \int_{t_1}^{t_2} \|x(t)\|_Q^2 \, dt$. The properties (i)-(ii) are due to the continuity of the state, and the property (iii) is due to Lemma 6.6.3.

Thus, as long as $x(t) \neq 0$, $J_T^*(t)$ will be strictly decreasing with the evolution of time, which shows that $J_T^*(t)$ can be Lyapunov function for asymptotic stability.

A more strict proof involves the details of stability theory of continuous-time systems, which are omitted here. $\qquad\square$

Chapter 7

State feedback synthesis approaches

The deficiency of classical MPC is that the uncertainty is not explicitly considered. Here, "explicitly" means that the uncertainty is considered in the optimization problem of MPC. Apparently, classical MPC (DMC, MAC, GPC, etc.) adopts a linear model for prediction, and optimizes a nominal performance cost. The uncertainty of the model is overcome by feedback correction or on-line refreshment of the model. Since a nominal model is adopted, the existing stability analysis is mainly for the nominal systems.

The studies on the robustness of MPC and robust MPC can be classified as robustness analysis (corresponding to adopting nominal model to design the controller, and analyzing stability when there is model-plant mismatch) and robustness synthesis (the uncertainty is directly considered in the controller design). This chapter introduces synthesis approaches of robust MPC for systems with polytopic description.

Sections 7.1 and 7.2 are referred to in [37]. Section 7.3 is referred to in [61]. Section 7.4 is referred to in [29]. Section 7.5 is referred to in [21], [25]. Section 7.6 is referred to in [21].

7.1 System with polytopic description, linear matrix inequality

Consider the following time-varying uncertain system

$$x(k+1) = A(k)x(k) + B(k)u(k), \quad [A(k)\,|B(k)] \in \Omega \qquad (7.1.1)$$

where $u \in \mathfrak{R}^m$ is the control input, $x \in \mathfrak{R}^n$ is the state. The input and state constraints are

$$-\bar{u} \le u(k+i) \le \bar{u}, \quad -\bar{\psi} \le \Psi x(k+i+1) \le \bar{\psi}, \ \forall i \ge 0 \qquad (7.1.2)$$

where $\bar{u} := [\bar{u}_1, \bar{u}_2, \cdots, \bar{u}_m]^T$, $\bar{u}_j > 0$, $j \in \{1, \ldots, m\}$, $\bar{\psi} := [\bar{\psi}_1, \bar{\psi}_2, \cdots, \bar{\psi}_q]^T$, $\bar{\psi}_s > 0$, $s \in \{1, \ldots, q\}$, $\Psi \in \Re^{q \times n}$. Note that state constraint has different starting time from input constraint. This is because the current state cannot be affected by the current and future input.

Suppose the matrix pair $[A(k) | B(k)] \in \Omega$, where Ω is defined as the following "polytope":

$$\Omega = Co\left\{[A_1 | B_1], [A_2 | B_2], \cdots, [A_L | B_L]\right\}, \ \forall k \geq 0,$$

i.e., there exist L nonnegative coefficients $\omega_l(k)$, $l \in \{1, \ldots, L\}$ such that

$$\sum_{l=1}^{L} \omega_l(k) = 1, \ [A(k) | B(k)] = \sum_{l=1}^{L} \omega_l(k) [A_l | B_l] \tag{7.1.3}$$

where $[A_l | B_l]$ is called the vertex of the polytope. Denote $[\hat{A} | \hat{B}] \in \Omega$ as the nominal model which is the "closest" to the actual system, such as $[\hat{A} | \hat{B}] = 1/L \sum_{l=1}^{L} [A_l | B_l]$.

Polytope (also known as multi-model) can be obtained by two different manners. At different operating points, or at different time periods, input/output data can be obtained for the same system (possibly nonlinear). From each data set, we develop a linear model (suppose the same state vector is selected for all the data sets). If each linear model is seen as a vertex, then polytope is obtained. Obviously, it is reasonable to suppose the analysis and design results for the system (7.1.1) are also suitable for all linear models.

Alternatively, for a nonlinear discrete time-varying system $x(k + 1) = f(x(k), u(k), k)$, one can suppose the Jacobian matrix pair $\begin{bmatrix} \partial f / \partial x & \partial f / \partial u \end{bmatrix}$ lies in the polytope Ω. Any possible dynamic behavior of the nonlinear system is contained in the possible dynamics of polytopic system, i.e., for any initial condition of the nonlinear system there exists a time-varying system belongs to Ω, which has the same dynamic response with the nonlinear system.

Linear matrix inequality (LMI) is especially suitable in the analysis and design works based on the polytopic description. LMI is a matrix inequality of the following form

$$F(v) = F_0 + \sum_{i=1}^{l} v_i F_i > 0, \tag{7.1.4}$$

where $v_1, v_2, \cdots, v_l$ are the variables, F_i is a given symmetric matrix, $F(v) > 0$ means that $F(v)$ is positive-definite. Usually, variables are in the form of matrices. Hence LMI is usually not written in the uniform form of (7.1.4). For transforming an inequality into an LMI, Schur complement is often applied.

Schur complements: For $Q(v) = Q(v)^T$, $R(v) = R(v)^T$ and $S(v)$, the following three groups of inequalities are equivalent:

(i) $\begin{bmatrix} Q(v) & S(v) \\ S(v)^T & R(v) \end{bmatrix} > 0;$

(ii) $R(v) > 0, Q(v) - S(v)R(v)^{-1}S(v)^T > 0$;

(iii) $Q(v) > 0, R(v) - S(v)^T Q(v)^{-1}S(v) > 0$.

Remark 7.1.1. In Schur complement, it is not required (i) to be LMI. Denote $F(v) = \begin{bmatrix} Q(v) & S(v) \\ S(v)^T & R(v) \end{bmatrix}$. If $F(v)$ can be expressed in the form of (7.1.4), then (i) is LMI. In controller synthesis, one often meets with matrix inequalities in the form of (ii)-(iii), which cannot be expressed in the form of (7.1.4) (i.e., often (ii)-(iii) are not LMIs). Thus, by applying Schur complement, the inequalities in the form of (ii)-(iii) can be equivalently transformed into LMI in the form of (i).

In MPC synthesis, one often meets minimization of a linear function satisfying a group of LMIs, i.e.,

$$\min_v c^T v, \text{ s.t. } F(v) > 0. \tag{7.1.5}$$

This is a convex optimization problem, which can be solved by polynomial-time algorithms [3].

7.2 On-line approach based on min-max performance cost: case zero-horizon

Consider the following quadratic performance index:

$$J_\infty(k) = \sum_{i=0}^{\infty} [\|x(k+i|k)\|_W^2 + \|u(k+i|k)\|_R^2],$$

where $W > 0$ and $R > 0$ are both the symmetric weighting matrices. MPC adopting this performance cost is called infinite-horizon MPC. Our target is to solve the following optimization problem:

$$\min_{u(k+i|k),i\geq 0} \max_{[A(k+i)|B(k+i)]\in\Omega,i\geq 0} J_\infty(k), \tag{7.2.1}$$

$$\text{s.t.} -\bar{u} \leq u(k+i|k) \leq \bar{u}, \quad -\bar{\psi} \leq \Psi x(k+i+1|k) \leq \bar{\psi}, \tag{7.2.2}$$

$$x(k+i+1|k) = A(k+i)x(k+i|k) + B(k+i)u(k+i|k), \quad x(k|k) = x(k). \tag{7.2.3}$$

Problem (7.2.1)-(7.2.3) is a "min-max" optimization problem. The "max" operation is finding $[A(k+i)|B(k+i)] \in \Omega$ based on which the largest $J_\infty(k)$ (or, called the worst value of $J_\infty(k)$) is found. Then, this worst value is minimized over control moves $u(k+i|k)$. If the finite-horizon optimization, rather than the infinite-horizon one, is encountered, then this "min-max" optimization problem is convex (i.e., a unique optimal solution exists), but is computationally intractable (i.e., finding the optimal solution in finite time is not guaranteed).

7.2.1 Performance cost handling and unconstrained MPC

To simplify the solution process of (7.2.1)-(7.2.3), we will derive the upper bound of the performance index, and then minimize this upper bound by adopting the following control law:

$$u(k+i|k) = Fx(k+i|k), \ i \geq 0. \tag{7.2.4}$$

To define the upper bound, firstly define quadratic function $V(x) = x^T Px, \ P > 0$ and impose the following constraint:

$$V(x(k+i+1|k)) - V(x(k+i|k)) \leq -[\|x(k+i|k)\|_W^2 + \|u(k+i|k)\|_R^2]. \tag{7.2.5}$$

For the sake of boundedness of the performance cost, $x(\infty|k) = 0$ and $V(x(\infty|k)) = 0$ has to be satisfied. Summing (7.2.5) from $i = 0$ to $i = \infty$, we get

$$\max_{[A(k+i)|B(k+i)]\in\Omega, i\geq 0} J_\infty(k) \leq V(x(k|k)),$$

where $V(x(k|k))$ gives an upper bound on the performance cost.

Thus MPC algorithm is redefined as: at each time step k, find (7.2.4) to minimize $V(x(k|k))$, but only $u(k|k) = Fx(k|k)$ is implemented; at the next time step $k + 1$, $x(k + 1)$ is measured, and the optimization is repeated to re-compute F.

Define a scalar $\gamma > 0$, and let

$$V(x(k|k)) \leq \gamma. \tag{7.2.6}$$

Then the minimization of $\max_{[A(k+i)|B(k+i)]\in\Omega, i\geq 0} J_\infty(k)$ is approximated by the minimization of γ satisfying (7.2.6). Define matrix $Q = \gamma P^{-1}$. By utilizing Schur complements, (7.2.6) is equivalent to the following LMI:

$$\begin{bmatrix} 1 & x(k|k)^T \\ x(k|k) & Q \end{bmatrix} \geq 0. \tag{7.2.7}$$

Substitute (7.2.4) into (7.2.5) yields

$$x(k+i|k)^T \{[A(k+i) + B(k+i)F]^T P[A(k+i) + B(k+i)F] - P \\ +F^T RF + W\}x(k+i|k) \leq 0, \tag{7.2.8}$$

which is satisfied, for all $i \geq 0$, if

$$[A(k+i) + B(k+i)F]^T P[A(k+i) + B(k+i)F] - P + F^T RF + W \leq 0. \tag{7.2.9}$$

Define $F = YQ^{-1}$. By substituting $P = \gamma Q^{-1}$ and $F = YQ^{-1}$ into (7.2.9), pre- and post-multiplying the obtained inequality by Q (congruence transformation), and using Schur complements, it is shown that (7.2.9) is equivalent to the following LMI:

$$
\begin{bmatrix}
Q & * & * & * \\
A(k+i)Q + B(k+i)Y & Q & * & * \\
W^{1/2}Q & 0 & \gamma I & * \\
R^{1/2}Y & 0 & 0 & \gamma I
\end{bmatrix} \geq 0. \tag{7.2.10}
$$

Since an LMI is symmetric, "$*$" in any LMI always denotes the blocks in the symmetric position. Eq. (7.2.10) is affine in $[\ A(k+i)\quad B(k+i)\ ]$ (the linear superposition principle is satisfied). Hence (7.2.10) is satisfied if and only if

$$
\begin{bmatrix}
Q & * & * & * \\
A_l Q + B_l Y & Q & * & * \\
W^{1/2}Q & 0 & \gamma I & * \\
R^{1/2}Y & 0 & 0 & \gamma I
\end{bmatrix} \geq 0, \ l \in \{1, \dots, L\}. \tag{7.2.11}
$$

Note that, strictly speaking, the variables γ, Q, P, F, Y in the above should be denoted by $\gamma(k)$, $Q(k)$, $P(k)$, $F(k)$, $Y(k)$. In the application, these variables can be time-varying. For $L = 1$, the optimal solution $F = YQ^{-1}$ for the discrete-time infinite-horizon unconstrained linear quadric regulator (LQR) is obtained by solving the following optimization problem:

$$
\min_{\gamma, Q, Y} \gamma, \text{ s.t. } (7.2.7), (7.2.11). \tag{7.2.12}
$$

The solution F has nothing to do with x, i.e., F is unique irrespective of the value of x.

When $L > 1$, it is apparent the (7.2.1)-(7.2.3) include the corresponding LQR as a special case, and is much more complex than LQR. When $L > 1$, by solving (7.2.12), the approximate solution of problem (7.2.1)-(7.2.3), without considering constraints, is obtained. This approximate solution is directly related to x, i.e., F varies with x. This shows that, even without considering hard constraints, receding horizon solving (7.2.12) can greatly improve the performance compared with adopting a single F.

7.2.2 Constraint handling

For constraint, we first consider the notion of invariant ellipsoidal set (invariant ellipsoid). Consider γ, Q, P, F, Y which have been defined in the previous section, and $\varepsilon = \{z | z^T Q^{-1} z \leq 1\} = \{z | z^T P z \leq \gamma\}$. Then ε is an ellipsoidal set. When (7.2.7) and (7.2.11) are satisfied, ε is an invariant ellipsoid, i.e.,

$$
x(k|k) \in \varepsilon \Rightarrow x(k+i|k) \in \varepsilon, \ \forall i \geq 1.
$$

Firstly, consider input constraints $-\bar{u} \leq u(k+i|k) \leq \bar{u}$ in (7.2.1)-(7.2.3). Since ε is an invariant ellipsoid, by considering the j-th element of u, denoting

ξ_j as the j-th row of the m-ordered identity matrix, we can make the following deduction:

$$\max_{i \geq 0} |\xi_j u(k+i|k)|^2 = \max_{i \geq 0} |\xi_j YQ^{-1}x(k+i|k)|^2$$

$$\leq \max_{z \in \varepsilon} |\xi_j YQ^{-1}z|^2 \leq \max_{z \in \varepsilon} \left\| \xi_j YQ^{-1/2} \right\|_2^2 \left\| Q^{-1/2}z \right\|_2^2$$

$$\leq \left\| \xi_j YQ^{-1/2} \right\|_2^2 = (YQ^{-1}Y^T)_{jj},$$

where $(\cdot)_{jj}$ is the j-th diagonal element of the square matrix, $\|\cdot\|_2$ the 2-norm. By using Schur complements, if there exists a symmetric matrix Z such that

$$\begin{bmatrix} Z & Y \\ Y^T & Q \end{bmatrix} \geq 0, \ Z_{jj} \leq \bar{u}_j^2, j \in \{1, \ldots, m\}, \qquad (7.2.13)$$

then $|u_j(k+i|k)| \leq \bar{u}_j, \ j \in \{1, 2, \ldots, m\}$.

Eq. (7.2.13) is a sufficient (not necessary) condition for satisfaction of input constraints. In general, utilizing (7.2.13) for handling input constraint is not very conservative, especially when the nominal case is addressed.

Then, consider state constraint $-\bar{\psi} \leq \Psi x(k+i+1|k) \leq \bar{\psi}$. Since ε is an invariant ellipsoid, by denoting ξ_s as the s-th row of the q-ordered identity matrix, we can make the following deduction:

$$\max_{i \geq 0} |\xi_s x(k+i+1|k)| = \max_{i \geq 0} |\xi_s [A(k+i) + B(k+i)F] x(k+i|k)|$$

$$= \max_{i \geq 0} \left| \xi_s [A(k+i) + B(k+i)F] Q^{1/2}Q^{-1/2}x(k+i|k) \right|$$

$$\leq \max_{i \geq 0} \left\| \xi_s [A(k+i) + B(k+i)F] Q^{1/2} \right\| \left\| Q^{-1/2}x(k+i|k) \right\|$$

$$\leq \max_{i \geq 0} \left\| \xi_s [A(k+i) + B(k+i)F] Q^{1/2} \right\|.$$

Thus, by using Schur complements, if there exists a symmetric matrix Γ such that

$$\begin{bmatrix} Q & * \\ \Psi[A(k+i)Q + B(k+i)Y] & \Gamma \end{bmatrix} \geq 0, \ \Gamma_{ss} \leq \bar{\psi}_s^2, \ l \in \{1, 2, \ldots, L\},$$
$$s \in \{1, 2, \ldots, q\}, \qquad (7.2.14)$$

then $|\xi_s x(k+i+1|k)| \leq \bar{\psi}_s, \ s \in \{1, 2, \ldots, q\}$.

Eq. (7.2.14) is affine with respect to $[\ A(k+i) \quad B(k+i)\ ]$ (superposition principle is satisfied). Therefore, (7.2.14) is satisfied if and only if

$$\begin{bmatrix} Q & * \\ \Psi(A_l Q + B_l Y) & \Gamma \end{bmatrix} \geq 0, \ \Gamma_{ss} \leq \bar{\psi}_s^2, \ l \in \{1, 2, \ldots, L\}, \ s \in \{1, 2, \ldots, q\}.$$
$$(7.2.15)$$

Now, problem (7.2.1)-(7.2.3) is approximately converted into the following optimization problem:

$$\min_{\gamma,Q,Y,Z,\Gamma} \gamma, \text{ s.t. } (7.2.7), (7.2.11), (7.2.13), (7.2.15). \qquad (7.2.16)$$

By the above deductions we obtain the following important property of predictive control:

Lemma 7.2.1. *(Feasibility) Any feasible solution of the optimization in (7.2.16) at time k is also feasible for all time $t > k$. Thus if the optimization problem in (7.2.16) is feasible at time k, then it is feasible for all times $t > k$.*

Proof. Let us suppose (7.2.16) is feasible at time k. The only LMI in (7.2.16) that depends explicitly on the measured state $x(k|k) = x(k)$ of the system is the following:

$$\begin{bmatrix} 1 & x(k|k)^T \\ x(k|k) & Q \end{bmatrix} \geq 0.$$

Thus, to prove the lemma, we need only prove that this LMI is feasible for all future measured states $x(k+i|k+i) = x(k+i)$.

In the above discussion we have shown that, when (7.2.7) and (7.2.11) are satisfied,

$$x(k+i|k)^T Q^{-1} x(k+i|k) < 1, \; i \geq 1.$$

Consider the state measured at $k+1$, then there is a $[A|B] \in \Omega$ such that

$$x(k+1|k+1) = x(k+1) = (A+BF)x(k|k). \qquad (7.2.17)$$

Eq. (7.2.17) is different from

$$x(k+1|k) = [A(k)+B(k)F]x(k|k) \qquad (7.2.18)$$

in that, $x(k+1|k)$ is uncertain while $x(k+1|k+1)$ is deterministic and measured. Apparently, $x(k+1|k)^T Q^{-1} x(k+1|k) \leq 1$ must lead to

$$x(k+1|k+1)^T Q^{-1} x(k+1|k+1) < 1.$$

Thus, the feasible solution of the optimization problem at time k is also feasible at time $k+1$. Hence the optimization (7.2.16) is feasible at time $k+1$.

Analogously, for $k+2, k+3, \cdots$, the same result as $k+1$ can be obtained. □

Theorem 7.2.1. *(Stability) Suppose the optimization in (7.2.16) is feasible at time $k = 0$. Then by receding-horizon implementation of $u(k) = F(k)x(k) = Y(k)Q(k)^{-1}x(k)$, the closed-loop system is robustly exponentially stable.*

Proof. Use $*$ to denote the optimal solution. To prove asymptotic stability, we shall establish that $\gamma^*(k)$ is strictly decreasing with the evolution of k.

First, let us suppose problem (7.2.16) is feasible at time $k = 0$. Lemma 7.2.1 then ensures feasibility of (7.2.16) at all $k > 0$. At each time k, problem (7.2.16) is convex and, therefore, has a unique minimum.

Since (7.2.5) is satisfied,

$$x^*(k+1|k)^T P^*(k)x^*(k+1|k) \le x(k|k)^T P^*(k)x(k|k) - \left[\|x(k)\|_W^2 + \|u^*(k|k)\|_R^2\|\right].$$
(7.2.19)

Since $x^*(k + 1|k)$ is the predicted state whereas $x(k + 1|k + 1)$ is the measured state, (7.2.19) must lead to

$$x(k + 1|k + 1)^T P^*(k)x(k + 1|k + 1) \le x(k|k)^T P^*(k)x(k|k)$$
$$- \left[\|x(k)\|_W^2 + \|u^*(k|k)\|_R^2\|\right].$$
(7.2.20)

Now, notice that

$$x(k + 1|k + 1)^T P(k + 1)x(k + 1|k + 1) \le \gamma(k + 1), \quad x(k|k)^T P^*(k)x(k|k)$$
$$\le \gamma^*(k).$$

According to (7.2.20), at time $k + 1$, it is feasible to choose $\gamma(k + 1) = \gamma^*(k) - \left[\|x(k)\|_W^2 + \|u^*(k|k)\|_R^2\|\right].$
This $\gamma(k + 1)$ is not necessarily optimal at time $k + 1$. Hence,

$$\gamma^*(k + 1) \le \gamma(k + 1),$$
(7.2.21)

and

$$\gamma^*(k + 1) - \gamma^*(k) \le - \left[\|x(k)\|_W^2 + \|u^*(k|k)\|_R^2\|\right] \le -\lambda_{\min}(W)\|x(k)\|^2.$$
(7.2.22)

Eq. (7.2.22) shows that $\gamma^*(k)$ is strictly decreasing with the evolution of k and, hence, can serve as Lyapunov function. Therefore, $\lim_{k\to\infty} x(k) = 0$ is concluded. $\qquad\square$

7.3 Off-line approach based on min-max performance cost: case zero-horizon

In this section, we design off-line MPC based on the notion of "asymptotically stable invariant ellipsoid." The so-called "off-line" means that all the optimizations are performed off-line. A series of control laws are optimized off-line, each corresponding to an ellipsoidal region of attraction. When the algorithm is implemented on-line, one only needs to find the ellipsoid in which the current state lies, and choose the control law corresponding to this ellipsoid.

Definition 7.3.1. *Consider a discrete-time dynamic system $x(k+1) = f(x(k))$ and a set $\varepsilon = \{x \in \Re^n | x^T Q^{-1} x \leq 1\}$. If*

$$x(k_1) \in \varepsilon \Rightarrow x(k) \in \varepsilon, \ \forall k \geq k_1, \ \lim_{k \to \infty} x(k) = 0,$$

then ε is said to be an asymptotically stable invariant ellipsoid.

Apparently, by solving (7.2.16) to obtain $\{F, Q\}$, then the ellipsoid ε corresponding to Q is an asymptotically stable invariant ellipsoid. If we off-line optimize a series of (e.g. a number of N; note that, at here N is not the switching horizon, but adopts the same notation as the switching horizon; the reader could think that this N has some relation with the switching horizon) ε, with each ε corresponding to its F, then on-line, we can choose F corresponding to the ellipsoid ε in which the state lies. Suppose we have chosen a "sufficiently large" number of ellipsoids, then it is easy to imagine that we can make off-line MPC approach to on-line MPC in the former chapter, to some extent.

Algorithm 7.1

Step 1. Off-line, choose states x_i, $i \in \{1, \dots, N\}$. Substitute $x(k|k)$ in (7.2.7) by x_i, and solve (7.2.16) to obtain the corresponding matrices $\{Q_i, Y_i\}$, ellipsoids $\varepsilon_i = \{x \in \Re^n | x^T Q_i^{-1} x \leq 1\}$ and feedback gains $F_i = Y_i Q_i^{-1}$. Notice that x_i should be chosen such that $\varepsilon_j \subset \varepsilon_{j-1}$, $\forall j \in \{2, \dots, N\}$. For each $i \neq N$, check if the following is satisfied:

$$Q_i^{-1} - (A_l + B_l F_{i+1})^T Q_i^{-1} (A_l + B_l F_{i+1}) > 0, \ l \in \{1, \dots, L\}. \quad (7.3.1)$$

Step 2. On-line, at each time k adopt the state feedback law:

$$u(k) = F(k)x(k) = \begin{cases} F(\alpha_i(k))x(k), & x(k) \in \varepsilon_i, x(k) \notin \varepsilon_{i+1}, \ i \neq N \\ F_N x(k), & x(k) \in \varepsilon_N \end{cases},$$

$$(7.3.2)$$

where $F(\alpha_i(k)) = \alpha_i(k)F_i + (1 - \alpha_i(k))F_{i+1}$, and

i) if (7.3.1) is satisfied, then $0 < \alpha_i(k) \leq 1$, $x(k)^T[\alpha_i(k)Q_i^{-1} + (1 - \alpha_i(k))Q_{i+1}^{-1}]x(k) = 1$;

ii) if (7.3.1) is not satisfied, then $\alpha_i(k) = 1$.

From the on-line approach, we know that the optimal control law and its corresponding asymptotically stable invariant ellipsoid depend on the state. Although the control law can be applied to all the states within the ellipsoid, it is not optimal (usually, we can only be sure that the feedback gain F_i is optimal for the state point x_i belonging to ε_i). So our off-line formulation sacrifices optimality somewhat while significantly reducing the on-line computational burden.

In the above algorithm, we also find that the selection of x_i is, to a large extent, arbitrary. However, in general we can select $x^{\max}$ as far as possible from $x = 0$. Then, we can select $x_i = \beta_i x^{\max}$, $\beta_1 = 1$, $1 > \beta_2 > \cdots > \beta_N > 0$.

Theorem 7.3.1. *(Stability) Suppose $x(0) \in \varepsilon_{x,1}$, then Algorithm 7.1 asymptotically stabilizes the closed-loop system. Further, if (7.3.1) is satisfied for all $i \neq N$, then the control law (7.3.2) in Algorithm 7.1 is a continuous function of the system state x.*

Proof. We only consider the case where (7.3.1) is satisfied for all $i \neq N$. Other cases are simpler.

The closed-loop system is given by

$$x(k+1) = \begin{cases} [A(k) + B(k)F(\alpha_i(k))]x(k), & x(k) \in \varepsilon_i, x(k) \notin \varepsilon_{i+1}, i \neq N \\ [A(k) + B(k)F_N]x(k), & x(k) \in \varepsilon_N \end{cases}.$$
(7.3.3)

For $x(k) \in \varepsilon_i \backslash \varepsilon_{i+1}$, denote

$$Q(\alpha_i(k))^{-1} = \alpha_i(k)Q_i^{-1} + (1 - \alpha_i(k))Q_{i+1}^{-1},$$
$$X(\alpha_i(k)) = \alpha_i(k)X_i + (1 - \alpha_i(k))X_{i+1}, \; X \in \{Z, \Gamma\}.$$

When (7.3.1) is satisfied, by considering the procedure of solving $\{Q_i, Y_i\}$ (i.e., $\{Q_i, Y_i\}$ satisfy stability constraint (7.2.11)), it is shown that

$$Q_i^{-1} - (A_l + B_l F(\alpha_i(k)))^T Q_i^{-1} (A_l + B_l F(\alpha_i(k))) > 0, \; l \in \{1, \dots, L\}.$$
(7.3.4)

Moreover, if both $\{Y_i, Q_i, Z_i, \Gamma_i\}$ and $\{Y_{i+1}, Q_{i+1}, Z_{i+1}, \Gamma_{i+1}\}$ satisfy (7.2.13) and (7.2.15), then

$$\begin{bmatrix} Q(\alpha_i(k))^{-1} & * \\ F(\alpha_i(k)) & Z(\alpha_i(k)) \end{bmatrix} \geq 0, Z(\alpha_i(k))_{jj} \leq \bar{u}_j^2, j \in \{1, \dots, m\}, \quad (7.3.5)$$

$$\begin{bmatrix} Q(\alpha_i(k))^{-1} & * \\ \Psi(A_l + B_l F(\alpha_i(k))) & \Gamma(\alpha_i(k)) \end{bmatrix} \geq 0,$$
$$l \in \{1, \dots, L\}, \Gamma(\alpha_i(k))_{ss} \leq \bar{\psi}_s^2, \; s \in \{1, \dots, q\}. \quad (7.3.6)$$

Eqs. (7.3.4))-(7.3.6) indicate that $u(k) = F(\alpha_i(k))x(k)$ will keep the state inside of ε_i and drive it towards ε_{i+1}, with the hard constraints satisfied. Finally, the state is converged to the origin by $u(k) = F_N x(k)$.

Consider two ring regions:

$$\mathbb{R}_{i-1} = \{x \in \mathfrak{R}^n | x^T Q_{i-1}^{-1} x \leq 1, x^T Q_i^{-1} x > 1\},$$
$$\mathbb{R}_i = \{x \in \mathfrak{R}^n | x^T Q_i^{-1} x \leq 1, x^T Q_{i+1}^{-1} x > 1\}.$$

Firstly, within $\mathbb{R}_i$, the solution of $x^T(\alpha_i Q_i^{-1} + (1 - \alpha_i)Q_{i+1}^{-1})x = 1$ is

$$\alpha_i = \frac{1 - x^T Q_{i+1}^{-1} x}{x^T (Q_i^{-1} - Q_{i+1}^{-1})x}.$$

Therefore, within $\mathbb{R}_i$, α_i is a continuous function of x, and so is $F(\alpha_i)$. The same argument holds for the region $\mathbb{R}_{i-1}$, where $\alpha_{i-1} = (1 - x^T Q_i^{-1} x)/(x^T (Q_{i-1}^{-1} - Q_i^{-1})x)$.

Secondly, when $x \in \mathbb{R}_i$, $x^T Q_i^{-1} x \to 1 \Rightarrow \alpha_i \to 1$. Thus on the boundary between $\mathbb{R}_i$ and $\mathbb{R}_{i-1}$,

$$\lim_{\alpha_{i-1} \to 0} F(\alpha_{i-1}) = \lim_{\alpha_i \to 1} F(\alpha_i) = F_i,$$

which establishes the continuity of $F(k)$ in (7.3.2) on the boundary between $\mathbb{R}_i$ and $\mathbb{R}_{i-1}$.

So, it can be concluded that $F(k)$ is a continuous function of x. $\square$

7.4   Off-line approach based on min-max performance cost: case varying-horizon

In Algorithm 7.1, when F_i is calculated, F_j, $\forall j > i$ are not considered. However, for ε_j, $\forall j > i$, F_j is the better choice than F_i. In the following, we select $\{Q_N, F_N, \gamma_N\}$ as in Algorithm 7.1, but the selection of $\{Q_j, F_j, \gamma_j, \forall j \le N - 1\}$ is different from Algorithm 7.1. For x_{N-h}, $\forall h \ge 1$, we select $\{Q_{N-h}, F_{N-h}\}$ such that, when $x(k) \in \varepsilon_{N-h}$, $x(k+i|k) \in \varepsilon_{N-h+i} \subset \varepsilon_{N-h}$, $1 \le i \le h$ and, inside of ε_{N-h+i}, F_{N-h+i} is adopted. For convenience, firstly define

$$J_{\text{tail}}(k) = \sum_{i=1}^{\infty} \left[\|x(k + i|k)\|_W^2 + \|u(k + i|k)\|_R^2 \right]. \qquad (7.4.1)$$

1. *Calculation of Q_{N-1}, F_{N-1}*
Suppose Q_N, F_N have been obtained, and consider $x(k) \notin \varepsilon_N$. The following control law is adopted to solve problem (7.2.16):

$$u(k) = F_{N-1}x(k), u(k + i|k) = F_N x(k + i|k), \forall i \ge 1, \qquad (7.4.2)$$

which yields

$$\max_{[A(k+i)|B(k+i)] \in \Omega, i \ge 1} J_{\text{tail}}(k) \le x(k + 1|k)^T P_N x(k + 1|k) \le \gamma_N, \qquad (7.4.3)$$

where $P_N = \gamma_N Q_N^{-1}$. Thus, the optimization of $J_\infty(k)$ is transformed into the optimization of the following cost function:

$$
\begin{aligned}
\bar{J}_{N-1}(k) \triangleq \bar{J}(k) &= \|x(k)\|_W^2 + \|u(k)\|_R^2 + \|x(k+1|k)\|_{P_N}^2 , = x(k)^T \\
&\times \left\{ W + F_{N-1}^T R F_{N-1} + [A(k) + B(k)F_{N-1}]^T P_N [A(k) + B(k)F_{N-1}] \right\} \\
&\times x(k).
\end{aligned}
\tag{7.4.4}
$$

Define $\bar{J}_{N-1}(k) \le \gamma_{N-1}$. Introduce the slack variable P_{N-1} such that

$$
\gamma_{N-1} - x_{N-1}^T P_{N-1} x_{N-1} \ge 0,
\tag{7.4.5}
$$

$$
W + F_{N-1}^T R F_{N-1} + [A(k) + B(k)F_{N-1}]^T P_N [A(k) + B(k)F_{N-1}] \le P_{N-1}.
\tag{7.4.6}
$$

Moreover, $u(k) = F_{N-1}x(k)$ should satisfy the input/state constraints

$$
-\bar{u} \le F_{N-1}x(k) \le \bar{u}, \quad -\bar{\psi} \le \Psi [A(k) + B(k)F_{N-1}] x(k) \le \bar{\psi}, \quad \forall x(k) \in \varepsilon_{N-1}
\tag{7.4.7}
$$

and the terminal constraint

$$
x(k+1|k) \in \varepsilon_N, \quad \forall x(k) \in \varepsilon_{N-1}.
\tag{7.4.8}
$$

Eq. (7.4.8) is equivalent to $[A(k) + B(k)F_{N-1}]^T Q_N^{-1} [A(k) + B(k)F_{N-1}] \le Q_{N-1}^{-1}$. By defining $Q_{N-1} = \gamma_{N-1} P_{N-1}^{-1}$ and $F_{N-1} = Y_{N-1} Q_{N-1}^{-1}$, (7.4.5), (7.4.6) and (7.4.8) can be transformed into the following LMIs:

$$
\begin{bmatrix} 1 & * \\ x_{N-1} & Q_{N-1} \end{bmatrix} \ge 0,
\tag{7.4.9}
$$

$$
\begin{bmatrix}
Q_{N-1} & * & * & * \\
A_l Q_{N-1} + B_l Y_{N-1} & \gamma_{N-1} P_N^{-1} & * & * \\
W^{1/2} Q_{N-1} & 0 & \gamma_{N-1} I & * \\
R^{1/2} Y_{N-1} & 0 & 0 & \gamma_{N-1} I
\end{bmatrix} \ge 0, \ l \in \{1, \dots, L\},
\tag{7.4.10}
$$

$$
\begin{bmatrix}
Q_{N-1} & * \\
A_l Q_{N-1} + B_l Y_{N-1} & Q_N
\end{bmatrix} \ge 0, \ l \in \{1, \dots, L\}.
\tag{7.4.11}
$$

Moreover, by satisfaction of the following LMIs, constraint (7.4.7) is satisfied:

$$
\begin{bmatrix} Z_{N-1} & Y_{N-1} \\ Y_{N-1}^T & Q_{N-1} \end{bmatrix} \ge 0, \ Z_{N-1,jj} \le \bar{u}_j^2, \ j \in \{1, \dots, m\},
\tag{7.4.12}
$$

$$
\begin{bmatrix}
Q_{N-1} & * \\
\Psi (A_l Q_{N-1} + B_l Y_{N-1}) & \Gamma_{N-1}
\end{bmatrix} \ge 0, \ \Gamma_{N-1,ss} \le \bar{\psi}_s^2,
$$

$$
l \in \{1, \dots, L\}, \ s \in \{1, \dots, q\}.
\tag{7.4.13}
$$

Thus, $\{Y_{N-1}, Q_{N-1}, \gamma_{N-1}\}$ can be obtained by solving the following optimization problem:

$$\min_{\gamma_{N-1}, Y_{N-1}, Q_{N-1}, Z_{N-1}, \Gamma_{N-1}} \gamma_{N-1}, \text{ s.t. } (7.4.9) - (7.4.13). \qquad (7.4.14)$$

2. *Calculation of Q_{N-h}, F_{N-h}, $\forall h \geq 2$*
Suppose Q_{N-h+1}, F_{N-h+1}, $\cdots$, Q_N, F_N have been obtained, and consider $x(k) \notin \varepsilon_{N-h+1}$. The following control law is adopted to solve problem (7.2.16):

$$u(k+i|k) = F_{N-h+i}x(k+i|k), \ i=\{0,\ldots,h-1\}; \ u(k+i|k)$$
$$= F_N x(k+i|k), \ \forall i \geq h. \qquad (7.4.15)$$

Consider (7.4.3) with $h = 2, 3, \cdots$ By induction, the following is obtained:

$$\max_{[A(k+i)|B(k+i)] \in \Omega, i \geq 1} J_{\text{tail}}(k) \leq x(k+1|k)^T P_{N-h+1} x(k+1|k) \leq \gamma_{N-h+1},$$
$$(7.4.16)$$

where $P_{N-h+1} = \gamma_{N-h+1} Q_{N-h+1}^{-1}$. Thus, the optimization of $J_\infty(k)$ is transformed into the optimization of the following cost function:

$$\bar{J}_{N-h}(k) \triangleq \bar{J}(k) = \|x(k)\|_W^2 + \|u(k)\|_R^2 + \|x(k+1|k)\|_{P_{N-h+1}}^2. \qquad (7.4.17)$$

Introduce the slack variable $P_{N-h} = \gamma_{N-h} Q_{N-h}^{-1}$ and define $\bar{J}_{N-h}(k) \leq \gamma_{N-h}$, $F_{N-h} = Y_{N-h} Q_{N-h}^{-1}$ such that

$$\begin{bmatrix} 1 & * \\ x_{N-h} & Q_{N-h} \end{bmatrix} \geq 0, \qquad (7.4.18)$$

$$\begin{bmatrix} Q_{N-h} & * & * & * \\ A_l Q_{N-h} + B_l Y_{N-h} & \gamma_{N-h} P_{N-h+1}^{-1} & * & * \\ W^{1/2} Q_{N-h} & 0 & \gamma_{N-h} I & * \\ R^{1/2} Y_{N-h} & 0 & 0 & \gamma_{N-h} I \end{bmatrix} \geq 0, l \in \{1,\ldots,L\}.$$
$$(7.4.19)$$

Moreover, $u(k|k) = F_{N-h}x(k|k)$ should satisfy

$$\begin{bmatrix} Q_{N-h} & * \\ A_l Q_{N-h} + B_l Y_{N-h} & Q_{N-h+1} \end{bmatrix} \geq 0, \ l \in \{1,\ldots,L\}, \qquad (7.4.20)$$

$$\begin{bmatrix} Z_{N-h} & Y_{N-h} \\ Y_{N-h}^T & Q_{N-h} \end{bmatrix} \geq 0, \ Z_{N-h,jj} \leq \bar{u}_j^2, \ j \in \{1,\ldots,m\}, \qquad (7.4.21)$$

$$\begin{bmatrix} Q_{N-h} & * \\ \Psi(A_l Q_{N-h} + B_l Y_{N-h}) & \Gamma_{N-h} \end{bmatrix} \geq 0, \ \Gamma_{N-h,ss} \leq \bar{\psi}_s^2, \ l \in \{1,\ldots,L\},$$
$$s \in \{1,\ldots,q\}. \qquad (7.4.22)$$

Thus, $\{Y_{N-h}, Q_{N-h}, \gamma_{N-h}\}$ can be obtained by solving the following optimization problem

$$\min_{\gamma_{N-h}, Y_{N-h}, Q_{N-h}, Z_{N-h}, \Gamma_{N-h}} \gamma_{N-h}, \text{ s.t. } (7.4.18) - (7.4.22). \qquad (7.4.23)$$

Algorithm 7.2 (Varying horizon off-line robust MPC)

Step 1. Off-line, select state points x_i, $i \in \{1, \ldots, N\}$. Substitute $x(k|k)$ in
(7.2.7) by x_N, and solve (7.2.16) to obtain Q_N, Y_N, γ_N, ellipsoid ε_N
and feedback gain $F_N = Y_N Q_N^{-1}$. For x_{N-h}, let h gradually increase,
from 1 to $N-1$, and solve (7.4.23) to obtain Q_{N-h}, Y_{N-h}, γ_{N-h},
ellipsoid ε_{N-h} and feedback gain $F_{N-h} = Y_{N-h} Q_{N-h}^{-1}$. Notice that
the selection of x_{N-h}, $h \in \{0, \ldots, N-1\}$ should satisfy $\varepsilon_j \supset \varepsilon_{j+1}$,
$\forall j \in \{1, \ldots, N-1\}$.

Step 2. On-line, at each time k, the following is adopted

$$F(k) = \begin{cases} F(\alpha_{N-h}), & x(k) \in \varepsilon_{N-h}, \ x(k) \notin \varepsilon_{N-h+1} \\ F_N, & x(k) \in \varepsilon_N \end{cases}, \qquad (7.4.24)$$

where $F(\alpha_{N-h}) = \alpha_{N-h} F_{N-h} + (1 - \alpha_{N-h}) F_{N-h+1}$,
$x(k)^T \left[\alpha_{N-h} Q_{N-h}^{-1} + (1 - \alpha_{N-h}) Q_{N-h+1}^{-1} \right] x(k) = 1, 0 \le \alpha_{N-h} \le 1.$

In Algorithm 7.2, $\alpha_{N-h}(k)$ is simplified as α_{N-h}, since $x(k)$ can only stay
in ε_{N-h} once.

Suppose, at time k, F_{N-h} is adopted and the control law in (7.4.15) is
considered. Since the same control law is adopted for all states satisfying
the same conditions, and the uncertain system is considered, it is usually
impossible to exactly satisfy $x(k+i|k) \in \varepsilon_{N-h+i}$, $x(k+i|k) \notin \varepsilon_{N-h+i+1}$,
$\forall i \in \{0, \ldots, h-1\}$. In the real applications, it is usually impossible to exactly
satisfy $x(k+i) \in \varepsilon_{N-h+i}$, $x(k+i) \notin \varepsilon_{N-h+i+1}$, $\forall i \in \{0, \ldots, h-1\}$. However,
when one considers F_{N-h+i} $(i > 1)$, it is more suitable to $\varepsilon_{N-h+i} \cdots \varepsilon_N$ than
F_{N-h}, hence, the optimality of off-line MPC can be improved by adopting
(7.4.15).

Notice that it is not bounded that the above rationale can improve optimality. Compared with Algorithm 7.1, Algorithm 7.2 utilizes (7.4.20), which
is an extra constraint. Adding any constraint can degrade the performance
with respect to feasibility and optimality.

Moreover, in (7.4.15), not a single F_{N-h}, but a sequence of control laws
F_{N-h}, F_{N-h+1}, F_N, is adopted. In the real implementation, applying (7.4.24)
implies application of the control law sequence $F(\alpha_{N-h})$, F_{N-h+1}, $\cdots$, F_N
where, however, only the current control law $F(\alpha_{N-h})$ is implemented. This
implies that varying horizon MPC is adopted, where the control horizon
changes within $\{N-1, \ldots, 0\}$ (the control horizon for the algorithm in section
7.2 is 0).

Theorem 7.4.1. *Given $x(0) \in \varepsilon_1$, by adopting Algorithm 7.2, the closed-loop system is asymptotically stable. Further, the control law (7.4.24) in Algorithm 7.2 is a continuous function of the system state x.*

Proof. For $h \neq 0$, if $x(k)$ satisfies $\|x(k)\|_{Q_{N-h}^{-1}}^2 \leq 1$ and $\|x(k)\|_{Q_{N-h+1}^{-1}}^2 \geq 1$, then let

$$Q(\alpha_{N-h})^{-1} = \alpha_{N-h}Q_{N-h}^{-1} + (1 - \alpha_{N-h})Q_{N-h+1}^{-1},$$
$$Z(\alpha_{N-h}) = \alpha_{N-h}Z_{N-h} + (1 - \alpha_{N-h})Z_{N-h+1},$$
$$\Gamma(\alpha_{N-h}) = \alpha_{N-h}\Gamma_{N-h} + (1 - \alpha_{N-h})\Gamma_{N-h+1}.$$

By linear interpolation, we obtain

$$\begin{bmatrix} Z(\alpha_{N-h}) & * \\ F(\alpha_{N-h})^T & Q(\alpha_{N-h})^{-1} \end{bmatrix} \geq 0, \quad \begin{bmatrix} Q(\alpha_{N-h})^{-1} & * \\ \Psi\,(A_l + B_l F(\alpha_{N-h})) & \Gamma(\alpha_{N-h}) \end{bmatrix} \geq 0.$$
$$(7.4.25)$$

Eq. (7.4.25) implies that $F(\alpha_{N-h})$ can satisfy the input and state constraints. For $\forall x(0) \in \varepsilon_{N-h+1}$, F_{N-h+1} is a stabilizing feedback law. Hence,

$$\begin{bmatrix} Q_{N-h+1}^{-1} & * \\ A_l + B_l F_{N-h+1} & Q_{N-h+1} \end{bmatrix} \geq 0.$$

Moreover, in (7.4.20), by multiplying both sides of
$\begin{bmatrix} Q_{N-h} & * \\ A_l Q_{N-h} + B_l Y_{N-h} & Q_{N-h+1} \end{bmatrix}$ by $\begin{bmatrix} Q_{N-h}^{-1} & 0 \\ 0 & I \end{bmatrix}$, it is shown that (7.4.20) is equivalent to

$$\begin{bmatrix} Q_{N-h}^{-1} & * \\ A_l + B_l F_{N-h} & Q_{N-h+1} \end{bmatrix} \geq 0.$$

Thus, applying linear interpolation yields

$$\begin{bmatrix} Q(\alpha_{N-h})^{-1} & * \\ A_l + B_l F(\alpha_{N-h}) & Q_{N-h+1} \end{bmatrix} \geq 0. \qquad (7.4.26)$$

Since $x(k) \in \varepsilon_{N-h,\alpha_{N-h}} = \{x \in \Re^n | x^T Q(\alpha_{N-h})^{-1} x \leq 1\}$, (7.4.26) indicates that $u(k) = F(\alpha_{N-h})x(k)$ can guarantee to drive $x(k+1)$ into ε_{N-h+1}, with the constraints satisfied. The continued proof is the same as the last section.   □

3. Numerical example

Consider $\begin{bmatrix} x^{(1)}(k+1) \\ x^{(2)}(k+1) \end{bmatrix} = \begin{bmatrix} 0.8 & 0.2 \\ \beta(k) & 0.8 \end{bmatrix} \begin{bmatrix} x^{(1)}(k) \\ x^{(2)}(k) \end{bmatrix} + \begin{bmatrix} 1 \\ 0 \end{bmatrix} u(k)$ (since x_i is the state point in off-line MPC, we use $x^{(i)}$ to denote the state element), where $\beta(k)$ satisfies $0.5 \leq \beta(k) \leq 2.5$, which is an uncertain parameter. The true state is generated by $\beta(k) = 1.5 + \sin(k)$. The constraint is $|u(k+i|k)| \leq 2$, $\forall i \geq 0$.

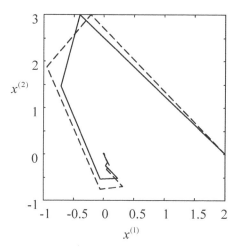

Figure 7.4.1: State trajectories of the closed-loop system.

Choose the weighting matrices as $W = I$ and $R = 1$. Choose $x_{N-h} = \begin{bmatrix} 2 - 0.01(N - h - 1) & 0 \end{bmatrix}^T$ and $x_N = \begin{bmatrix} 1 & 0 \end{bmatrix}^T$ (for simplicity more concrete partition is not given). The initial state lies at $x(0) = \begin{bmatrix} 2 & 0 \end{bmatrix}^T$. Adopt Algorithms 7.1 and 7.2. The state trajectories, state responses and the control input signal by adopting the two algorithms are shown in Figures 7.4.1, 7.4.2 and 7.4.3. The solid line refers to Algorithm 7.2, and the dotted line refers to Algorithm 7.1. It can be conceived from the figures that, in a single simulation, the state does not necessarily stay in every ellipsoid and, rather, the state can jump over some ellipsoids and only stay in part of the ellipsoids. If a series of simulations are performed, then each of the ellipsoids will be useful.

Further, denote $\hat{J} = \sum_{i=0}^{\infty} \left[\|x(i)\|_W^2 + \|u(i)\|_R^2 \right]$. Then, for Algorithm 7.1, $\hat{J}^* = 24.42$, and for Algorithm 7.2, $\hat{J}^* = 21.73$. The simulation results show that Algorithm 7.2 is better for optimality.

7.5   Off-line approach based on nominal performance cost: case zero-horizon

As mentioned earlier, since the problem (7.2.16) can involve huge computational burden, the corresponding off-line MPC is given such that problem (7.2.16) is performed off-line. However, while off-line MPC greatly reduce the on-line computational burden, its feasibility and optimality is largely degraded as compared with on-line MPC. The algorithms in this and the next sections are to compensate two deficiencies of problem (7.2.16):

- The "worst-case" is adopted so that there are L LMIs in (7.2.11).

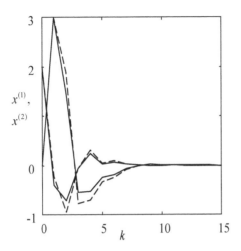

Figure 7.4.2: State responses of the closed-loop system.

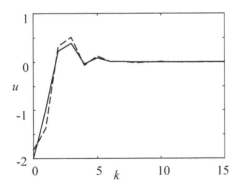

Figure 7.4.3: Control input signal.

- Problem (7.2.16) only adopts feedback MPC with control horizon $M = 0$ (see section 7.4; about feedback MPC, more details are given in Chapter 9).

These two deficiencies restrict the optimality and feasibility. Therefore, in this section we will adopt nominal performance cost in synthesizing off-line MPC, while in the next section we will further combine MPC based on the nominal performance cost with the varying horizon approach.

1. *Basic algorithm*

Using nominal performance cost, let us solve the following problem at each time k:

$$\min_{u(k+i|k)=F(k)x(k+i|k),P(k)} J_{n,\infty}(k) = \sum_{i=0}^{\infty} \left[\|\hat{x}(k+i|k)\|_W^2 + \|u(k+i|k)\|_R^2 \right],$$
(7.5.1)

s.t. $\hat{x}(k+i+1|k) = \hat{A}\hat{x}(k+i|k) + \hat{B}u(k+i|k),$

$\hat{x}(k|k) = x(k), \ \forall i \geq 0, \ (7.2.2), (7.2.3),$ (7.5.2)

$\|x(k+i+1|k)\|_{P(k)}^2 - \|x(k+i|k)\|_{P(k)}^2 < 0, \ \forall i \geq 0, \ P(k) > 0,$ (7.5.3)

$\|\hat{x}(k+i+1|k)\|_{P(k)}^2 - \|\hat{x}(k+i|k)\|_{P(k)}^2 \leq - \|\hat{x}(k+i|k)\|_W^2$

$\qquad - \|u(k+i|k)\|_R^2, \ \forall k, i \geq 0,$ (7.5.4)

where $\hat{x}$ denotes nominal state; (7.5.3) is for guaranteeing stability; (7.5.4) is for cost monotonicity. Since $[\hat{A}|\hat{B}] \in \Omega$, (7.5.3)-(7.5.4) are less restrictive than (7.2.5). Hence, compared with (7.2.1)-(7.2.3)+(7.2.5) ("+" indicates satisfaction of (7.2.5) when solving (7.2.1)-(7.2.3)), (7.5.1)-(7.5.4) is easier to be feasible, i.e., (7.5.1)-(7.5.4) can be utilized to a wider class of systems.

For stable closed-loop system, $\hat{x}(\infty|k) = 0$. Hence, summing (7.5.4) from $i = 0$ to $i = \infty$ obtains $J_{n,\infty}(k) \leq \|x(k)\|_{P(k)}^2 \leq \gamma$, where the same notation γ as in the former sections is adopted that should not induce confusion. It is easy to show that constraint (7.5.4) is equivalent to:

$$(\hat{A} + \hat{B}F(k))^T P(k)(\hat{A} + \hat{B}F(k)) - P(k) \leq -W - F(k)^T RF(k).$$
(7.5.5)

Define $Q = \gamma P(k)^{-1}$, $F(k) = YQ^{-1}$, then (7.5.5) and (7.5.3) can be transformed into the following LMIs:

$$\begin{bmatrix} Q & * & * & * \\ \hat{A}Q + \hat{B}Y & Q & * & * \\ W^{1/2}Q & 0 & \gamma I & * \\ R^{1/2}Y & 0 & 0 & \gamma I \end{bmatrix} \geq 0,$$
(7.5.6)

$$\begin{bmatrix} Q & * \\ A_l Q + B_l Y & Q \end{bmatrix} > 0, \ l \in \{1, \ldots, L\}.$$
(7.5.7)

For treating input and state constraints, (7.5.7) and (7.2.11) will have the same role, i.e., (7.2.7) and (7.5.7) also lead to $x(k+i|k)^T Q^{-1} x(k+i|k) \leq 1,$

$\forall i \geq 0$. Therefore, with (7.2.7) and (7.5.7) satisfied, (7.2.13) and (7.2.15) guarantee satisfaction of (7.2.2). Thus, problem (7.5.1)-(7.5.4) can be approximated by

$$\min_{\gamma, Q, Y, Z, \Gamma} \gamma, \text{ s.t. } (7.5.6) - (7.5.7), (7.2.7), (7.2.13) \text{ and } (7.2.15). \qquad (7.5.8)$$

Eq. (7.5.6) is a necessary condition of (7.2.11), (7.5.7) a part of (7.2.11). Hence, (7.5.8) is easier to be feasible than (7.2.16).

Algorithm 7.3 (Type I off-line MPC adopting nominal performance cost)

Stage 1. Off-line, choose states x_i, $i \in \{1, \cdots, N\}$. Substitute $x(k)$ in (7.2.7) by x_i, and solve (7.5.8) to obtain the corresponding matrices $\{Q_i, Y_i\}$, ellipsoids $\varepsilon_i = \{x \in \mathfrak{R}^n | x^T Q_i^{-1} x \leq 1\}$ and feedback gains $F_i = Y_i Q_i^{-1}$. Notice that x_i should be chosen such that $\varepsilon_{i+1} \subset \varepsilon_i$, $\forall i \neq N$. For each $i \neq N$, check if (7.3.1) is satisfied.

Stage 2. On-line, at each time k adopt the state feedback law (7.3.2).

Theorem 7.5.1. *(Stability) Suppose $x(0) \in \varepsilon_1$. Then, Algorithm 7.3 asymptotically stabilizes the closed-loop system. Moreover, if (7.3.1) is satisfied for all $i \neq N$, then the control law (7.3.2) in Algorithm 7.3 is a continuous function of the system state x.*

Proof. For $x(k) \in \varepsilon_i$, since $\{Y_i, Q_i, Z_i, \Gamma_i\}$ satisfy (7.2.7), (7.2.13), (7.2.15) and (7.5.7), F_i is feasible and stabilizing. For $x(k) \in \varepsilon_i \backslash \varepsilon_{i+1}$, denote $Q(\alpha_i(k))^{-1} = \alpha_i(k)Q_i^{-1} + (1 - \alpha_i(k))Q_{i+1}^{-1}$, $X(\alpha_i(k)) = \alpha_i(k)X_i + (1 - \alpha_i(k))X_{i+1}$, $X \in \{Z, \Gamma\}$. If (7.3.1) is satisfied and $\{Y_i, Q_i\}$ satisfies (7.5.7), then

$$Q_i^{-1} - (A_l + B_l F(\alpha_i(k)))^T Q_i^{-1} (A_l + B_l F(\alpha_i(k))) > 0, \; l \in \{1, \ldots, L\}. \qquad (7.5.9)$$

Moreover, if both $\{Y_i, Q_i, Z_i, \Gamma_i\}$ and $\{Y_{i+1}, Q_{i+1}, Z_{i+1}, \Gamma_{i+1}\}$ satisfy (7.2.13) and (7.2.15), then

$$\begin{bmatrix} Q(\alpha_i(k))^{-1} & * \\ F(\alpha_i(k)) & Z(\alpha_i(k)) \end{bmatrix} \geq 0, \; Z(\alpha_i(k))_{jj} \leq \bar{u}_j^2, \; j \in \{1, \ldots, m\}, \qquad (7.5.10)$$

$$\begin{bmatrix} Q(\alpha_i(k))^{-1} & * \\ \Psi (A_l + B_l F(\alpha_i(k))) & \Gamma(\alpha_i(k)) \end{bmatrix} \geq 0, \; l \in \{1, \ldots, L\}, \; \Gamma(\alpha_i(k))_{ss} \leq \bar{\psi}_s^2,$$

$$s \in \{1, \ldots, q\}. \qquad (7.5.11)$$

Eqs. (7.5.9)-(7.5.11) indicate that $u(k) = F(\alpha_i(k))x(k)$ will keep the state inside of ε_i and drive it towards ε_{i+1}, with the hard constraints satisfied. More details are referred to Theorem 7.3.1. $\qquad \square$

2. *Algorithm utilizing variable G*

In order to improve optimality, define $F(k) = YG^{-1}$ (rather than $F(k) = YQ^{-1}$). Then, (7.5.5) and (7.5.3) can be transformed into the following LMIs (where the fact that $(G-Q)^T Q^{-1}(G-Q) \geq 0 \Rightarrow G^T + G - Q \leq G^T Q^{-1} G$ is utilized):

$$
\begin{bmatrix}
G + G^T - Q & * & * & * \\
\hat{A}G + \hat{B}Y & Q & * & * \\
W^{1/2}G & 0 & \gamma I & * \\
R^{1/2}Y & 0 & 0 & \gamma I
\end{bmatrix} \geq 0,
\tag{7.5.12}
$$

$$
\begin{bmatrix}
G + G^T - Q & * \\
A_l G + B_l Y & Q
\end{bmatrix} > 0, \ l \in \{1, \dots, L\}.
\tag{7.5.13}
$$

Further, (7.2.2) is satisfied by satisfaction of the following LMIs:

$$
\begin{bmatrix}
Z & Y \\
Y^T & G + G^T - Q
\end{bmatrix} \geq 0, \ Z_{jj} \leq \bar{u}_j^2, \ j \in \{1, \dots, m\},
\tag{7.5.14}
$$

$$
\begin{bmatrix}
G + G^T - Q & * \\
\Psi (A_l G + B_l Y) & \Gamma
\end{bmatrix} \geq 0, \ \Gamma_{ss} \leq \bar{\psi}_s^2, \ l \in \{1, \dots, L\}, \ s \in \{1, \dots, q\}.
\tag{7.5.15}
$$

Thus, optimization problem (7.5.1)-(7.5.4) is approximated by

$$
\min_{\gamma, Q, Y, G, Z, \Gamma} \gamma, \text{ s.t. } (7.2.7), (7.5.12) - (7.5.15).
\tag{7.5.16}
$$

Let $G = G^T = Q$, then (7.5.16) becomes into (7.5.8). Since an extra variable G is introduced, the degree of freedom for optimization is increased and, in general, the optimality of $F(k)$ is enhanced.

Algorithm 7.4 (Type II off-line MPC adopting nominal performance cost)

Stage 1.   Off-line, choose states x_i, $i \in \{1, \dots, N\}$. Substitute $x(k)$ in (7.2.7) by x_i, and solve (7.5.16) to obtain the corresponding matrices $\{G_i, Q_i, Y_i\}$, ellipsoids $\varepsilon_i = \{x \in \mathfrak{R}^n | x^T Q_i^{-1} x \leq 1\}$ and feedback gains $F_i = Y_i G_i^{-1}$. Note that x_i should be chosen such that $\varepsilon_{i+1} \subset \varepsilon_i$, $\forall i \neq N$. For each $i \neq N$, check if (7.3.1) is satisfied.

Stage 2.   On-line, at each time k adopt the state feedback law (7.3.2).

Theorem 7.5.2. *(Stability) Suppose $x(0) \in \varepsilon_1$. Then, by applying Algorithm 7.4 the closed-loop system is asymptotically stable. Further, if (7.3.1) is satisfied for all $i \neq N$, then the control law (7.3.2) is a continuous function of the system state x.*

3. *Numerical example*
Consider

$$
\begin{bmatrix}
x^{(1)}(k+1) \\
x^{(2)}(k+1)
\end{bmatrix} =
\begin{bmatrix}
1 & 0 \\
K(k) & 1
\end{bmatrix}
\begin{bmatrix}
x^{(1)}(k) \\
x^{(2)}(k)
\end{bmatrix} +
\begin{bmatrix}
1 \\
0
\end{bmatrix} u(k),
$$

where $K(k)$ is an uncertain parameter. Take $W = I$, $R = 1$ and the input constraint as $|u| \leq 1$. The initial state is $x(0) = [7, 80]^T$. Consider the following two cases:

CASE 1: Take $K(k) \in [1, K_M]$ and vary K_M. When $K_M \geq 49.4$ Algorithm 7.1 becomes infeasible. However, Algorithms 7.3 and 7.4 are still feasible when $K_M = 56.9$. This shows that adopting nominal performance cost improves feasibility.

CASE 2: $K(k) \in [0.1, 2.9]$. Take $x_i^{(1)} = x_i^{(2)} = \xi_i$, $\xi_1 = 10$, $\xi_2 = 8$, $\xi_3 = 7$, $\xi_4 = 6$, $\xi_5 = 5$, $\xi_6 = 4$, $\xi_7 = 3$, $\xi_8 = 2.5$, $\xi_9 = 2$, $\xi_{10} = 1.5$. When the state are generated with different methods, by adopting Algorithms 7.1, 7.4 and the on-line algorithm in this chapter, respectively, we obtain the cost values $J_{\text{true},\infty}^*$ shown in Table 7.5.1, where $J_{\text{true},\infty} = \sum_{i=0}^{\infty} \left[\|x(i)\|_W^2 + \|u(i)\|_R^2 \right]$. These results show that, even for some "extreme" cases, Algorithm 7.4 can still improve optimality. When the state is generated by $K(k) = 1.5 + 1.4\sin(k)$, by adopting Algorithms 7.1 and 7.4, respectively, the obtained closed-loop state trajectories are shown in Figure 7.5.1, where the dotted line refers to Algorithm 7.1 and solid line to Algorithm 7.4.

Table 7.5.1: The cost values $J_{\text{true},\infty}^*$ adopting different state generation methods and different Algorithms.

State generation formula	Algorithm 7.1	Algorithm 7.4	On-line Method
$K(k) = 1.5 + 1.4\sin(k)$	353123	290041	342556
$K(k) = 1.5 + (-1)^k 1.4$	331336	272086	321960
$K(k) = 0.1$	653110	527367	526830
$K(k) = 2.9$	454295	364132	453740
$K(k) = 0.1 + 2.8\text{rand}(0, 1)$	330616	268903	319876

where: $\text{rand}(0, 1)$ is a random number in the interval $[0, 1]$.
For the three algorithms, the sequences of random numbers are identical.

7.6 Off-line approach based on nominal performance cost: case varying-horizon

Algorithms 7.1, 7.3 and 7.4 have a common feature, i.e., for any initial state $x(k) = x_{N-h}$, the following law is adopted

$$u(k + i|k) = F_{N-h}x(k + i|k), \quad \forall i \geq 0. \tag{7.6.1}$$

However, after applying F_{N-h} in ε_{N-h} ($h > 0$), the state may have been driven into the smaller ellipsoid ε_{N-h+1} in which F_{N-h+1} is more appropriate than F_{N-h}. Hence, we can substitute (7.6.1) by

$$\begin{aligned} u(k + i|k) &= F_{N-h+i}x(k + i|k), i \in \{0, \ldots, h - 1\}, \\ u(k + i|k) &= F_N x(k + i|k), \quad \forall i \geq h. \end{aligned} \tag{7.6.2}$$

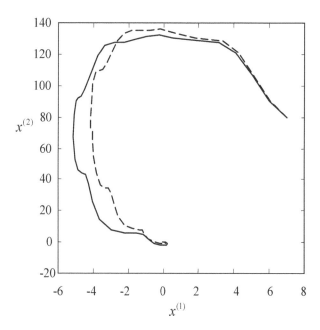

Figure 7.5.1: Closed-loop state trajectories by adopting Algorithms 7.1 and 7.4.

If

$$x(k) \in \varepsilon_{N-h}, \ x(k) \notin \varepsilon_{N-h+1} \Rightarrow x(k+i) \in \varepsilon_{N-h+i}, \ \forall i \in \{0,\ldots,h\}, \quad (7.6.3)$$

then adopting (7.6.2) is apparently better than adopting (7.6.1). Section 7.4 has adopted this observation to give the controller. In this section, we heuristically combine this observation with adopting nominal performance cost.

1. *Calculating* $\{Q_N, \ F_N\}$ *and* $\{Q_{N-h}, F_{N-h}\}$, $h = 1$

Firstly, Q_N and F_N are determined by Algorithm 7.3, by which $P_N = \gamma_N Q_N^{-1}$ is obtained. Then, considering an $x(k) = x_{N-1} \notin \varepsilon_N$, we select the control laws as (7.4.2).

If $x(k+1|k) \in \varepsilon_N$, then according to the procedure for calculating Q_N and F_N, it follows that

$$\sum_{i=1}^{\infty} \left[\|\hat{x}(k+i|k)\|_W^2 + \|u(k+i|k)\|_R^2 \right] \leq \|\hat{x}(k+1|k)\|_{P_N}^2,$$

$$J_{n,\infty}(k) \leq \bar{J}_{n,\infty}(N-1,k) = \|x(k)\|_W^2 + \|u(k)\|_R^2 + \|\hat{x}(k+1|k)\|_{P_N}^2.$$

Let

$$\gamma_{N-1} - x(k)^T P(k) x(k) \geq 0,$$

$$W + F_{N-1}^T R F_{N-1} + (\hat{A} + \hat{B} F_{N-1})^T P_N (\hat{A} + \hat{B} F_{N-1}) \leq P(k). \quad (7.6.4)$$

Then by applying (7.4.2) it follows that $\bar{J}_{n,\infty}(N-1,k) \leq \gamma_{N-1}$.

Let us consider, instead of (7.5.1)-(7.5.4), the following optimization problem:

$$\min_{u(k)=F_{N-1}x(k),P(k),\gamma_{N-1}} \gamma_{N-1},$$

$$\text{s.t. } (7.2.2),(7.2.3),(7.5.3) \text{ and } (7.6.4), \text{ where } i = 0. \qquad (7.6.5)$$

Define $Q_{N-1} = \gamma_{N-1}P(k)^{-1}$ and $F_{N-1} = Y_{N-1}Q_{N-1}^{-1}$, then (7.6.4) can be transformed into (7.4.18) and the following LMI:

$$\begin{bmatrix} Q_{N-h} & * & * & * \\ \hat{A}Q_{N-h} + \hat{B}Y_{N-h} & \gamma_{N-h}P_{N-h+1}^{-1} & * & * \\ W^{1/2}Q_{N-h} & 0 & \gamma_{N-h}I & * \\ R^{1/2}Y_{N-h} & 0 & 0 & \gamma_{N-h}I \end{bmatrix} \geq 0. \qquad (7.6.6)$$

In addition, (7.5.3) for $i = 0$ is transformed into

$$\begin{bmatrix} Q_{N-h} & * \\ A_lQ_{N-h} + B_lY_{N-h} & Q_{N-h} \end{bmatrix} > 0, \; l \in \{1,\dots,L\}, \qquad (7.6.7)$$

while (7.2.2) for $i = 0$ is guaranteed by (7.4.21)-(7.4.22).

Thus, problem (7.6.5) can be solved by

$$\min_{\gamma_{N-1},Y_{N-1},Q_{N-1},Z_{N-1},\Gamma_{N-1}} \gamma_{N-1}, \text{ s.t. } (7.4.18),(7.6.6)$$

$$-(7.6.7) \text{ and } (7.4.21)-(7.4.22), \qquad (7.6.8)$$

where $h = 1$.

Notice that imposing (7.6.6)-(7.6.7) cannot guarantee $x(k+1|k) \in \varepsilon_N$. Hence, (7.6.8) for calculating $\{Q_{N-1}, F_{N-1}\}$ is heuristic.

2. *Calculating* $\{Q_{N-h}, F_{N-h}\}$, $h \in \{2,\dots,N-1\}$

The procedure for calculating Q_{N-1}, F_{N-1} can be generalized. Considering an $x(k) = x_{N-h} \notin \varepsilon_{N-h+1}$, we select the control law (7.6.2), where F_{N-h+1}, F_{N-h+2}, $\cdots$, F_N have been obtained in earlier time.

By induction, if $x(k+j|k) \in \varepsilon_{N-h+j}$ for all $j \in \{1,\cdots,h\}$, then according to the procedure for calculating Q_{N-h+j} and F_{N-h+j}, it follows that

$$\sum_{i=1}^{\infty} \left[\|\hat{x}(k+i|k)\|_W^2 + \|u(k+i|k)\|_R^2 \right] \leq \|\hat{x}(k+1|k)\|_{P_{N-h+1}}^2,$$

$$J_{n,\infty}(k) \leq \bar{J}_{n,\infty}(N-h,k) = \|x(k)\|_W^2 + \|u(k)\|_R^2 + \|\hat{x}(k+1|k)\|_{P_{N-h+1}}^2,$$

where $P_{N-h+1} = \gamma_{N-h+1}Q_{N-h+1}^{-1}$. Let

$$\gamma_{N-h} - x(k)^T P(k)x(k) \geq 0,$$

$$W + F_{N-h}^T R F_{N-h} + (\hat{A} + \hat{B}F_{N-h})^T P_{N-h+1}(\hat{A} + \hat{B}F_{N-h}) \leq P(k). \qquad (7.6.9)$$

Then by applying (7.6.2) it follows that $\bar{J}_{n,\infty}(N-h,k) \leq \gamma_{N-h}$.

Let us solve, instead of (7.5.1)-(7.5.4), the following optimization problem:

$$\min_{u(k)=F_{N-h}x(k),P(k),\gamma_{N-h}} \gamma_{N-h},$$

s.t. (7.2.2), (7.2.3), (7.5.3) and (7.6.9), where $i = 0$.   (7.6.10)

By defining $Q_{N-h} = \gamma_{N-h}P(k)^{-1}$ and $F_{N-h} = Y_{N-h}Q_{N-h}^{-1}$, problem (7.6.10) can be solved by

$$\min_{\gamma_{N-h},Y_{N-h},Q_{N-h},Z_{N-h},\Gamma_{N-h}} \gamma_{N-h},$$

s.t. (7.4.18), (7.6.6) $-$ (7.6.7) and (7.4.21) $-$ (7.4.22).   (7.6.11)

Since $x(k+j|k) \in \varepsilon_{N-h+j}$ for all $j \in \{1,\ldots,h\}$ cannot be guaranteed, (7.6.11) for calculating $\{Q_{N-h}, F_{N-h}\}$ is heuristic.

Algorithm 7.5 (Varying horizon off-line MPC adopting nominal performance cost)

Stage 1. Off-line, generate states x_i, $i \in \{1,\ldots,N\}$. Substitute $x(k)$ in (7.2.7) by x_N and solve (7.5.8) to obtain the matrices $\{Q_N, Y_N\}$, ellipsoid ε_N and feedback gain $F_N = Y_N Q_N^{-1}$. For x_{N-h}, let h gradually increase, from 1 to $N-1$, and solve (7.6.11) to obtain $\{Q_{N-h}, Y_{N-h}\}$, ellipsoid ε_{N-h} and feedback gain $F_{N-h} = Y_{N-h}Q_{N-h}^{-1}$. Notice that x_i should be chosen such that $\varepsilon_{i+1} \subset \varepsilon_i$, $\forall i \neq N$. For each $i \neq N$, check if (7.3.1) is satisfied.

Stage 2. On-line, at each time k adopt the state feedback law (7.3.2).

Remark 7.6.1. The Algorithm in section 7.4 adopts "worst-case" performance cost and control law in the form of (7.6.2), and the following constraint is explicitly imposed:

$$x(k) \in \varepsilon_{N-h}, \ x(k) \notin \varepsilon_{N-h+1} \Rightarrow x(k+i|k) \in \varepsilon_{N-h+i}, \ \forall i \in \{0,\ldots,h\},$$
(7.6.12)

which means that the state in a larger ellipsoid will be driven into the neighboring smaller ellipsoid in one step. The adverse effect of imposing (7.6.12) is that the number of ellipsoids tends to be very large in order to attain feasibility. Without imposing (7.6.12), (7.6.11) is much easier to be feasible, and N of Algorithm 7.5 can be dramatically reduced. There are still underlying reasons. Since we are dealing with uncertain systems and ellipsoidal domains of attraction, imposing (7.6.12) is very conservative for guaranteeing (7.6.3). Hence, by imposing (7.6.12), it is nearly impossible to gain

$$x(k) \in \varepsilon_{N-h}, \ x(k) \notin \varepsilon_{N-h+1} \Rightarrow x(k+i) \in \varepsilon_{N-h+i},$$
$$x(k+i) \notin \varepsilon_{N-h+i+1}, \forall i \in \{1,\ldots,h-1\}, \ x(k+h) \in \varepsilon_N.$$
(7.6.13)

It is more likely that (7.6.13) can be achieved by properly constructing smaller number of ellipsoids.

Remark 7.6.2. It is better to satisfy (7.6.13) by appropriately spacing x_i. With F_{i+1} known, (7.3.1) can be transformed into LMI and be incorporated into calculating $\{\varepsilon_i, F_i\}$. In such a way, it is much easier to satisfy (7.3.1). However, loss of optimality may result.

Theorem 7.6.1. *(Stability). Suppose $x(0) \in \varepsilon_1$. Then, by applying Algorithm 7.5 the closed-loop system is asymptotically stable. Further, if (7.3.1) is satisfied for all $i \neq N$, then the control law (7.3.2) in Algorithm 7.5 is a continuous function of the system state x.*

Proof. The proof is based on the same rationale used for proving Theorem 7.5.1. In Algorithm 7.5, if $\gamma_{N-h} P_{N-h+1}^{-1}$ in (7.6.6) is replaced by Q_{N-h}, $h \in \{1, \cdots, N-1\}$, then Algorithm 7.3 is retrieved. Hence, the only difference between Algorithms 7.3 and 7.5 lies in (7.5.12) and (7.6.6). We do not use (7.5.12) and (7.6.6) for proving stability of the algorithms. □

The complexity of solving LMI optimization problem (7.2.16), (7.5.8) or (7.6.11) is polynomial-time, which (regarding the interior-point algorithms) is proportional to $\mathfrak{K}^3 \mathfrak{L}$, where $\mathfrak{K}$ is the number of scalar variables in LMIs and $\mathfrak{L}$ the number of rows (see [31]). For (7.2.16), (7.5.8) and (7.6.11), $\mathfrak{K} = 1 + \frac{1}{2}(n^2 + n) + mn + \frac{1}{2}(m^2 + m) + \frac{1}{2}(q^2 + q)$; for (7.2.16), $\mathfrak{L} = (4n + m + q)L + 2n + 2m + q + 1$; for (7.5.8) and (7.6.11), $\mathfrak{L} = (3n + q)L + 5n + 3m + q + 1$.

3. *Numerical example*

Consider

$$\begin{bmatrix} x^{(1)}(k+1) \\ x^{(2)}(k+1) \end{bmatrix} = \begin{bmatrix} 1 - \beta & \beta \\ K(k) & 1 - \beta \end{bmatrix} \begin{bmatrix} x^{(1)}(k) \\ x^{(2)}(k) \end{bmatrix} + \begin{bmatrix} 1 \\ 0 \end{bmatrix} u(k),$$

where $K(k) \in [0.5, \ 2.5]$ is an uncertain parameter. The constraint is $|u| \leq 2$. Choose $\hat{A} = \begin{bmatrix} 1 - \beta & \beta \\ 1.5 & 1 - \beta \end{bmatrix}$, $\hat{B} = \begin{bmatrix} 1 \\ 0 \end{bmatrix}$. Take $W = I$, $R = 1$. The true state is generated by $K(k) = 1.5 + \sin(k)$.

Simulate Algorithms 7.1, 7.3 and 7.5. For $i \in \{1, \ldots, N\}$, choose $x_{N-i+1} = [0.6 + d(i-1), \ 0]^T$, where d represents the spacing of the ellipsoids. Denote $x_1^{(1),\max}$ as the maximum value such that, when $x_1 = [x_1^{(1),\max}, \ 0]^T$, the corresponding optimization problem remains feasible.

By varying d, we find that,

 i) for $\beta = 0$, Algorithm 7.3 is much easier to be feasible than Algorithm 7.1;

 ii) for $\beta = \pm 0.1$, Algorithm 7.5 gives smaller cost value than Algorithm 7.3;

 iii) for $\beta = 0$, either Algorithm 7.5 is easier to be feasible, or it gives smaller cost value, than Algorithm 7.3.

Tables 7.6.1- 7.6.2 lists the simulation results for four typical cases:

A. $\beta = -0.1$, $d = 0.1$, $N = 16$, $x(0) = [2.1, \ 0]^T$;

B. $\beta = 0$, $d = 0.02$, $N = 76$, $x(0) = [2.1, \ 0]^T$;

C. $\beta = 0$, $d = 5$, $N = 12$, $x(0) = [55.6, \ 0]^T$;

D. $\beta = 0.1$, $d = 0.1$, $N = 39$, $x(0) = [4.4, \ 0]^T$.

For simplicity, we have spaced x_i equally. By unequal spacing, it can give lower cost value $J_{\text{true},\infty}$, result into larger $x_1^{(1),\text{max}}$, and render satisfaction of (7.3.1) for all $i \neq N$ (especially for Algorithm 7.5).

Table 7.6.1: The simulation results by Algorithm 7.1 (A7.1), Algorithm 7.3 (A7.3) and Algorithm 7.5 (A7.5) under four typical cases.

	$x_1^{(1),\text{max}}$			The set of i for which (7.3.1) is not satisfied		
	A7.1	A7.3	A7.5	A7.1	A7.3	A7.5
Case A	2.1	2.1	2.1	{1}	{2,1}	{5,4,3,2,1}
Case B	59.28	74.66	72.46	{}	{68,67,64}	{75,71,64,61,59}
Case C	55.6	70.6	90.6	{11,10}	{11, 10}	{11,10,9,8}
Case D	4.4	4.4	4.4	{}	{38}	{38, 35}

Table 7.6.2: Table 7.6.1 continued ($a(b)$ indicates that $M = a$ is repeated for b times).

	$J_{\text{true},\infty}$			Control horizon M for Algorithm 7.5
	A7.1	A7.3	A7.5	for $k = 0, 1, 2, 3, \cdots$, $M =$
Case A	60.574	57.932	54.99	15, 11, 10, 6, 0, 0, 0, $\cdots$
Case B	37.172	34.687	34.485	75, 47, 35, 0, 0, 0, $\cdots$
Case C	64407677	61469235	64801789	11(7), 10(10), 9(15), 8(14), 7(11), 6(7), 5(6), 4(5), 3(14), 2(11), 1(15), 0, 0, 0, $\cdots$
Case D	575	552	542	38, 30, 29, 26, 23, 19, 16, 11, 2, 0, 0, 0, $\cdots$

Chapter 8

Synthesis approaches with finite switching horizon

For MPC with switching horizon N, we mainly consider two approaches. One is called standard approach. By standard approach we mean the usual approach, i.e., in industrial MPC and most of the synthesis approaches of MPC, an finite-horizon optimization problem is solved and, in synthesis approaches, in general the three ingredients are selected off-line. Another approach is on-line approach. In on-line approaches, one, two or all (usually all) of the three ingredients are served as the decision variables of the on-line optimization problem.

Especially in MPC for uncertain systems, on-line approach is often applied, the merits of which include enlarging the region of attraction and enhancing the optimality.

Section 8.1 is referred to in [43]. Section 8.2 is referred to in [60]. Section 8.3 is referred to in [2]. Section 8.4 is referred to in [17], [16]. Section 8.5 is referred to in [58]. Section 8.6 is referred to in [55], [62].

8.1 Standard approach for nominal systems

The following time-invariant discrete-time linear system will be considered:

$$x(k+1) = Ax(k) + Bu(k), \qquad (8.1.1)$$

where $u \in \Re^m$ is the control input, $x \in \Re^n$ is the state. The input and state constraints are

$$-\underline{u} \leq u(k+i) \leq \bar{u}, \quad -\underline{\psi} \leq \Psi x(k+i+1) \leq \bar{\psi}, \ \forall i \geq 0 \qquad (8.1.2)$$

where $\underline{u} := [\underline{u}_1, \underline{u}_2, \cdots, \underline{u}_m]^T$, $\bar{u} := [\bar{u}_1, \bar{u}_2, \cdots, \bar{u}_m]^T$, $\underline{u}_j > 0$, $\bar{u}_j > 0$, $j \in \{1, \ldots, m\}$; $\underline{\psi} := [\underline{\psi}_1, \underline{\psi}_2, \cdots, \underline{\psi}_q]^T$, $\bar{\psi} := [\bar{\psi}_1, \bar{\psi}_2, \cdots, \bar{\psi}_q]^T$, $\underline{\psi}_s > 0$, $\bar{\psi}_s > 0$,

$s \in \{1,\ldots,q\}$; $\Psi \in \mathfrak{R}^{q \times n}$. The output constraint can be expressed as the above form. Constraint (8.1.2) can also be expressed as

$$u(k+i) \in \mathcal{U}, \ x(k+i+1) \in \mathcal{X}. \tag{8.1.3}$$

Suppose (8.1.2) and (8.1.3) are equivalent.

We define the following finite-horizon cost function:

$$J(x(k)) = \sum_{i=0}^{N-1} \left[\|x(k+i|k)\|_W^2 + \|u(k+i|k)\|_R^2 \right] + \|x(k+N|k)\|_P^2, \tag{8.1.4}$$

where $W > 0$, $R > 0$, $P > 0$ are symmetric weighting matrices; N is the switching horizon. We suppose P satisfies the following inequality condition:

$$P \geq (A + BF)^T P(A + BF) + W + F^T RF, \tag{8.1.5}$$

where F is the feedback gain which will be explained later.

Define $X = P^{-1}$, $F = YX^{-1}$. By applying Schur complements, it is shown that (8.1.5) is equivalent to the following LMI:

$$\begin{bmatrix} X & * & * & * \\ AX + BY & X & * & * \\ W^{1/2}X & 0 & I & * \\ R^{1/2}Y & 0 & 0 & I \end{bmatrix} \geq 0. \tag{8.1.6}$$

Based on Chapter 7, we can easily obtain the following conclusion.

Lemma 8.1.1. *Suppose that the symmetric matrix* $X = P^{-1}$, $\{Z, \Gamma\}$ *and matrix* Y *satisfy (8.1.6) and*

$$\begin{bmatrix} Z & Y \\ Y^T & X \end{bmatrix} \geq 0, \ Z_{jj} \leq \bar{u}_{j,\inf}^2, \ j \in \{1,\ldots,m\}, \tag{8.1.7}$$

$$\begin{bmatrix} X & * \\ \Psi(AX+BY) & \Gamma \end{bmatrix} \geq 0, \ \Gamma_{ss} \leq \bar{\psi}_{s,\inf}^2, \ s \in \{1,\ldots,q\}, \tag{8.1.8}$$

where $u_{j,\inf} = \min\{\underline{u}_j, \bar{u}_j\}$, $\psi_{s,\inf} = \min\{\underline{\psi}_s, \bar{\psi}_s\}$; Z_{jj} (Γ_{ss}) *is the* j-th (s-th) *diagonal element of* Z (Γ). *Then, when* $x(k+N) \in \varepsilon_P = \{z \in \mathfrak{R}^n | z^T Pz \leq 1\}$ *and* $u(k+i+N) = YX^{-1}x(k+i+N)$, $i \geq 0$ *is adopted, the closed-loop system is exponentially stable,* $x(k+i+N)$, $i \geq 0$ *always remains in the region* ε_P *and constraint (8.1.2) is satisfied for all* $i \geq N$.

The standard approach of MPC considers the following optimization problem at each time k:

$$\min_{u(k|k),\cdots,u(k+N-1|k)} J(x(k)), \tag{8.1.9}$$

s.t. $-\underline{u} \leq u(k+i|k) \leq \bar{u}, \ -\underline{\psi} \leq \Psi x(k+i+1|k) \leq \bar{\psi}, \ i \in \{0,1,\ldots,N-1\},$

$$\tag{8.1.10}$$

$$\|x(k+N|k)\|_P^2 \leq 1. \tag{8.1.11}$$

Define the feasible set of initial states

$$\mathcal{F}(P,N) \;=\; \{x(0) \in \Re^n | \exists u(i) \in \mathcal{U},$$
$$i = 0 \ldots N - 1, \text{ s.t. } x(i+1) \in \mathcal{X}, \; x(N) \in \varepsilon_P\} \text{ (8.1.12)}$$

in which problem (8.1.9)-(8.1.11) exists a feasible solution. The optimization problem (8.1.9)-(8.1.11) has the following property.

Theorem 8.1.1. *(Feasibility) Suppose that $x(k) \in \mathcal{F}(P,N)$. Then, there exist $\kappa > 0$ and $u(k+i|k) \in \mathcal{U}$, $i \in \{0, 1, \ldots, N-1\}$ such that $|u(k+i|k)|^2 \leq \kappa |x(k)|^2$, $x(k+i+1|k) \in \mathcal{X}$, $i \in \{0, 1, \ldots, N-1\}$ and $x(k+N|k) \in \varepsilon_P$.*

Proof. We consider the case that $x(k) \neq 0$ because $x(k) = 0$ gives the trivial solution $u(k+i|k) = 0$. Let $B(\gamma)$ be a closed ball with a radius $\gamma > 0$ such that $B(\gamma) \subset \mathcal{F}$. If $x(k) \in B(\gamma)$, define $\alpha(x(k))$ $(1 \leq \alpha(x(k)) < \infty)$ such that $\alpha(x(k))x(k) \in \partial B(\gamma)$, where $\partial B(\gamma)$ denotes the boundary of $B(\gamma)$. Otherwise, define $\alpha(x(k))$ $(1 \leq \alpha(x(k)) < \infty)$ such that $\alpha(x(k))x(k) \in \mathcal{F} - B(\gamma)$.

According to the definition of (8.1.12), $\hat{u}(k+i|k) \in \mathcal{U}$ exists that drives $\alpha(x(k))x(k)$ into the ellipsoid ε_P in N steps while satisfying the state constraint. Because the system is linear, $1/\alpha(x(k))\hat{u}(k+i|k) \in \mathcal{U}$ drives $x(k)$ into ε_P while satisfying the state constraint.

Denoting $u_{\sup} = \max_j \max \{\underline{u}_j, \bar{u}_j\}$, we obtain $\|\hat{u}(k+i|k)\|_2^2 \leq m u_{\sup}^2$. Hence, it holds that

$$\left\| \frac{1}{\alpha(x(k))} \hat{u}(k+i|k) \right\|_2^2 \leq \left(\frac{1}{\alpha(x(k))} \right)^2 m u_{\sup}^2 \leq \left(\frac{m u_{\sup}^2}{\gamma^2} \right) \|x(k)\|_2^2.$$

Apparently, let $u(k+i|k) = 1/\alpha(x(k))\hat{u}(k+i|k)$ and $\kappa = m u_{\sup}^2/\gamma^2$ then we obtain the final conclusion. $\square$

According to the definition (8.1.12), Theorem 8.1.1 is almost straightforward. The significance of the above proof is to construct a κ in order to pave the way for selecting Lyapunov function in stability proof.

Theorem 8.1.2. *(Stability) Suppose (8.1.6)-(8.18) are satisfied and $x(0) \in \mathcal{F}(P,N)$. Then, (8.1.9)-(8.1.11) is always feasible for all $k \geq 0$. Further, by receding horizon implementation of $u^*(k|k)$, the closed-loop system is exponentially stable.*

Proof. Suppose the optimal solution $u^*(k+i|k)$ at the current time k exists, and let $F = YX^{-1}$. Then, according to Lemma 8.1.1, at the next time step $k+1$,

$$u(k+i|k+1) = u^*(k+i|k), \; i \in \{1, \ldots, N-1\}, \; u(k+N|k+1) = Fx(k+N|k+1)$$
$$(8.1.13)$$

is a feasible solution. Thus, by induction, we observe that (8.1.9)-(8.1.11) is always feasible for all $k \geq 0$.

In order to show the exponential stability, we need to show that a, b, c $(0 < a, b, c < \infty)$ exist such that

$$a \left\| x(k) \right\|_2^2 \leq J^*(x(k)) \leq b \left\| x(k) \right\|_2^2, \quad \Delta J^*(x(k+1)) < -c \left\| x(k) \right\|_2^2,$$

where $\Delta J^*(x(k+1)) = J^*(x(k+1)) - J^*(x(k))$. When the above condition is satisfied, $J^*(x(k))$ serves as Lyapunov functional for proving exponential stability.

Firstly, it is apparent that

$$J^*(x(k)) \geq x(k)^T W x(k) \geq \lambda_{\min}(W) \left\| x(k) \right\|_2^2,$$

i.e., one can select $a = \lambda_{\min}(W)$.

According to Theorem 8.1.1, we can obtain that

$$\begin{aligned}
J^*(x(k)) &\leq \sum_{i=0}^{N-1} \left[\left\| x(k+i|k) \right\|_W^2 + \left\| u(k+i|k) \right\|_R^2 \right] + \left\| x(k+N|k) \right\|_P^2 \\
&\leq \left[(N+1)\mathcal{A}^2 (1 + N \left\| B \right\| \sqrt{\kappa})^2 \cdot \max \left\{ \lambda_{\max}(W), \lambda_{\max}(P) \right\} \right. \\
&\quad \left. + N\kappa\lambda_{\max}(R) \right] \left\| x(k) \right\|_2^2,
\end{aligned}$$

where $\mathcal{A} := \max_{i \in \{0,1,\ldots,N\}} \left\| A^i \right\|$, i.e., we can choose

$$b = (N+1)\mathcal{A}^2 (1 + N \left\| B \right\| \sqrt{\kappa})^2 \cdot \max \left\{ \lambda_{\max}(W), \lambda_{\max}(P) \right\} + N\kappa\lambda_{\max}(R).$$

Let $\bar{J}(x(k+1))$ be the cost value when the control sequence (8.1.13) is implemented at the next time step $k+1$. Then

$$J^*(x(k)) \geq \left\| x(k) \right\|_W^2 + \left\| u(k) \right\|_R^2 + \bar{J}(x(k+1)) \geq \left\| x(k) \right\|_W^2 + \left\| u(k) \right\|_R^2 + J^*(x(k+1))$$

which shows that

$$\Delta J^*(x(k+1)) \leq - \left\| x(k) \right\|_W^2 - \left\| u(k) \right\|_R^2 \leq -\lambda_{\min}(W) \left\| x(k) \right\|_2^2,$$

i.e., one can select $c = \lambda_{\min}(W)$.

Therefore, $J^*(x(k))$ serves as a Lyapunov functional for exponential stability. $\qquad\square$

8.2 Optimal solution to infinite-horizon constrained linear quadratic control utilizing synthesis approach of MPC

Consider the system (8.1.1)-(8.1.3). The infinite-horizon cost function is adopted,

$$J(x(k)) = \sum_{i=0}^{\infty} \left[\left\| x(k+i|k) \right\|_W^2 + \left\| u(k+i|k) \right\|_R^2 \right]. \tag{8.2.1}$$

Suppose (A, B) is stabilizable, $(A, W^{1/2})$ detectable. Define $\pi = \{u(k|k), u(k+1|k), \cdots\}$.

Now we give the three relevant problems.

Problem 8.1 Infinite-horizon unconstrained LQR:

$$\min_{\pi} J(x(k)), \text{ s.t. } (8.2.2).$$

$$x(k+i+1|k) = Ax(k+i|k) + Bu(k+i|k), \ i \geq 0. \qquad (8.2.2)$$

Problem 8.1 has been solved by Kalman (see Chapter 1), with the following solution:

$$u(k) = -Kx(k), \qquad (8.2.3)$$

where the state feedback gain K is represented as

$$K = (R+B^T PB)^{-1}B^T PA, \ P = W + A^T PA - A^T PB(R+B^T PB)^{-1}B^T PA. \qquad (8.2.4)$$

Problem 8.2 Infinite-horizon constrained LQR:

$$\min_{\pi} J(x(k)), \text{ s.t. } (8.2.2), (8.2.5).$$

$$-\underline{u} \leq u(k+i|k) \leq \bar{u}, \ -\underline{\psi} \leq \Psi x(k+i+1|k) \leq \bar{\psi}, \ i \geq 0. \qquad (8.2.5)$$

Problem 8.2 is a direct generalization of Problem 8.1, with constraints included. Since Problem 8.2 involves an infinite number of decision variables and infinite number of constraints, it is impossible to solve it directly.

Problem 8.3 MPC problem:

$$\min_{\pi} J(x(k)), \text{ s.t. } (8.2.2), (8.1.10), (8.2.6).$$

$$u(k+i|k) = Kx(k+i|k), \ i \geq N. \qquad (8.2.6)$$

Problem 8.3 involves a finite number of decision variables, which can be solved by the quadratic programming. However, the effect of Problem 8.3 is to help the exact solution of Problem 8.2.

Define a set $\mathfrak{X}_K$ as, whenever $x(k) \in \mathfrak{X}_K$, with $u(k+i|k) = Kx(k+i|k), \ i \geq 0$ applied (8.2.5) can be satisfied. Define $\mathfrak{P}_N(x(k))$ as the set of π satisfying (8.1.10) and (8.2.6), and $\mathfrak{P}(x(k))$ as the set of π satisfying (8.2.5). Thus, $\mathfrak{P}(x(k))$ is the limit of $\mathfrak{P}_N(x(k))$ when $N \to \infty$. Define $\mathfrak{W}_N$ as the set of $x(k)$ in which Problem 8.3 is feasible, and $\mathfrak{W}$ the set of $x(k)$ in which Problem 8.2 is feasible.

Applying the above definitions, we can define Problems 8.2 and 8.3 via the following manners.

Problem 8.2 Infinite-horizon constrained LQR: given $x(k) \in \mathfrak{W}$, find π^* such that

$$J^*(x(k)) = \min_{\pi \in \mathfrak{P}(x(k))} J(x(k))$$

where $J^*(x(k))$ is the optimum.

Problem 8.3 MPC problem: given a finite N and $x(k) \in \mathfrak{W}_N$, find π_N^* (the optimum of $\pi_N = \{u(k|k), u(k+1|k), \cdots, u(k+N-1|k), -Kx(k+N|k), -Kx(k+N+1|k), \cdots\}$) such that

$$J_N^*(x(k)) = \min_{\pi \in \mathfrak{P}_N(x(k))} J_N(x(k))$$

where $J_N^*(x(k))$ is the optimum,

$$J_N(x(k)) = \sum_{i=0}^{N-1} \left[\|x(k+i|k)\|_W^2 + \|u(k+i|k)\|_R^2 \right] + \|x(k+N|k)\|_P^2 .$$

The theme of this section is to show that, in a finite time, the optimal solution (rather than the suboptimal solution) of Problem 8.2 can be found. This optimal solution is obtained by solving Problem 8.3.

Denote the optimal solution to Problem 8.3 as $u^0(k+i|k)$, that to Problem 8.2 as $u^*(k+i|k)$; $x(k+i|k)$ generated by the optimal solution to Problem 8.3 is denoted as $x^0(k+i|k)$, by that of Problem 8.2 as $x^*(k+i|k)$.

Lemma 8.2.1. $x^*(k+i|k) \in \mathfrak{X}_K \Leftrightarrow u^*(k+i|k) = Kx^*(k+i|k), \ \forall i \geq 0.$

Proof. "⇒": According to the definition of $\mathfrak{X}_K$, when $x(k+i|k) \in \mathfrak{X}_K$, the optimal solution to the constrained LQR is $u(k+i|k) = Kx(k+i|k), \ \forall i \geq 0$. According to Bellman's optimality principle, $\{u^*(k|k), u^*(k+1|k), \cdots\}$ is the overall optimal solution.

"⇐": (by contradiction) According to the definition of $\mathfrak{X}_K$, when $x(k+i|k) \notin \mathfrak{X}_K$, if $u^*(k+i|k) = Kx^*(k+i|k), \ \forall i \geq 0$ is adopted then (8.2.5) cannot be satisfied. However, the overall optimal solution $\{u^*(k|k), u^*(k+1|k), \cdots\}$ satisfies (8.2.5)! Hence, $u^*(k+i|k) = Kx^*(k+i|k), \ \forall i \geq 0$ implies $x^*(k+i|k) \in \mathfrak{X}_K$. □

Theorem 8.2.1. *(Optimality) When $x(k) \in \mathfrak{W}$, there exists a finite N_1 such that, whenever $N \geq N_1$, $J^*(x(k)) = J_N^*(x(k))$ and $\pi^* = \pi_N^*$.*

Proof. According to (8.1.3), $\mathfrak{X}_K$ will contain a neighborhood of the origin. The optimal solution π^* will drive the state to the origin. Therefore, there exists a finite N_1 such that $x^*(k+N_1|k) \in \mathfrak{X}_K$. According to Lemma 8.2.1, for any $N \geq N_1$, $\pi^* \in \mathfrak{P}_N(x(k))$, i.e., $\pi^* \in \mathfrak{P}(x(k)) \bigcap \mathfrak{P}_N(x(k))$. Since π_N^* optimizes $J_N(x(k))$ inside of $\mathfrak{P}_N(x(k))$, when $N \geq N_1$, $J^*(x(k)) = J_N^*(x(k))$ and, correspondingly, $\pi^* = \pi_N^*$. □

We can conclude finding the optimal solution to Problem 8.2 as the following algorithm.

Algorithm 8.1 (solution to the constrained LQR)

Step 1. Choose an initial (finite number) N_0. Take $N = N_0$.

Step 2. Solve Problem 8.3.

Step 3. If $x^0(k + N|k) \in \mathfrak{X}_K$, then turn to Step 5.

Step 4. Increase N and turn to Step 2.

Step 5. Implement $\pi^* = \pi_N^*$.

Finiteness of N_1 guarantees that Algorithm 8.1 will terminate in finite time.

8.3 On-line approach for nominal systems

Consider the system (8.1.1)-(8.1.3). Adopt the cost function (8.2.1). For handling this infinite-horizon cost function, we split it into two parts:

$$J_1(k) = \sum_{i=0}^{N-1} \left[\|x(k + i|k)\|_W^2 + \|u(k + i|k)\|_R^2 \right],$$

$$J_2(k) = \sum_{i=N}^{\infty} \left[\|x(k + i|k)\|_W^2 + \|u(k + i|k)\|_R^2 \right].$$

For $J_2(k)$, introduce the following stability constraint (see Chapter 7):

$$V(x(k+i+1|k)) - V(x(k+i|k)) \leq - \|x(k + i|k)\|_W^2 - \|u(k + i|k)\|_R^2, \quad (8.3.1)$$

where $V(x(k + i|k)) = \|x(k + i|k)\|_P^2$. Summing (8.3.1) from $i = N$ to $i = \infty$ yields

$$J_2(k) \leq V(x(k + N|k)) = \|x(k + N|k)\|_P^2.$$

Hence, it is easy to know that

$$J(x(k)) \leq \bar{J}(x(k)) = \sum_{i=0}^{N-1} \left[\|x(k + i|k)\|_W^2 + \|u(k + i|k)\|_R^2 \right] + \|x(k + N|k)\|_P^2.$$

$$(8.3.2)$$

We will transform the optimization problem corresponding to $J(x(k))$ into that corresponding to $\bar{J}(x(k))$.

If P and F (strictly P, F should be written as $P(k)$, $F(k)$) were not solved at each sampling time, then there is no intrinsic difference between the methods in this section and section 8.1. On-line MPC of this section solves the following optimization problem at each time k:

$$\min_{\gamma, u(k|k), \cdots, u(k+N-1|k), F, P} J_1(x(k)) + \gamma, \tag{8.3.3}$$

$$\text{s.t.} \; -\underline{u} \le u(k+i|k) \le \bar{u}, \; -\underline{\psi} \le \Psi x(k+i+1|k) \le \bar{\psi}, \; i \ge 0, \tag{8.3.4}$$

$$\|x(k+N|k)\|_P^2 \le \gamma, \tag{8.3.5}$$

$$(8.3.1), \; u(k+i|k) = Fx(k+i|k), \; i \ge N. \tag{8.3.6}$$

The so-called "on-line" indicates that P and F (involving the three ingredients for stability, which are referred to in Chapter 6) are solved at each sampling instant. Apparently, this involves a heavier computational burden than when P and F are fixed (see the standard approach in section 8.1). However, since P and F are also decision variables, the optimization problem is easier to be feasible (with larger region of attraction), and the cost function can be better optimized, such that the performance of the overall closed-loop system is enhanced.

An apparent question is, why not adopt the optimal solution of LQR as in section 8.2? By applying Algorithm 8.1, the finally determined N can be very large, so that (especially for high dimensional system) the computational burden will be increased. By taking a smaller N, the computational burden can be greatly decreased, while the lost optimality can be partially compensated by the receding horizon optimization. This is also the important difference between MPC and the traditional optimal control.

Define $Q = \gamma P^{-1}$ and $F = YQ^{-1}$. Then, by applying Schur complement, it is known that (8.3.1) and (8.3.5) are equivalent to the following LMIs:

$$\begin{bmatrix} Q & * & * & * \\ AQ + BY & Q & * & * \\ W^{1/2}Q & 0 & \gamma I & * \\ R^{1/2}Y & 0 & 0 & \gamma I \end{bmatrix} \ge 0, \tag{8.3.7}$$

$$\begin{bmatrix} 1 & * \\ x(k+N|k) & Q \end{bmatrix} \ge 0. \tag{8.3.8}$$

The input/state constraints after the switching horizon N can be guaranteed by the following LMIs:

$$\begin{bmatrix} Z & Y \\ Y^T & Q \end{bmatrix} \ge 0, Z_{jj} \le \bar{u}_{j,\text{inf}}^2, j \in \{1, \ldots, m\}, \tag{8.3.9}$$

$$\begin{bmatrix} Q & * \\ \Psi(AQ + BY) & \Gamma \end{bmatrix} \ge 0, \Gamma_{ss} \le \bar{\psi}_{s,\text{inf}}^2, s \in \{1, \ldots, q\}. \tag{8.3.10}$$

Now, we wish to transform the whole optimization problem (8.3.3)-(8.3.6) into LMI optimization problem. For this reason, consider the state predictions,

$$
\begin{bmatrix} x(k+1|k) \\ x(k+2|k) \\ \vdots \\ x(k+N|k) \end{bmatrix} = \begin{bmatrix} A \\ A^2 \\ \vdots \\ A^N \end{bmatrix} x(k)
$$

$$
+ \begin{bmatrix} B & 0 & \cdots & 0 \\ AB & B & \ddots & \vdots \\ \vdots & \ddots & \ddots & 0 \\ A^{N-1}B & \cdots & AB & B \end{bmatrix} \begin{bmatrix} u(k|k) \\ u(k+1|k) \\ \vdots \\ u(k+N-1|k) \end{bmatrix}.
$$

$$(8.3.11)$$

We simply denote these state predictions as

$$
\begin{bmatrix} \tilde{x}(k+1|k) \\ x(k+N|k) \end{bmatrix} = \begin{bmatrix} \tilde{A} \\ A^N \end{bmatrix} x(k) + \begin{bmatrix} \tilde{B} \\ \tilde{B}_N \end{bmatrix} \tilde{u}(k|k). \qquad (8.3.12)
$$

Let $\tilde{W}$ be the block diagonal matrix with the diagonal element being W, and $\tilde{R}$ be the block diagonal matrix with the diagonal element being R. It is easy to know that

$$
\bar{J}(x(k)) \le \|x(k)\|_W^2 + \left\| \tilde{A}x(k) + \tilde{B}\tilde{u}(k|k) \right\|_{\tilde{W}}^2 + \|\tilde{u}(k|k)\|_{\tilde{R}}^2 + \gamma. \qquad (8.3.13)
$$

Define

$$
\left\| \tilde{A}x(k) + \tilde{B}\tilde{u}(k|k) \right\|_{\tilde{W}}^2 + \|\tilde{u}(k|k)\|_{\tilde{R}}^2 \le \gamma_1. \qquad (8.3.14)
$$

By applying Schur complement, (8.3.14) can be transformed into the following LMI:

$$
\begin{bmatrix} \gamma_1 & * & * \\ \tilde{A}x(k) + \tilde{B}\tilde{u}(k|k) & \tilde{W}^{-1} & * \\ \tilde{u}(k|k) & 0 & \tilde{R}^{-1} \end{bmatrix} \ge 0. \qquad (8.3.15)
$$

Thus, by also considering Lemma 8.1.1, problem (8.3.3)-(8.3.6) is approximately transformed into the following LMI optimization problem:

$$
\min_{u(k|k),\cdots,u(k+N-1|k),\gamma_1,\gamma,Q,Y,Z,\Gamma} \gamma_1 + \gamma,
$$

$$
\text{s.t. } -\underline{u} \le u(k+i|k) \le \bar{u}, \quad -\underline{\psi} \le \Psi x(k+i+1|k) \le \bar{\psi},
$$

$$
i \in \{0,1,\ldots,N-1\}, \ (8.3.7) - (8.3.10), (8.3.15), \qquad (8.3.16)
$$

where, when the optimization problem is being solved, $x(k+N|k)$ in (8.3.8) should be substituted with $A^N x(k) + \tilde{B}_N \tilde{u}(k|k)$.

Notice that,

(i) since the ellipsoidal confinement is invoked for handling the input/state constraints after the switching horizon N, which is relatively conservative, (8.3.16) is only an approximation of (8.3.3)-(8.3.6);

(ii) $x(k+i+1|k)$ in the state constraint should be substituted with (8.3.11).

At each time k, solve (8.3.16), but only the obtained $u^*(k|k)$ is implemented. At the next sampling instant $k + 1$, based on the new measurement $x(k+1)$, the optimization is re-done such that $u^*(k+1|k+1)$ is obtained. If, after implementation of $u^*(k|k)$, the input/state constraints are still active ("active" means affecting the optimal solution of (8.3.16)), then re-optimization at time $k + 1$ can improve the performance. The basic reason is that, the handling of the constraints by (8.3.9)-(8.3.10) has its conservativeness, which can be reduced by "receding horizon optimization."

Theorem 8.3.1. *(Stability) Suppose (8.3.16) is feasible at the initial time $k = 0$. Then (8.3.16) is feasible for any $k \geq 0$. Further, by receding horizon implementing the optimal $u^*(k|k)$, the closed-loop system is exponentially stable.*

Proof. Suppose (8.3.16) is feasible at time k (the solution is denoted by $*$). Then the following is a feasible solution at time $k + 1$:

$$u(k+i|k+1) = u^*(k+i|k), i \in \{1,\ldots,N-1\}; \ u(k+N|k+1) = F^*(k)x^*(k+N|k).$$
$$(8.3.17)$$

By applying (8.3.17), the following is easily obtained:

$$
\begin{aligned}
\bar{J}(x(k+1)) &= \sum_{i=0}^{N-1} \left[\|x(k+i+1|k+1)\|_W^2 + \|u(k+i+1|k+1)\|_R^2 \right] \\
&\quad + \|x(k+N+1|k+1)\|_{P(k+1)}^2 \\
&= \sum_{i=1}^{N-1} \left[\|x^*(k+i|k)\|_W^2 + \|u^*(k+i|k)\|_R^2 \right] \\
&\quad + \left[\|x^*(k+N|k)\|_W^2 + \|F^*(k)x^*(k+N|k)\|_R^2 \right] \\
&\quad + \|x^*(k+N+1|k)\|_{P^*(k)}^2 .
\end{aligned}
\tag{8.3.18}
$$

Since (8.3.1) is applied, the following is satisfied:

$$
\begin{aligned}
\|x^*(k+N+1|k)\|_{P^*(k)}^2 &\leq \|x^*(k+N|k)\|_{P^*(k)}^2 \\
&\quad - \|x^*(k+N|k)\|_W^2 - \|F^*(k)x^*(k+N|k)\|_R^2 .
\end{aligned}
\tag{8.3.19}
$$

Substituting (8.3.19) into (8.3.18) yields

$$\bar{J}(x(k+1)) \leq \sum_{i=1}^{N-1} \left[\|x^*(k+i|k)\|_W^2 + \|u^*(k+i|k)\|_R^2 \right] + \|x^*(k+N|k)\|_{P^*(k)}^2$$

$$= \bar{J}^*(x(k)) - \|x(k)\|_W^2 - \|u^*(k|k)\|_R^2. \tag{8.3.20}$$

Now, notice that
$$\bar{J}^*(x(k)) \leq \eta^*(k) := \|x(k)\|_W^2 + \gamma_1^*(k) + \gamma^*(k)$$
and
$$\bar{J}(x(k+1)) \leq \eta(k+1) := \|x^*(k+1|k)\|_W^2 + \gamma_1(k+1) + \gamma(k+1).$$
According to (8.3.20), at time $k+1$ it is feasible to choose

$$\gamma_1(k+1) + \gamma(k+1) = \gamma_1^*(k) + \gamma^*(k) - \|u^*(k|k)\|_R^2 - \|x^*(k+1|k)\|_W^2.$$

Since at time $k+1$ the optimization is re-done, it must lead to $\gamma_1^*(k+1) + \gamma^*(k+1) \leq \gamma_1(k+1) + \gamma(k+1)$, which means that

$$\eta^*(k+1) - \eta^*(k) \leq -\|x(k)\|_W^2 - \|u^*(k|k)\|_R^2. \tag{8.3.21}$$

Therefore, $\eta^*(k)$ can be Lyapunov function for proving exponential stability. $\qquad\square$

The difference between (8.1.6) and (8.3.7) should be noted. One includes γ, the other does not include γ. Correspondingly, the definitions X and Q are different. One contains γ, the other does not contain γ. These two different manners can have different effects on the control performance. It is noted that the two different manners are not the criterion for distinguishing between standard approach and on-line approach.

8.4 Quasi-optimal solution to the infinite-horizon constrained linear time-varying quadratic regulation utilizing synthesis approach of MPC

In order to emphasize the speciality of the studied problem, this section adopts some slightly different notations. We study the following linear time-varying discrete-time system:

$$x(t+1) = A(t)x(t) + B(t)u(t), \tag{8.4.1}$$

where $x(t) \in \mathfrak{R}^n$ and $u(t) \in \mathfrak{R}^m$ are the measurable state and input, respectively. We aim at solving the constrained linear time-varying quadratic regulation (CLTVQR), by finding $u(0), u(1), \cdots, u(\infty)$ that minimizes the following infinite-horizon performance cost:

$$\Phi(x(0)) = \sum_{i=0}^{\infty} \left[x(i)^T \Pi x(i) + u(i)^T R u(i) \right], \tag{8.4.2}$$

and satisfies the following constraints:

$$x(i+1) \in \mathfrak{X}, \ \ u(i) \in \mathfrak{U}, \ \forall i \geq 0, \tag{8.4.3}$$

where $\Pi > 0$, $R > 0$ are symmetric weighting matrices, $\mathfrak{U} \subset \mathfrak{R}^m$, $\mathfrak{X} \subset \mathfrak{R}^n$ are compact, convex and contain the origin as interior point.

Differently from the linear time-invariant systems, it is usually difficult to find the optimal solution to CLTVQR, except for some special $[A(t)|B(t)]$ such as periodic time-varying systems, etc. The method developed in this section is not for optimal solution, but for finding suboptimal solution such that the corresponding suboptimal cost value can be arbitrarily close (although may be not equal) to the theoretically optimal cost value. The so-called theoretically optimal cost value is the one such that (8.4.2) is "absolutely" minimized.

In order to make the suboptimal arbitrarily close to the optimal, the infinite-horizon cost index is divided into two parts. The second part is formulated as an infinite-horizon min-max LQR based on polytopic inclusion of the time-varying dynamics in a neighborhood of the origin. Outside this neighborhood, the first part calculates the finite-horizon control moves by solving a finite-horizon optimization problem. For sufficiently large switching horizon, the resultant overall controller achieves desired closed-loop optimality. Denote

$$u_i^j := \left[u(i|0)^T, u(i+1|0)^T, \cdots, u(j|0)^T \right].$$

8.4.1   Overall idea

We suppose $[A(t), B(t)]$ is bounded, uniformly stabilizable and

$$[A(i)|B(i)] \in \Omega = Co\left\{ [A_1|B_1], \ [A_2|B_2], \ \cdots, \ [A_L|B_L] \right\}, \ \forall i \geq N_0. \tag{8.4.4}$$

With the solution of CLTVQR applied, the state will converge to the origin. Therefore, (8.4.4) actually defines a polytope of dynamics (8.4.1) in a neighborhood of the origin. CLTVQR with (8.4.2)-(8.4.3) is implemented by

$$\min_{u_0^\infty} \Phi(x(0|0)) = \sum_{i=0}^{\infty} \left[\|x(i|0)\|_{\Pi}^2 + \|u(i|0)\|_R^2 \right], \tag{8.4.5}$$

$$\text{s.t. } x(i+1|0) = A(i)x(i|0) + B(i)u(i|0), \ x(0|0) = x(0), \ i \geq 0, \tag{8.4.6}$$

$$- \underline{u} \leq u(i|0) \leq \bar{u}, \ i \geq 0, \tag{8.4.7}$$

$$- \underline{\psi} \leq \Psi x(i+1|0) \leq \bar{\psi}, \ i \geq 0, \tag{8.4.8}$$

where u_0^∞ are the decision variables.

Then we state the four relevant control problems in the following.

Problem 8.4 CLTVQR of infinite-horizon (optimal CLTVQR):

$$\Phi^* = \min_{u_0^\infty} \Phi(x(0|0)), \text{ s.t. } (8.4.6) - (8.4.8). \qquad (8.4.9)$$

The key idea is to find a suboptimal solution of Problem 8.4, say Φ^f, such that

$$\frac{\left(\Phi^f - \Phi^*\right)}{\Phi^*} \leq \delta, \qquad (8.4.10)$$

where $\delta > 0$ is a pre-specified scalar. In the sense that δ can be chosen arbitrarily small, (8.4.10) means that the suboptimal solution can be arbitrarily close to the optimal solution. For achieving (8.4.10), $\Phi(x(0|0))$ is divided into two parts,

$$\Phi\left(x(0|0)\right) = \sum_{i=0}^{N-1} \left[\|x(i|0)\|_{\Pi}^2 + \|u(i|0)\|_R^2\right] + \Phi_{\text{tail}}\left(x(N|0)\right), \qquad (8.4.11)$$

$$\Phi_{\text{tail}}\left(x(N|0)\right) = \sum_{i=N}^{\infty} \left[\|x(i|0)\|_{\Pi}^2 + \|u(i|0)\|_R^2\right], \qquad (8.4.12)$$

where $N \geq N_0$. For (8.4.12), the optimization is still of infinite-horizon, hence without general guarantee of finding the optimal solution. For this reason, we turn to solve the following problem:

Problem 8.5 Min-max CLQR:

$$\min_{u_N^\infty} \max_{[A(j)|B(j)]\in\Omega, \; j\geq N} \Phi_{\text{tail}}(x(N|0)), \text{ s.t. } (8.4.6) - (8.4.8), \; i \geq N. \qquad (8.4.13)$$

Problem 8.5 has been solved in Chapter 7. By defining control law in the following form:

$$u(i|0) = Fx(i|0), \; \forall i \geq N, \qquad (8.4.14)$$

the bound on (8.4.12) can be deduced as

$$\Phi_{\text{tail}}\left(x(N|0)\right) \leq x(N|0)^T P_N x(N|0) \leq \gamma, \qquad (8.4.15)$$

where $P_N > 0$ is a symmetric weighting matrix. Hence,

$$\Phi\left(x(0|0)\right) \leq \sum_{i=0}^{N-1} \left[\|x(i|0)\|_{\Pi}^2 + \|u(i|0)\|_R^2\right] + x(N|0)^T P_N x(N|0) = \bar{\Phi}_{P_N}\left(x(0|0)\right). \qquad (8.4.16)$$

For $P_N = 0$, $\bar{\Phi}_{P_N}\left(x(0|0)\right) = \bar{\Phi}_0\left(x(0|0)\right)$. Denote $\mathfrak{X}(N|0) \subset \mathfrak{R}^n$ as the set of $x(N|0)$ in which Problem 8.5 exists a feasible solution of the form (8.4.14). The other two problems of interests are as follows.

Problem 8.6 CLTVQR of finite-horizon without terminal weighting:

$$\bar{\Phi}_0^* = \min_{u_0^{N-1}} \bar{\Phi}_0\left(x(0|0)\right), \text{ s.t. } (8.4.6) - (8.4.8), \ i \in \{0, 1, \ldots, N-1\}. \quad (8.4.17)$$

Problem 8.7 CLTVQR of finite-horizon with terminal weighting:

$$\bar{\Phi}_{P_N}^* = \min_{u_0^{N-1}} \bar{\Phi}_{P_N}\left(x(0|0)\right), \text{ s.t. } (8.4.6)-(8.4.8), i \in \{0, 1, \ldots, N-1\}. \quad (8.4.18)$$

The resultant terminal state by the optimal solution of Problem 8.6 is denoted as $x^0(N|0)$; the resultant terminal state by the optimal solution of Problem 8.7 is denoted as $x^*(N|0)$. In the next section, we will achieve the optimality requirement (8.4.10) through solving Problems 8.5-8.7.

8.4.2   Solution to the min-max constrained linear quadratic control

Define $Q = \gamma P_N^{-1}, F = YQ^{-1}$. Based on what has been described in the previous section, we can transform Problem 8.5 into the following optimization problem:

$$\min_{\gamma, Q, Y, Z, \Gamma} \gamma, \text{ s.t. } (8.4.20), (8.3.9), (8.4.21) \text{ and } \begin{bmatrix} 1 & * \\ x(N|0) & Q \end{bmatrix} \geq 0, \quad (8.4.19)$$

$$\begin{bmatrix} Q & * & * & * \\ A_l Q + B_l Y & Q & * & * \\ \Pi^{1/2} Q & 0 & \gamma I & * \\ R^{1/2} Y & 0 & 0 & \gamma I \end{bmatrix} \geq 0, \ \forall l \in \{1, \ldots, L\}, \quad (8.4.20)$$

$$\begin{bmatrix} Q & * \\ \Psi \left(A_l Q + B_l Y\right) & \Gamma \end{bmatrix} \geq 0, \ \Gamma_{ss} \leq \bar{\psi}_{s,\text{inf}}^2, \ \forall l \in \{1, \ldots, L\}, \ s \in \{1, \ldots, q\}. \quad (8.4.21)$$

Lemma 8.4.1. *(Determination of the upper bound) Any feasible solution of problem (8.4.19) defines a set* $\mathfrak{X}(N|0) = \{x \in \mathfrak{R}^n | x^T Q^{-1} x \leq 1\}$ *in which a local controller* $Fx = YQ^{-1}x$ *exists such that the closed-loop cost value over the infinite-horizon beginning from* N *is bounded by*

$$\Phi_{\text{tail}}\left(x(N|0)\right) \leq x(N|0)^T \gamma Q^{-1} x(N|0). \quad (8.4.22)$$

Proof. The proof is similar to Chapter 7. Since $\mathfrak{X}(N|0) = \{x|x^T Q^{-1}x \leq 1\}$, $\begin{bmatrix} 1 & * \\ x(N|0) & Q \end{bmatrix} \geq 0$ means $x(N|0) \in \mathfrak{X}(N|0)$. $\square$

8.4.3   Case finite-horizon without terminal weighting

Define

$$\tilde{x} = \left[x(0|0)^T, x(1|0)^T, \cdots, x(N-1|0)^T \right]^T, \tag{8.4.23}$$

$$\tilde{u} = \left[u(0|0)^T, u(1|0)^T, \cdots, u(N-1|0)^T \right]^T. \tag{8.4.24}$$

Then

$$\tilde{x} = \tilde{A}\tilde{x} + \tilde{B}\tilde{u} + \tilde{x}_0, \tag{8.4.25}$$

where $\tilde{A} = \begin{bmatrix} 0 & \cdots & \cdots & & \cdots & 0 \\ A(0) & \ddots & \ddots & & \ddots & \vdots \\ 0 & A(1) & \ddots & & \ddots & \vdots \\ \vdots & \ddots & \ddots & & \ddots & \vdots \\ 0 & \cdots & 0 & A(N-2) & 0 \end{bmatrix}$,

$$\tilde{B} = \begin{bmatrix} 0 & \cdots & \cdots & & \cdots & 0 \\ B(0) & \ddots & \ddots & & \ddots & \vdots \\ 0 & B(1) & \ddots & & \ddots & \vdots \\ \vdots & \ddots & \ddots & & \ddots & \vdots \\ 0 & \cdots & 0 & B(N-2) & 0 \end{bmatrix},$$

$$\tilde{x}_0 = \left[x(0|0)^T, 0, \cdots, 0 \right]^T. \tag{8.4.26}$$

Equation (8.4.25) can be rewritten as

$$\tilde{x} = \tilde{W}\tilde{u} + \tilde{V}_0 \tag{8.4.27}$$

where

$$\tilde{W} = (I - \tilde{A})^{-1}\tilde{B}, \quad \tilde{V}_0 = (I - \tilde{A})^{-1}\tilde{x}_0. \tag{8.4.28}$$

Thus, the cost function of Problem 8.6 can be represented by

$$\bar{\Phi}_0\left(x(0|0) \right) = \|\tilde{x}\|_{\tilde{\Pi}}^2 + \|\tilde{u}\|_{\tilde{R}}^2 = \tilde{u}^T W \tilde{u} + W_v \tilde{u} + V_0 \le \eta^0, \tag{8.4.29}$$

where η^0 is a scalar, $\tilde{\Pi} = \begin{bmatrix} \Pi & 0 & \cdots & 0 \\ 0 & \Pi & \ddots & \vdots \\ \vdots & \ddots & \ddots & 0 \\ 0 & \cdots & 0 & \Pi \end{bmatrix}$, $\tilde{R} = \begin{bmatrix} R & 0 & \cdots & 0 \\ 0 & R & \ddots & \vdots \\ \vdots & \ddots & \ddots & 0 \\ 0 & \cdots & 0 & R \end{bmatrix}$,

$$W = \tilde{W}^T \tilde{\Pi} \tilde{W} + \tilde{R}, \quad W_v = 2\tilde{V}_0^T \tilde{\Pi} \tilde{W}, \quad V_0 = \tilde{V}_0^T \tilde{\Pi} \tilde{V}_0. \tag{8.4.30}$$

Equation (8.4.29) can be represented by the following LMI:

$$\begin{bmatrix} \eta^0 - W_v \tilde{u} - V_0 & * \\ W^{1/2}\tilde{u} & I \end{bmatrix} \ge 0. \tag{8.4.31}$$

Further, define $\tilde{x}^+ = \left[x(1|0)^T, \cdots, x(N-1|0)^T, x(N|0)^T \right]^T$, then $\tilde{x}^+ = \tilde{A}^+\tilde{x} + \tilde{B}^+\tilde{u}$, where

$$
\tilde{A}^+ = \begin{bmatrix} A(0) & 0 & \cdots & 0 \\ 0 & A(1) & \ddots & \vdots \\ \vdots & \ddots & \ddots & 0 \\ 0 & \cdots & 0 & A(N-1) \end{bmatrix},
$$

$$
\tilde{B}^+ = \begin{bmatrix} B(0) & 0 & \cdots & 0 \\ 0 & B(1) & \ddots & \vdots \\ \vdots & \ddots & \ddots & 0 \\ 0 & \cdots & 0 & B(N-1) \end{bmatrix}. \tag{8.4.32}
$$

The constraints in Problem 8.6 are transformed into

$$
-\underline{\tilde{u}} \le \tilde{u} \le \bar{\tilde{u}}, \tag{8.4.33}
$$

$$
-\underline{\tilde{\psi}} \le \tilde{\Psi}(\tilde{A}^+\tilde{W}\tilde{u} + \tilde{B}^+\tilde{u} + \tilde{A}^+\tilde{V}_0) \le \bar{\tilde{\psi}}, \tag{8.4.34}
$$

where

$$
\underline{\tilde{u}} = \left[\underline{u}^T, \underline{u}^T \cdots \underline{u}^T \right]^T \in \mathfrak{R}^{mN}, \quad \bar{\tilde{u}} = \left[\bar{u}^T, \bar{u}^T \cdots \bar{u}^T \right]^T \in \mathfrak{R}^{mN}, \tag{8.4.35}
$$

$$
\tilde{\Psi} = \begin{bmatrix} \Psi & 0 & \cdots & 0 \\ 0 & \Psi & \ddots & \vdots \\ \vdots & \ddots & \ddots & 0 \\ 0 & \cdots & 0 & \Psi \end{bmatrix} \in \mathfrak{R}^{qN \times nN}, \quad \underline{\tilde{\psi}} = \left[\underline{\psi}^T, \underline{\psi}^T \cdots \underline{\psi}^T \right]^T \in \mathfrak{R}^{qN},
$$

$$
\bar{\tilde{\psi}} = \left[\bar{\psi}^T, \bar{\psi}^T \cdots \bar{\psi}^T \right]^T \in \mathfrak{R}^{qN}. \tag{8.4.36}
$$

Problem 8.6 is transformed into

$$
\min_{\eta^0, \tilde{u}} \eta^0, \text{ s.t. } (8.4.31), (8.4.33) - (8.4.34). \tag{8.4.37}
$$

Denote the optimal $\tilde{u}$ by solving (8.4.37) as $\tilde{u}^0$.

8.4.4   Case finite-horizon with terminal weighting

Problem 8.7 can be solved similarly to Problem 8.6. The cost function of Problem 8.7 can be represented by

$$
\bar{\Phi}_{P_N}(x(0|0)) = \|\tilde{x}\|_{\tilde{\Pi}}^2 + \|\tilde{u}\|_{\tilde{R}}^2 + \|\mathcal{A}_{N,0}x(0|0) + \bar{B}\tilde{u}\|_{P_N}^2
$$

$$
= \left(\tilde{u}^T W \tilde{u} + W_v \tilde{u} + V_0 \right) + \|\mathcal{A}_{N,0}x(0|0) + \bar{B}\tilde{u}\|_{P_N}^2 \le \eta, \tag{8.4.38}
$$

where η is a scalar, $\mathcal{A}_{j,i} = \prod_{l=i}^{j-1} A(j-1+i-l)$,

$$\bar{B} = [\mathcal{A}_{N,1}B(0), \cdots, \mathcal{A}_{N,N-1}B(N-2),\ B(N-1)]. \tag{8.4.39}$$

Equation (8.4.38) can be represented by the following LMI:

$$\begin{bmatrix} \eta - W_v\tilde{u} - V_0 & * & * \\ W^{1/2}\tilde{u} & I & * \\ \mathcal{A}_{N,0}x(0|0) + \bar{B}\tilde{u} & 0 & P_N^{-1} \end{bmatrix} \geq 0. \tag{8.4.40}$$

Problem 8.7 is transformed into

$$\min_{\eta,\tilde{u}} \eta,\ \text{s.t. } (8.4.40), (8.4.33) - (8.4.34). \tag{8.4.41}$$

Denote the optimal $\tilde{u}$ by solving (8.4.41) as $\tilde{u}^*$.

8.4.5   Quasi-optimality, algorithm and stability

Let us firstly give the following conclusion.

Lemma 8.4.2. *The design requirement* (8.4.10) *is satisfied if* $\bar{\Phi}_{P_N}^*$ *serves as* Φ^f *and*

$$\frac{\bar{\Phi}_{P_N}^* - \bar{\Phi}_0^*}{\bar{\Phi}_0^*} \leq \delta. \tag{8.4.42}$$

Proof. By the optimality principle and notations, $\bar{\Phi}_0^* \leq \Phi^* \leq \bar{\Phi}_{P_N}^*$. If $\bar{\Phi}_{P_N}^*$ serves as Φ^f and (8.4.42) is satisfied, then

$$\frac{\Phi^f - \Phi^*}{\Phi^*} = \frac{\bar{\Phi}_{P_N}^* - \Phi^*}{\Phi^*} \leq \frac{\bar{\Phi}_{P_N}^* - \bar{\Phi}_0^*}{\bar{\Phi}_0^*} \leq \delta.$$

The above formula shows that (8.4.10) is satisfied.   □

Algorithm 8.2

Step 1. Choose initial (large) $x(N|0) = \hat{x}(N|0)$ satisfying $\|\hat{x}(N|0)\| > \Delta$, where Δ is a pre-specified scalar. Notice that N is unknown at this step.

Step 2. Solve (8.4.19) to obtain γ^*, Q^*, F^*.

Step 3. If (8.4.19) is infeasible and $\|x(N|0)\| > \Delta$, then decrease $x(N|0)$ $(x(N|0) \leftarrow rx(N|0)$, where r is a pre-specified scalar satisfying $0 < r < 1)$, and return to Step 2. However, if (8.4.19) is infeasible and $\|x(N|0)\| \leq \Delta$, then mark the overall Algorithm as INFEASIBLE and STOP.

Step 4. $P_N = \gamma^* Q^{*-1}$, $\mathfrak{X}(N|0) = \{x | x^T Q^{*-1} x \leq 1\}$.

Step 5. Choose initial $N \geq N_0$.

Step 6. Solve (8.4.41) to obtain $\tilde{u}^*$, $x^*(N|0) = \mathcal{A}_{N,0} x(0) + \bar{B} \tilde{u}^*$.

Step 7. If $x^*(N|0) \notin \mathfrak{X}(N|0)$, then increase N and return to Step 6.

Step 8. Choose $\tilde{u} = \tilde{u}^*$ in Step 6 as initial solution and solve (8.4.37) to obtain $\tilde{u}^0$ and $x^0(N|0) = \mathcal{A}_{N,0} x(0) + \bar{B} \tilde{u}^0$.

Step 9. If $x^0(N|0) \notin \mathfrak{X}(N|0)$, then increase N and return to Step 6.

Step 10. If $\left(\bar{\Phi}^*_{P_N} - \bar{\Phi}^*_0 \right) / \bar{\Phi}^*_0 > \delta$, then increase N and return to Step 6.

Step 11. Implement $\tilde{u}^*, F^*$.

Remark 8.4.1. How to choose r in Algorithm 8.2 depends on the system dynamics. However, given the initial choice $x(N|0) = \hat{x}(N|0)$, r should satisfy $r^{M_1} \leq \Delta / \|\hat{x}(N|0)\| \leq r^{M_2}$, where M_1 and M_2 are the maximum and the minimum allowable iterations between Step 2 and Step 3. In case (8.4.19) is feasible, then the iteration between Step 2 and Step 3 will not stop within M_2 times, and will not continue after M_1 times. Usually, we can choose M_0 satisfying $M_2 \leq M_0 \leq M_1$, and then choose $r = \sqrt[M_0]{\Delta / \|\hat{x}(N|0)\|}$.

Remark 8.4.2. For the same N, if $x^0(N|0) \in \mathfrak{X}(N|0)$ then $x^*(N|0) \in \mathfrak{X}(N|0)$; however, $x^*(N|0) \in \mathfrak{X}(N|0)$ does not mean $x^0(N|0) \in \mathfrak{X}(N|0)$. This is because problem (8.4.41) incorporates terminal cost. The terminal cost can suppress the evolution of the terminal state, such that it is easier for the terminal state to enter $\mathfrak{X}(N|0)$. Therefore, in Algorithm 8.2, if (8.4.37) is solved prior to (8.4.41), the computational burden can be reduced.

Denote $\mathfrak{X}(0|0) \subset \mathfrak{R}^n$ as the set of state $x(0|0)$ in which Problem 8.4 exists a feasible solution. Then the following result shows the feasibility and stability of the suboptimal CLTVQR.

Theorem 8.4.1. *(Stability) By applying Algorithm 8.2 , if problem (8.4.19) exists a feasible solution for a suitable $x(N|0)$, then for all $x(0|0) \in \mathfrak{X}(0|0)$ there exists a finite N and feasible $\tilde{u}^*$ such that the design requirement (8.4.10) is satisfied, and the closed-loop system is asymptotically stable.*

Proof. For sufficiently large N, $x^*(N|0)$ moves sufficiently close to the origin and $\left(\bar{\Phi}^*_{P_N} - \bar{\Phi}^*_0 \right)$ becomes sufficiently small such that (8.4.42) can be satisfied. Moreover, for sufficiently large N, $x^*(N|0) \in \mathfrak{X}(N|0)$. Inside of $\mathfrak{X}(N|0)$, (8.4.19) gives stable feedback law F.   □

However, as long as CLTVQR is substituted by min-max CLQR in the neighborhood of the origin, there is a possibility that the suboptimal solution does not exist for $x(0|0) \in \mathfrak{X}(0|0)$. Whether or not this will happen depends on the concrete system.

Remark 8.4.3. The method for finding the suboptimal solution to CLTVQR can be easily generalized to the nonlinear systems.

8.4.6   Numerical example

The following model is adopted:

$$
\begin{bmatrix} x_1(t+1) \\ x_2(t+1) \\ x_3(t+1) \\ x_4(t+1) \end{bmatrix} = \begin{bmatrix} 1 & 0 & 0.1 & 0 \\ 0 & 1 & 0 & 0.1 \\ -0.1\frac{K(t)}{m_1} & 0.1\frac{K(t)}{m_1} & 1 & 0 \\ 0.1\frac{K(t)}{m_2} & -0.1\frac{K(t)}{m_2} & 0 & 1 \end{bmatrix} \begin{bmatrix} x_1(t) \\ x_2(t) \\ x_3(t) \\ x_4(t) \end{bmatrix} + \begin{bmatrix} 0 \\ 0 \\ \frac{0.1}{m_1} \\ 0 \end{bmatrix} u(t)
$$

$$(8.4.43)$$

which is modified from the model of the two-mass (m_1, m_2) spring system, where $m_1 = m_2 = 1$, $K(t) = 1.5 + 2e^{-0.1t}(1 + \sin t) + 0.973 \sin(t\pi/11)$. The initial state is $x(0) = \alpha \times [5, 5, 0, 0]^T$ where α is a constant, the weighting matrices $Q = I$, $R = 1$ and the input constraint $|u(t)| \le 1$.

Let us first consider Algorithm 8.2. The control objective is to find a sequence of control input signals such that (8.4.10) is satisfied with $\delta \le 10^{-4}$. At $t = 50$, $2e^{-0.1t} \approx 0.0135$. Hence, $0.527 \le K(t) \le 2.5$, $\forall t \ge 50$. We choose $N_0 = 50$,

$$
[A_1|B_1] = \begin{bmatrix} 1 & 0 & 0.1 & 0 & 0 \\ 0 & 1 & 0 & 0.1 & 0 \\ -0.0527 & 0.0527 & 1 & 0 & 0.1 \\ 0.0527 & -0.0527 & 0 & 1 & 0 \end{bmatrix},
$$

$$
[A_2|B_2] = \begin{bmatrix} 1 & 0 & 0.1 & 0 & 0 \\ 0 & 1 & 0 & 0.1 & 0 \\ -0.25 & 0.25 & 1 & 0 & 0.1 \\ 0.25 & -0.25 & 0 & 1 & 0 \end{bmatrix}. \qquad (8.4.44)
$$

Choose $\hat{x}(N|0) = 0.02 \times [1, 1, 1, 1]^T$, then problem (8.4.19) exists feasible solution $F = \begin{bmatrix} -8.7199 & 6.7664 & -4.7335 & -2.4241 \end{bmatrix}$. Algorithm 8.2 exists a feasible solution whenever $\alpha \le 23.0$. Choose $\alpha = 1$, $N = 132$, then $\bar{\Phi}^*_{P_{132}} = 1475.91$, $\bar{\Phi}^*_0 = 1475.85$ and the desired optimality requirement (8.4.10) is achieved. Figure 8.4.1 shows the state responses of the closed-loop system. Figure 8.4.2 shows the control input signal.

8.4.7   A comparison with another approach

By introducing some extra assumptions, the on-line approach in Chapter 7 can be utilized to find the suboptimal solution of CLTVQR (the design requirement (8.4.10) is ignored). Suppose

$$[A(t + i)|B(t + i)] \in \Omega(t), \; \forall t \ge 0, \; \forall i \ge 0, \qquad (8.4.45)$$

where

$$\Omega(t) := Co\{[A_1(t)|B_1(t)], [A_2(t)|B_2(t)], \cdots, [A_L(t)|B_L(t)]\}, \qquad (8.4.46)$$

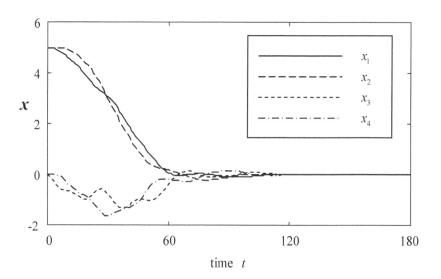

Figure 8.4.1: Closed-loop state responses of CLTVQR.

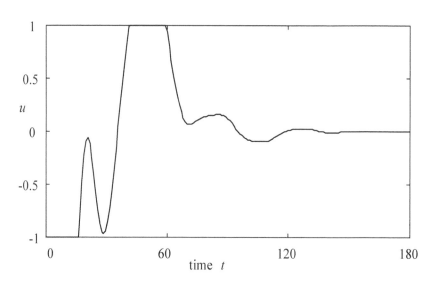

Figure 8.4.2: Control input signal of CLTVQR.

i.e., $\forall t, i \geq 0$ there exist L nonnegative coefficients $\omega_l(t+i,t)$, $l \in \{1, 2, \ldots, L\}$ such that

$$\sum_{l=1}^{L} \omega_l(t+i,t) = 1, \quad [A(t+i)|B(t+i)] = \sum_{l=1}^{L} \omega_l(t+i,t) [A_l(t)|B_l(t)]. \quad (8.4.47)$$

Equations (8.4.45)-(8.4.47) define a time-varying polytopic inclusion of the dynamic system (8.4.1). For $t = N_0$, (8.4.45)-(8.4.47) reduce to (8.4.4). Define the following optimization problem:

$$\min_{\gamma(t),Q(t),Y(t),Z(t),\Gamma(t)} \gamma(t), \text{ s.t. } \begin{bmatrix} 1 & * \\ x(t) & Q(t) \end{bmatrix} \geq 0 \text{ and } (8.4.20),$$

$$(8.3.9), (8.4.21), \text{by substituting } \{Q, Y, Z, \Gamma, \gamma, A_l, B_l\}$$
$$\text{with } \{Q(t), Y(t), Z(t), \Gamma(t), \gamma(t), A_l(t), B_l(t)\}. \quad (8.4.48)$$

According to the theoretical results in Chapter 7, if at time $t = 0$, (8.4.48) is solvable, then by receding horizon implementation of (8.4.48), the control sequence $u(t) = Y(t)Q(t)^{-1}x(t)$, $t \geq 0$ asymptotically stabilizes the system (8.4.1).

Remark 8.4.4. Notice that, in Chapter 7, only the time-invariant polytopic inclusion, i.e., $[A(t+i)|B(t+i)] \in \Omega$, $\forall t \geq 0$, $\forall i \geq 0$, is considered. However, by the time-varying polytopic inclusion (8.4.45)-(8.4.47), $\Omega(t+1) \subseteq \Omega(t)$, $\forall t \geq 0$. Due to this reason, stability property of Chapter 7 is suitable to technique (8.4.48).

Let us firstly compare Algorithm 8.2 and the technique based on (8.4.48) with respect to the computational burden.

In finding the feasible solution of LMI, the interior point algorithm is often applied, which is a polynomial time algorithm, i.e., the complexity for solving the feasibility problem can be represented by a polynomial. This complexity is proportional to $\mathfrak{K}^3\mathfrak{L}$, where $\mathfrak{K}$ is the number of scalar LMI variables and $\mathfrak{L}$ the number of rows (referring to scalar row) of the total LMI system; see [31]. For problem (8.4.37), $\mathfrak{K}_1(N) = N+1$ and $\mathfrak{L}_1(N) = (2m+2q+1)N+1$; for problem (8.4.41), $\mathfrak{K}_2(N) = N+1$ and $\mathfrak{L}_2(N) = (2m+2q+1)N+n+1$; for problem (8.4.19), $\mathfrak{K}_3 = \frac{1}{2}(n^2+n+m^2+m+q^2+q)+mn+1$ and $\mathfrak{L}_3 = (4n+m+q)L+2n+2m+q+1$.

The computational burden of Algorithm 8.2 mainly comes from solving LMI optimization problem. Denote $\mathfrak{N}_2$ ($\mathfrak{N}_1$) as the set of temporary and final switching horizons in implementing Step 6 (Step 8) of Algorithm 8.2, and $\bar{M}_0$ as the repeating times between Step 2 and Step 3. Then, the computational burden of Algorithm 8.2 is proportional to

$$\sum_{N\in\mathfrak{N}_1} \mathfrak{K}_1(N)^3\mathfrak{L}_1(N) + \sum_{N\in\mathfrak{N}_2} \mathfrak{K}_2(N)^3\mathfrak{L}_2(N) + \bar{M}_0\mathfrak{K}_3^3\mathfrak{L}_3.$$

The main source of computational burden for the method based on (8.4.48) also comes from solving LMI optimization problem. During $0 \leq t \leq N - 1$, this computational burden is proportional to $N \mathfrak{K}_3^3 \mathfrak{L}_3$. In general, the computational burden involved in Algorithm 8.2 is larger than that involved in the method based on (8.4.48). However, Algorithm 8.2 can give suboptimal solution arbitrarily close to the theoretically optimal solution of CLTVQR (degree of closeness is pre-specified by δ), while the method based on (8.4.48) cannot achieve the same target.

Consider the model (8.4.43). By applying the method based on (8.4.48), one can obtain a sequence of suboptimal control moves to stabilize (8.4.43). Then, in (8.4.45)-(8.4.47),

$$[A_1(t)|B_1(t)] = [A_1|B_1],$$

$$[A_2(t)|B_2(t)] = \begin{bmatrix} 1 & 0 & 0.1 & 0 & 0 \\ 0 & 1 & 0 & 0.1 & 0 \\ -0.1\left(2.473 + 4e^{-0.1t}\right) & 0.1\left(2.473 + 4e^{-0.1t}\right) & 1 & 0 & 0.1 \\ 0.1\left(2.473 + 4e^{-0.1t}\right) & -0.1\left(2.473 + 4e^{-0.1t}\right) & 0 & 1 & 0 \end{bmatrix}.$$

Problem (8.4.48) exists a feasible solution whenever $\alpha \leq 21.6$ (this result is worse than that of Algorithm 8.2). Choose $\alpha = 1$, then by solving problem (8.4.48) in a receding horizon way, Figure 8.4.3 shows the state responses of the closed-loop system; Figure 8.4.4 shows the control input signal. The cost value $\Phi(x(0)) = 3914.5$ is much larger than the theoretically minimum one (which lies between $\bar{\Phi}^*_{P_{132}} = 1475.91$ and $\bar{\Phi}^*_0 = 1475.85$).

In the simulation, we have utilized LMI Toolbox of Matlab 5.3 on our laptop (1.5G Pentium IV CPU, 256 M Memory); it takes $9\frac{2}{3}$ minutes for calculating $\bar{\Phi}^*_{P_{132}}$, $7\frac{1}{3}$ minutes for calculating $\bar{\Phi}^*_0$, $1\frac{1}{3}$ minutes for calculating $\Phi(x(0)) = 3914.5$ (by solving (8.4.48) for 280 sampling intervals).

8.5 On-line approach for systems with poly-topic description

Consider the following time-varying uncertain system:

$$x(k + 1) = A(k)x(k) + B(k)u(k), \quad [A(k)|B(k)] \in \Omega. \tag{8.5.1}$$

Suppose

$$[A(k)|B(k)] \in \Omega = Co\left\{[A_1|B_1], [A_2|B_2], \cdots, [A_L|B_L]\right\}, \quad \forall k \geq 0.$$

The constraints are as in (8.1.2)-(8.1.3).

Define the performance cost (8.2.1), introduce stability constraint (8.3.1), and substitute the optimization problem based on the performance cost (8.2.1) with the optimization problem for (8.3.2). These are the same as in section 8.3. Since the uncertain system is considered, the control performance will

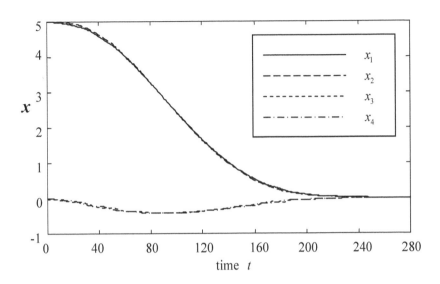

Figure 8.4.3: Closed-loop state responses of the technique based on (8.4.48).

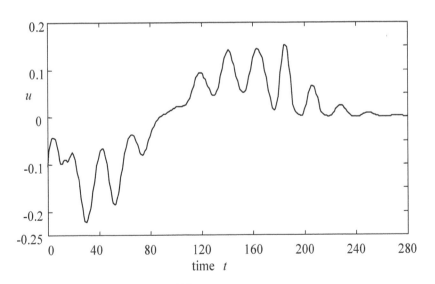

Figure 8.4.4: Control input signal of the technique based on (8.4.48).

get worse by adopting a standard approach. Therefore, we directly apply the on-line approach.

MPC here solves the following optimal problem at each time k:

$$\min_{\{u(k|k),\cdots u(k+N-1|k),F,P\}} \max_{[A(k+i)|B(k+i)]\in\Omega,i\geq0} \bar{J}(x(k)), \text{ s.t. (8.3.4)} - \text{(8.3.6)}.$$
$$(8.5.2)$$

It is easily seen that (8.5.2) is different from (8.3.3)-(8.3.6) in that "min" problem is substituted with "min-max" problem.

Define $Q = \gamma P^{-1}$, $F = YQ^{-1}$. Using Schur complements for a convex hull, (8.3.1) is equivalent to LMI

$$\begin{bmatrix} Q & * & * & * \\ A_l Q + B_l Y & Q & * & * \\ W^{1/2}Q & 0 & \gamma I & * \\ R^{1/2}Y & 0 & 0 & \gamma I \end{bmatrix} \geq 0, \ \forall l \in \{1,\ldots,L\}. \tag{8.5.3}$$

The input/state constraints after the switching horizon N can be guaranteed by (8.3.9) and (8.4.21).

For solving problem (8.5.2), i.e., transforming (8.5.2) into LMI optimization, we need the state predictions $x(k+i|k)$. Although the future state prediction is uncertain, we can determine the set to include this prediction.

Lemma 8.5.1. *Define the set $S(k+i|k)$ as*

$$S(k+i|k) = Co\{v_{l_{i-1}\cdots l_1 l_0}(k+i|k), \ l_0, l_1, \cdots l_{i-1} = 1 \ldots L\}, \ S(k|k) = \{x(k)\},$$
$$(8.5.4)$$

where $i \geq 0$. Suppose $x(k+i|k) \in S(k+i|k)$. If $v_{l_i \cdots l_1 l_0}(k+i+1|k), \ l_0, l_1, \cdots l_i \in \{1,\ldots,L\}$ satisfies

$$v_{l_i \cdots l_1 l_0}(k+i+1|k) = A_{l_i} v_{l_{i-1} \cdots l_1 l_0}(k+i|k) + B_{l_i} u(k+i|k),$$

then $S(k+i+1|k)$ is the tightest set that contains all possible $x(k+i+1|k)$.

Proof. (By induction) For $i = 0$ the result is trivial. For $i > 1$ the prediction equation is given by

$$x(k+i+1|k) = A(k+i)x(k+i|k) + B(k+i)u(k+i|k). \tag{8.5.5}$$

Note that, from the definition of a convex hull and with $\omega_l \geq 0$,

$$A(k+i) = \sum_{l=1}^{L} \omega_l A_l, \ B(k+i) = \sum_{l=1}^{L} \omega_l B_l, \ \sum_{l=1}^{L} \omega_l = 1. \tag{8.5.6}$$

Suppose

$$x(k+i|k) = \sum_{l_0 l_1 \cdots l_{i-1}=1}^{L} \left(\left(\prod_{h=0}^{i-1} \omega_{l_h} \right) v_{l_{i-1} \cdots l_1 l_0}(k+i|k) \right), \tag{8.5.7}$$

where $\sum_{l_0 l_1 \cdots l_{i-1}=1}^{L} \left(\prod_{h=0}^{i-1} \omega_{l_h} \right) = 1$, $\sum_{l_0=1}^{L} \cdots \sum_{l_i=1}^{L} (\cdots)$ is simplified as $\sum_{l_0 \cdots l_i=1}^{L} (\cdots)$. Substitution of (8.5.7) and (8.5.6) into (8.5.5) gives (where $\omega_l = \omega_{l_i}$)

$$
\begin{aligned}
x(k+i+1|k) &= \sum_{l_i=1}^{L} \omega_{l_i} A_{l_i} \sum_{l_0 l_1 \cdots l_{i-1}=1}^{L} \left(\left(\prod_{h=0}^{i-1} \omega_{l_h} \right) v_{l_{i-1} \cdots l_1 l_0}(k+i|k) \right) \\
&+ \sum_{l_i=1}^{L} \omega_{l_i} B_{l_i} u(k+i|k) \\
&= \sum_{l_0 l_1 \cdots l_i=1}^{L} \left(\prod_{h=0}^{i} \omega_{l_h} \right) v_{l_i \cdots l_1 l_0}(k+i|k).
\end{aligned}
\tag{8.5.8}
$$

Hence, $x(k+i+1|k) \in S(k+i+1|k) = Co\{v_{l_i \cdots l_1 l_0}(k+i+1|k), l_0, l_1, \cdots l_i = 1 \ldots L\}$.

Furthermore, there is no tighter set that contains all possible $x(k+i+1|k)$, since it is generated by a convex hull.   □

Define

$$
\|x(k+i|k)\|_W^2 + \|u(k+i|k)\|_R^2 \le \gamma_i.
\tag{8.5.9}
$$

Then $\bar{J}(x(k)) \le \sum_{i=0}^{N-1} \gamma_i + \gamma$.

For MPC in this section, a key is to take

$$
\begin{aligned}
u(k+i|k) &= F(k+i|k)x(k+i|k) + c(k+i|k), \\
F(\cdot|0) &= 0, \ i \in \{0, 1, \ldots, N-1\},
\end{aligned}
\tag{8.5.10}
$$

where $F(k+i|k)$, $k > 0$, $i \in \{0, 1, \ldots, N-1\}$ are supposed known, their values being carried over from the previous sampling instant, i.e., when $k > 0$, $F(k+i|k) = F(k+i|k-1)$, $i \in \{0, 1, \ldots, N-2\}$, $F(k+N-1|k) = F(k-1)$. For the significance of selecting F's in this manner, one is referred to Chapter 9 for more details.

In (8.5.10), c is the perturbation item. In the final optimization problem, c, rather than u, will be the decision variable. By utilizing (8.5.10) and Lemma 8.5.1, (8.5.9) is converted into LMI

$$
\begin{bmatrix}
\gamma_0 & * & * \\
x(k) & W^{-1} & * \\
F(k|k)x(k) + c(k|k) & 0 & R^{-1}
\end{bmatrix} \ge 0,
$$

$$
\begin{bmatrix}
\gamma_i & * & * \\
v_{l_{i-1} \cdots l_1 l_0}(k+i|k) & W^{-1} & * \\
F(k+i|k)v_{l_{i-1} \cdots l_1 l_0}(k+i|k) + c(k+i|k) & 0 & R^{-1}
\end{bmatrix} \ge 0,
$$

$$
l_0, l_1, \cdots l_{i-1} \in \{1, 2, \ldots, L\}, \ i \in \{1, \ldots, N-1\}.
\tag{8.5.11}
$$

Note that in (8.5.11), $v_{l_{i-1}\cdots l_1 l_0}(k+i|k)$ should be represented as the function of $c(k|k), c(k+1|k), \cdots, c(k+i-1|k)$ (by using Lemma 8.5.1 and (8.5.10)). In the first LMI of (8.5.11), one can remove the rows and columns corresponding to W^{-1}; since these rows and columns do not include LMI variables, this removal will not affect the feasibility and optimality of the control algorithm.

By using (8.5.10) and Lemma 8.5.1, (8.3.5) is converted into LMI

$$\begin{bmatrix} 1 & * \\ v_{l_{N-1}\cdots l_1 l_0}(k+N|k) & Q \end{bmatrix} \geq 0, \ l_0, l_1, \cdots l_{N-1} \in \{1, 2, \ldots, L\}. \quad (8.5.12)$$

Note that in (8.5.12), $v_{l_{N-1}\cdots l_1 l_0}(k+N|k)$ should be represented as the function of $c(k|k), c(k+1|k), \cdots, c(k+N-1|k)$.

Moreover, the satisfaction of input/state constraints before the switching horizon can be guaranteed by imposing

$$-\underline{u} \leq F(k|k)x(k) + c(k|k) \leq \bar{u},$$
$$-\underline{u} \leq F(k+i|k)v_{l_{i-1}\cdots l_1 l_0}(k+i|k) + c(k+i|k) \leq \bar{u},$$
$$l_0, l_1, \cdots l_{i-1} \in \{1, 2, \ldots, L\}, \ i \in \{1, \ldots, N-1\}, \quad (8.5.13)$$
$$-\underline{\psi} \leq \Psi v_{l_{i-1}\cdots l_1 l_0}(k+i|k) \leq \bar{\psi}, \ l_0, l_1, \cdots l_{i-1} \in \{1, 2, \ldots, L\}, \ i \in \{1, 2, \ldots, N\}. \quad (8.5.14)$$

So, problem (8.5.2) is converted into LMI optimization problem

$$\min_{c(k|k), \cdots, c(k+N-1|k), \gamma_i, \gamma, Q, Y, Z, \Gamma} \sum_{i=0}^{N-1} \gamma_i$$
$$+\gamma, \ \text{s.t.} \ (8.3.9), (8.4.21), (8.5.3), (8.5.11) - (8.5.14). \quad (8.5.15)$$

Although (8.5.15) is solved at each time k, only $c^*(k|k)$ among the decided values is implemented. At the next time $k + 1$, based on the newly measured $x(k)$, the optimization is re-performed to obtain $c^*(k + 1|k + 1)$. If, after implementing $c^*(k|k)$, input/state constraints are still active (here "active" means affecting the optimum of (8.5.15)), then the re-performed optimization at time $k + 1$ can significantly improve the control performance (if the input/state constraints are not active, receding horizon optimization still improves performance).

Theorem 8.5.1. *(Stability) Suppose* (8.5.15) *is feasible at time* $k = 0$. *Then* (8.5.15) *is feasible for all* $k \geq 0$. *Further, by receding horizon implementing the optimum* $c^*(k|k)$, *the closed-loop system is exponentially stable.*

Proof. Suppose (8.5.15) is feasible at time $k = 0$ (the solution is denoted by $*$), then the following is feasible at time $k + 1$:

$$c(k+i|k+1) = c^*(k+i|k), \ i \in \{1, 2, \ldots, N-1\},$$
$$u(k+N|k+1) = F^*(k)x^*(k+N|k). \quad (8.5.16)$$

Adopting (8.5.16), by analogy to Theorem 8.3.1 we can obtain (8.3.20). Now, notice that

$$\bar{J}^*(x(k)) \leq \eta^*(k) := \sum_{i=0}^{N-1} \gamma_i^*(k) + \gamma^*(k)$$

and

$$\bar{J}(x(k+1)) \leq \eta(k+1) := \sum_{i=0}^{N-1} \gamma_i(k+1) + \gamma(k+1).$$

According to (8.3.20), at time $k+1$ it is feasible to choose

$$\sum_{i=0}^{N-1} \gamma_i(k+1) + \gamma(k+1) = \sum_{i=0}^{N-1} \gamma_i^*(k) + \gamma^*(k) - \left[\|x(k)\|_W^2 + \|u^*(k|k)\|_R^2 \right].$$

Since at time $k+1$ the optimization is re-done, it must lead to $\sum_{i=0}^{N-1} \gamma_i^*(k+1) + \gamma^*(k+1) \leq \sum_{i=0}^{N-1} \gamma_i(k+1) + \gamma(k+1)$, which means that

$$\eta^*(k+1) - \eta^*(k) \leq -\|x(k)\|_W^2 - \|u^*(k|k)\|_R^2.$$

Therefore, $\eta^*(k)$ can be Lyapunov function for proving exponential stability. $\qquad\square$

8.6 Parameter-dependent on-line approach for systems with polytopic description

The problem description is the same as section 8.5. For $N \geq 2$, we take

$$u(k+i|k) = \sum_{l_0 \cdots l_{i-1}=1}^{L} \left(\left(\prod_{h=0}^{i-1} \omega_{l_h}(k+h) \right) u^{l_{i-1} \cdots l_0}(k+i|k) \right),$$

$$\sum_{l_0 \cdots l_{i-1}=1}^{L} \left(\prod_{h=0}^{i-1} \omega_{l_h}(k+h) \right) = 1, \ i \in \{1, \ldots, N-1\}. \qquad (8.6.1)$$

At each time k we solve the following optimization problem

$$\min_{\{\tilde{u}(k), F, P\}} \max_{[A(k+i)|B(k+i)] \in \Omega, i \geq 0} J(x(k)), \ \text{s.t.} \ (8.3.4) - (8.3.6), (8.6.1) \quad (8.6.2)$$

where

$$\tilde{u}(k) := \{u(k|k), u^{l_0}(k+1|k), \cdots, u^{l_{N-2} \cdots l_0}(k+N-1|k)|l_j = 1 \ldots L,$$
$$j = 0 \ldots N-2\}$$

is the collection of the "vertex control moves" $u^{l_{i-1} \cdots l_0}(k+i|k)$ (which amounts to taking a different control move for each corner of the uncertainty evolution). After solving problem (8.6.2), only $u(k) = u(k|k)$ is implemented. Then, problem (8.6.2) is solved again at $k+1$.

Note that $u(k+i|k)$, $i \in \{1, \ldots, N-1\}$ (or $c(k+i|k)$, $i \in \{1, \ldots, N-1\}$) in the former sections and chapters are single, whereas in this section they are parameter-dependent (i.e., dependent on the unknown parameters). Hence, differently from robust MPC in section 8.5, when $N > 1$, (8.6.2) can only be implemented in a receding-horizon manner. MPC in Chapter 7 can also be implemented in a non-receding-horizon manner; on-line MPC in section 8.5 can also be implemented in a non-receding-horizon manner; by these non-receding-horizon controllers, the closed-loop systems are still stable.

Define

$$\gamma_1 \geq \|u(k|k)\|_R^2 + \sum_{i=1}^{N-1} \left[\|x(k+i|k)\|_W^2 + \|u(k+i|k)\|_R^2 \right],$$

$$1 \geq \|x(k+N|k)\|_{Q^{-1}}^2. \tag{8.6.3}$$

According to the deductions in section 8.5, one can approximate problem (8.6.2) by the following LMI optimization problem:

$$\min_{\tilde{u}(k),\gamma_1,\gamma,Q,Y,Z,\Gamma} \quad \max_{[A(k+i)|B(k+i)]\in\Omega, i\in\{0,\ldots,N-1\}} \gamma_1 + \gamma,$$

s.t. (8.3.4), $i \in \{0, \ldots, N-1\}$, (8.3.9), (8.4.21), (8.5.3), (8.6.1), (8.6.3).
$$\tag{8.6.4}$$

The state predictions before the switching horizon can be expressed as

$$
\begin{bmatrix} x(k+1|k) \\ x(k+2|k) \\ \vdots \\ x(k+N|k) \end{bmatrix} = \sum_{l_0\cdots l_{N-1}=1}^{L} \left(\prod_{h=0}^{N-1} \omega_{l_h}(k+h) \right) \left(\begin{bmatrix} x^{l_0}(k+1|k) \\ x^{l_1 l_0}(k+2|k) \\ \vdots \\ x^{l_{N-1}\cdots l_1 l_0}(k+N|k) \end{bmatrix} \right),
$$

$$
\begin{bmatrix} x^{l_0}(k+1|k) \\ x^{l_1 l_0}(k+2|k) \\ \vdots \\ x^{l_{N-1}\cdots l_1 l_0}(k+N|k) \end{bmatrix} = \begin{bmatrix} A_{l_0} \\ A_{l_1} A_{l_0} \\ \vdots \\ \prod_{i=0}^{N-1} A_{l_{N-1-i}} \end{bmatrix} x(k)
$$

$$
+ \begin{bmatrix} B_{l_0} & 0 & \cdots & 0 \\ A_{l_1} B_{l_0} & B_{l_1} & \ddots & \vdots \\ \vdots & \vdots & \ddots & 0 \\ \prod_{i=0}^{N-2} A_{l_{N-1-i}} B_{l_0} & \prod_{i=0}^{N-3} A_{l_{N-1-i}} B_{l_1} & \cdots & B_{l_{N-1}} \end{bmatrix}
$$

$$
\begin{bmatrix} u(k|k) \\ u^{l_0}(k+1|k) \\ \vdots \\ u^{l_{N-2}\cdots l_1 l_0}(k+N-1|k) \end{bmatrix}, \tag{8.6.5}
$$

where $x^{l_{i-1}\cdots l_1 l_0}(k+i|k)$, $i \in \{1, \ldots, N\}$ are "vertex state predictions." Since $\sum_{l_0\cdots l_{i-1}=1}^{L} \left(\prod_{h=0}^{i-1} \omega_{l_h}(k+h) \right) = 1$, the state predictions $x(k+i|k)$,

$i \in \{1, \ldots, N\}$ belong to the polytopes (see section 8.5). Notice that, within the switching horizon, (8.6.5) is consistent with Lemma 8.5.1. By applying Schur complement, (8.6.1) and the convexity of the set of state predictions, one can transform (8.6.3) into the following LMIs:

$$
\left[
\begin{array}{cccccccc}
\gamma_1 & * & * & \cdots & * & * & \cdots & * \\
u(k|k) & R^{-1} & * & \cdots & * & * & \cdots & * \\
u^{l_0}(k+1|k) & 0 & R^{-1} & \cdots & * & * & \cdots & * \\
\vdots & \vdots & \vdots & \ddots & \vdots & \vdots & & \vdots \\
u^{l_{N-2}\cdots l_1 l_0}(k+N-1|k) & 0 & 0 & \cdots & R^{-1} & * & \cdots & * \\
A_{l_0}x(k)+B_{l_0}u(k|k) & 0 & 0 & \cdots & 0 & W^{-1} & \cdots & * \\
\vdots & \vdots & \vdots & & \vdots & \vdots & \ddots & \vdots \\
\begin{array}{c}\prod_{i=0}^{N-2}A_{l_{N-2-i}}x(k)+\prod_{i=0}^{N-3}A_{l_{N-2-i}} \\ B_{l_0}u(k|k)+\cdots \\ +B_{l_{N-2}}u^{l_{N-3}\cdots l_1 l_0}(k+N-2|k)\end{array} & 0 & 0 & \cdots & 0 & 0 & \cdots & W^{-1}
\end{array}
\right] \geq 0,
$$

$$l_0, \cdots, l_{N-2} \in \{1, \ldots, L\}, \tag{8.6.6}$$

$$
\left[
\begin{array}{cc}
1 & * \\
\begin{array}{c}\prod_{i=0}^{N-1}A_{l_{N-1-i}}x(k)+\prod_{i=0}^{N-2}A_{l_{N-1-i}}B_{l_0}u(k|k) \\ +\cdots+B_{l_{N-1}}u^{l_{N-2}\cdots l_1 l_0}(k+N-1|k)\end{array} & Q
\end{array}
\right] \geq 0,
$$

$$l_0, \cdots, l_{N-1} \in \{1, \ldots, L\}. \tag{8.6.7}$$

For $i \in \{0, \ldots, N-1\}$, the hard constraint (8.3.4) can be transformed into the following LMI:

$$
-\underline{u} \leq u(k|k) \leq \bar{u}, \quad -\underline{u} \leq u^{l_{j-1}\cdots l_1 l_0}(k+j|k) \leq \bar{u}, \; l_0, \cdots, l_{j-1} \in \{1, \ldots, L\},
$$
$$j \in \{1, \ldots, N-1\}, \tag{8.6.8}$$

$$
-\left[\begin{array}{c}\psi \\ \psi \\ \vdots \\ \psi\end{array}\right] \leq \tilde{\Psi}\left[\begin{array}{c}A_{l_0} \\ \prod_{i=0}^{1}A_{l_{1-i}} \\ \vdots \\ \prod_{i=0}^{N-1}A_{l_{N-1-i}}\end{array}\right]x(k)
$$

$$
+\tilde{\Psi}\left[\begin{array}{cccc}B_{l_0} & 0 & \cdots & 0 \\ A_{l_1}B_{l_0} & B_{l_1} & \ddots & \vdots \\ \vdots & \vdots & \ddots & 0 \\ \prod_{i=0}^{N-2}A_{l_{N-1-i}}B_{l_0} & \prod_{i=0}^{N-3}A_{l_{N-1-i}}B_{l_1} & \cdots & B_{l_{N-1}}\end{array}\right]
$$

$$
\times\left[\begin{array}{c}u(k|k) \\ u^{l_0}(k+1|k) \\ \vdots \\ u^{l_{N-2}\cdots l_1 l_0}(k+N-1|k)\end{array}\right] \leq \left[\begin{array}{c}\bar{\psi} \\ \bar{\psi} \\ \vdots \\ \bar{\psi}\end{array}\right], \; l_0, \cdots, l_{N-1} \in \{1, \ldots, L\},
$$

$$\tag{8.6.9}$$

where $\tilde{\Psi} = \text{diag}\{\Psi, \cdots, \Psi\}$.

Thus, the optimization problem (8.6.4) is eventually approximated by the following LMI optimization problem:

$$\min_{\gamma_1, \gamma, \tilde{u}(k), Y, Q, Z, \Gamma} \gamma_1 + \gamma, \text{ s.t. } \quad (8.3.9), (8.4.21), (8.5.3), (8.6.6) - (8.6.9).$$

$$(8.6.10)$$

Theorem 8.6.1. *(Stability) Suppose (8.6.10) is feasible at the initial time $k = 0$. Then, (8.6.10) is feasible for any $k \geq 0$. Further, by receding-horizon implementing the optimal solution $u^*(k|k)$, the closed-loop system is exponentially stable.*

Proof. Suppose at time k there is a feasible solution $\{\tilde{u}(k)^*, Y(k)^*, Q(k)^*\}$, by which we obtain $\{x^{l_{i-1}(k)\cdots l_0(k)}(k+i|k)^*, i = 1 \ldots N, F(k)^*, P(k)^*\}$. Then, at time $k + 1$ the following is feasible:

$$u^{l_{i-2}(k+1)\cdots l_0(k+1)}(k+i|k+1) = \sum_{l_0(k)=1}^{L} \omega_{l_0(k)}(k) u^{l_{i-1}(k)\cdots l_0(k)}(k+i|k)^*,$$

$$i = 1 \ldots N - 1, \tag{8.6.11}$$

$$u^{l_{N-2}(k+1)\cdots l_0(k+1)}(k+N|k+1) = F(k)^* \tag{8.6.12}$$

$$\sum_{l_0(k)=1}^{L} \omega_{l_0(k)}(k) x^{l_{N-1}(k)\cdots l_0(k)}(k+N|k)^*, \tag{8.6.13}$$

$$u(k+i|k+1) = F(k)^* x(k+i|k)^*, \ i \geq N+1. \tag{8.6.14}$$

Applying (8.6.11) yields

$$u(k+i|k+1) = \sum_{l_0(k+1)\cdots l_{i-2}(k+1)=1}^{L} \left(\left(\prod_{h=0}^{i-2} \omega_{l_h(k+1)}(k+1+h) \right) \right.$$

$$\left. u^{l_{i-2}(k+1)\cdots l_0(k+1)}(k+i|k+1) \right)$$

$$= \sum_{l_0(k+1)\cdots l_{i-2}(k+1)=1}^{L} \left(\left(\prod_{h=0}^{i-2} \omega_{l_h(k+1)}(k+1+h) \right) \right.$$

$$\left. \left(\sum_{l_0(k)=1}^{L} \omega_{l_0(k)}(k) u^{l_{i-1}(k)\cdots l_0(k)}(k+i|k)^* \right) \right),$$

$$i \in \{1, \ldots, N-1\}.$$

Since $\omega_{l_{h+1}(k)}(k+1+h) = \omega_{l_h(k+1)}(k+1+h)$, we further obtain

$$
\begin{aligned}
u(k+i|k+1) &= \sum_{l_1(k)\cdots l_{i-1}(k)=1}^{L} \left(\left(\prod_{h=1}^{i-1} \omega_{l_h(k)}(k+h) \right) \right. \\
&\qquad \left. \left(\sum_{l_0(k)=1}^{L} \omega_{l_0(k)}(k) u^{l_{i-1}(k)\cdots l_0(k)}(k+i|k)^* \right) \right) \\
&= \sum_{l_0(k)\cdots l_{i-1}(k+1)=1}^{L} \left(\left(\prod_{h=0}^{i-1} \omega_{l_h(k)}(k+h) \right) u^{l_{i-1}(k+1)\cdots l_0(k)}(k+i|k)^* \right) \\
&= u(k+i|k)^*, \ i = 1 \ldots N-1.
\end{aligned}
$$

Analogously, applying (8.6.13) and $\omega_{l_{h+1}(k)}(k+1+h) = \omega_{l_h(k+1)}(k+1+h)$ yields

$$
u(k+N|k+1) = F(k)^* x(k+N|k)^*.
$$

Hence, (8.6.11)-(8.6.14) are equivalent to

$$
\begin{aligned}
u(k+i+1|k+1) &= u^*(k+i+1|k), \ i \in \{0, \ldots, N-2\}, \\
u(k+i|k+1) &= F^*(k)x^*(k+i|k), \ i \geq N.
\end{aligned}
$$

The continued proof is the same as Theorem 8.3.1. $\qquad\qquad\qquad\square$

Remark 8.6.1. In the algorithm of section 8.5, $\gamma_0, \gamma_1, \cdots, \gamma_{N-1}$ are adopted. However, in (8.6.10) a γ_1 is utilized to incorporate all γ_i's. These two paradigms are equivalent with respect to the feasibility. By adopting $\gamma_0, \gamma_1, \cdots, \gamma_{N-1}$, there are $N-1$ more decision variables. However, the dimensions of LMI are smaller, which simplifies coding. The algorithm in section 8.6 can also adopt $\gamma_0, \gamma_1, \cdots, \gamma_{N-1}$, while the algorithm in section 8.5 can also adopt a single γ_1.

Chapter 9

Open-loop optimization and closed-loop optimization in synthesis approaches

In this chapter, we continue the topic of MPC with switching horizon N. When $u(k|k), u(k+1|k), \cdots, u(k+N-1|k)$ are optimized and $N > 1$, one usually meets with MPC based on the open-loop optimization, i.e., open-loop MPC. MPC is always a closed-loop strategy when it is really implemented. When the states $x(k+2|k), x(k+3|k), \cdots, x(k+N|k)$ are being predicted, if the effect of closed-loop is not taken into consideration, then the corresponding optimization is called "open-loop optimization." The basic feature of open-loop MPC is "open-loop optimization, closed-loop control."

When $u(k|k), K(k+1|k), \cdots, K(k+N-1|k)$ (where K is the in-time state feedback gain) are optimized, one usually meets with MPC based on the closed-loop optimization, i.e., feedback MPC. When the states $x(k+2|k), x(k+3|k), \cdots, x(k+N|k)$ are being predicted, if the effect of closed-loop (i.e., the effect of feedback) is taken into consideration, then the corresponding optimization is called "closed-loop optimization." The basic feature of feedback MPC is "closed-loop optimization, closed-loop control."

Take the prediction of $x(k+2|k)$, based on the system $x(k+1) = Ax(k) + Bu(k)$, as an example. In the open-loop prediction,

$$x(k+2|k) = A^2 x(k) + ABu(k) + Bu(k+1|k);$$

in the closed-loop prediction,

$$x(k+2|k) = (A + BK(k+1|k))Ax(k) + (A + BK(k+1|k))Bu(k).$$

Apparently, if there is uncertainty, $K(k+1|k)$ can reduce the conservativeness in the predicted values.

For nominal systems, open-loop prediction and closed-loop prediction are equivalent. For uncertain systems, there is a large difference between the open-loop optimization and the closed-loop optimization. For $N > 1$, it is hard to directly solve MPC based on the closed-loop optimization. Usually, the partial closed-loop form is adopted, i.e., $u = Kx + c$ is defined where c is called the perturbation item.

Section 9.1 is referred to in [38]. Sections 9.2 and 9.3 are referred to in [10].

9.1   A simple approach based on partial closed-loop optimization

In Chapter 7, on-line approach for state feedback MPC has been given. It is actually MPC based on the closed-loop optimization with switching horizon 0. On-line approach incurs huge on-line computational burden. Although the corresponding off-line approach exists, the feasibility and optimality of off-line approach are greatly discounted.

This section introduces MPC with switching horizon $N \geq 1$, and before the switching horizon the control move is defined as $u = Kx + c$, where K is (off-line given) fixed state feedback gain.

9.1.1   Aim: achieving larger region of attraction

Consider the following time-varying polytopic uncertain system:

$$x(k+1) = A(k)x(k) + B(k)u(k), k \geq 0, \qquad (9.1.1)$$

where $u \in \mathfrak{R}^m$, $x \in \mathfrak{R}^n$ are input and measurable state, respectively. Suppose $[A(k)|B(k)]$ belongs to the convex hull of the set of extreme points $[A_l|B_l]$, $l \in \{1, \dots, L\}$:

$$[A(k)|B(k)] \in \Omega = Co\{[A_1|B_1],\ [A_2 B_2],\ \cdots,\ [A_L|B_L]\},\ \forall k \geq 0, \quad (9.1.2)$$

i.e., there exist L nonnegative coefficients $\omega_l(k)$, $l \in \{1, \dots, L\}$ such that

$$\sum_{l=1}^{L} \omega_l(k) = 1,\ \ [A(k)|B(k)] = \sum_{l=1}^{L} \omega_l(k)\,[A_l|B_l]. \qquad (9.1.3)$$

Differently from the former chapters we adopt the following constraints:

$$-\underline{g} \leq Gx(k) + Du(k) \leq \bar{g} \qquad (9.1.4)$$

where $\underline{g} := \left[\underline{g}_1, \underline{g}_2, \cdots, \underline{g}_q\right]^T$, $\bar{g} := [\bar{g}_1, \bar{g}_2, \cdots, \bar{g}_q]^T$, $\underline{g}_s > 0$, $\bar{g}_s > 0$, $s \in \{1, \dots, q\}$; $G \in \mathfrak{R}^{q \times n}$, $D \in \mathfrak{R}^{q \times m}$. Note that the state constraint $-\underline{\psi} \leq$

$\Psi x(k+i+1) \leq \bar{\psi}$ in the former two chapters can be expressed as (9.1.4), where

$$G = [A_1^T \Psi^T, A_2^T \Psi^T, \cdots, A_L^T \Psi^T]^T, \ D = [B_1^T \Psi^T, B_2^T \Psi^T, \cdots, B_L^T \Psi^T]^T;$$

$$g := \left[\psi^T, \psi^T, \cdots, \psi^T\right]^T, \ \bar{g} := \left[\bar{\psi}^T, \bar{\psi}^T, \cdots, \bar{\psi}^T\right]^T.$$

For on-line approach in Chapter 7, when (9.1.4) is considered one needs to substitute LMI for the constraints with

$$\begin{bmatrix} Q & * \\ GQ + DY & \Gamma \end{bmatrix} \geq 0, \ \Gamma_{ss} \leq g_{s,\text{inf}}^2, \ s \in \{1, \ldots, q\}, \tag{9.1.5}$$

where $g_{s,\text{inf}} = \min\{\underline{g}_s, \bar{g}_s\}$, Γ_{ss} is the s-th diagonal element of $\Gamma(k)$. Accordingly, on-line MPC solves the following optimization problem at each time k:

$$\min_{\gamma, Q, Y, \Gamma} \gamma, \ \text{s.t.} \ (9.1.5), (9.1.7) \ \text{and} \ (9.1.8), \tag{9.1.6}$$

$$\begin{bmatrix} Q & * & * & * \\ A_l Q + B_l Y & Q & * & * \\ W^{1/2} Q & 0 & \gamma I & * \\ R^{1/2} Y & 0 & 0 & \gamma I \end{bmatrix} \geq 0, \ l \in \{1, \ldots, L\}, \tag{9.1.7}$$

$$\begin{bmatrix} 1 & * \\ x(k) & Q \end{bmatrix} \geq 0, \tag{9.1.8}$$

and implement $u(k) = F(k)x(k) = YQ^{-1}x(k)$.

When the interior-point algorithm is adopted to solve (9.1.6), the computational burden is proportional to $\mathfrak{K}^3 \mathfrak{L}$, where $\mathfrak{K}$ is number of scalar variables in (9.1.6), and $\mathfrak{L}$ is the number of rows. For (9.1.6), $\mathfrak{K} = \frac{1}{2}(n^2 + n) + mn + \frac{1}{2}(q^2 + q) + 1$, $\mathfrak{L} = (3n + m)L + 2n + 2q + 1$. Hence, increasing L linearly increases the computational burden.

In on-line approach, (9.1.6) is solved at each time k. Due to this reason, on-line approach can only be applied on slow dynamics and low dimensional systems. Based on (9.1.6), we can easily obtain the corresponding off-line approach.

Algorithm 9.1 (Off-line MPC)

Stage 1. Off-line, choose states x_i, $i \in \{1, \ldots, N\}$. Substitute $x(k)$ in (9.1.8) by x_i, and solve (9.1.6) to obtain the corresponding matrices $\{Q_i, Y_i\}$, ellipsoids $\varepsilon_{x,i} = \{x \in \mathfrak{R}^n | x^T Q_i^{-1} x \leq 1\}$ and feedback gains $F_i = Y_i Q_i^{-1}$. Note that x_i should be chosen such that $\varepsilon_{x,j} \subset \varepsilon_{x,j-1}, \forall j \in \{2, \ldots, N\}$. For each $i \neq N$, check if the following is satisfied:

$$Q_i^{-1} - (A_l + B_l F_{i+1})^T Q_i^{-1} (A_l + B_l F_{i+1}) > 0, \ l \in \{1, \ldots, L\}. \tag{9.1.9}$$

Stage 2. On-line, at each time k, adopt the following state feedback law:

$$u(k) = F(k)x(k) = \begin{cases} F(\alpha(k))x(k), & x(k) \in \varepsilon_{x,i}, \ x(k) \notin \varepsilon_{x,i+1}, \ i \neq N \\ F_N x(k), & x(k) \in \varepsilon_{x,N} \end{cases}$$

$$(9.1.10)$$

where $F(\alpha(k)) = \alpha(k)F_i + (1 - \alpha(k))F_{i+1}$, and

i) if (9.1.9) is satisfied, then $0 < \alpha(k) \leq 1$, and $x(k)^T \left[\alpha(k)Q_i^{-1} + (1 - \alpha(k))Q_{i+1}^{-1}\right]x(k) = 1$;

ii) if (9.1.9) is not satisfied, then $\alpha(k) = 1$.

Now, suppose a fixed feedback gain K is obtained which asymptotically stabilizes (9.1.1) for the unconstrained case (for example, we can give an $x(k)$, and solve (9.1.6) to obtain $K = YQ^{-1}$). Based on K, we can find a matrix $Q_{x,\chi}$ larger than Q. The ellipsoidal set corresponding to $Q_{x,\chi}$ is the region of attraction of the new approach in this section.

9.1.2   Efficient algorithm

Define

$$u(k + i|k) = Kx(k + i|k) + c(k + i|k), \ c(k + n_c + i|k) = 0, \ \forall i \geq 0 \quad (9.1.11)$$

where n_c is the switching horizon of the new approach (although it is not denoted as N). Thus, the state predictions are represented by

$$x(k+i+1|k) = \mathcal{A}(k+i)x(k+i|k) + B(k+i)c(k+i|k), \ x(k|k) = x(k) \quad (9.1.12)$$

where $\mathcal{A}(k + i) = A(k + i) + B(k + i)K$.

Equation (9.1.12) is equivalent to the following autonomous state space model:

$$\chi(k + i + 1|k) = \Phi(k + i)\chi(k + i|k),$$

$$\Phi(k + i) = \begin{bmatrix} \mathcal{A}(k + i) & [B(k + i) & 0 & \cdots & 0] \\ 0 & & \Pi & \end{bmatrix},$$

$$(9.1.13)$$

$$\chi = \begin{bmatrix} x \\ f \end{bmatrix}, \ f(k + i|k) = \begin{bmatrix} c(k + i|k) \\ c(k + 1 + i|k) \\ \vdots \\ c(k + n_c - 1 + i|k) \end{bmatrix}, \ \Pi = \begin{bmatrix} 0_m & I_m & 0_m & \cdots & 0_m \\ 0_m & 0_m & I_m & \ddots & \vdots \\ 0_m & 0_m & 0_m & \ddots & 0_m \\ \vdots & \ddots & \ddots & \ddots & I_m \\ 0_m & \cdots & 0_m & 0_m & 0_m \end{bmatrix},$$

$$(9.1.14)$$

where 0_m is an m-ordered zero matrix and I_m is an m-ordered identity matrix. Consider the following ellipsoid:

$$\varepsilon_\chi = \left\{\chi \in \mathfrak{R}^{n+mn_c} | \chi^T Q_\chi^{-1} \chi \leq 1\right\}. \quad (9.1.15)$$

Denote $Q_\chi^{-1} = \begin{bmatrix} \hat{Q}_{11} & \hat{Q}_{21}^T \\ \hat{Q}_{21} & \hat{Q}_{22} \end{bmatrix}$, $\hat{Q}_{11} \in \Re^{n \times n}$, $\hat{Q}_{21} \in \Re^{mn_c \times n}$, $\hat{Q}_{22} \in \Re^{mn_c \times mn_c}$ and define

$$\varepsilon_{x,\chi} = \{x \in \Re^n | x^T Q_{x,\chi}^{-1} x \le 1\}, \quad Q_{x,\chi} = \left[\hat{Q}_{11} - \hat{Q}_{21}^T \hat{Q}_{22}^{-1} \hat{Q}_{21}\right]^{-1} = TQ_\chi T^T,$$
(9.1.16)

where T is defined by $x = T\chi$. The invariance of ε_χ ($\varepsilon_{x,\chi}$) is ensured by

$$\Phi_l^T Q_\chi^{-1} \Phi_l - Q_\chi^{-1} \le 0 \Leftrightarrow \begin{bmatrix} Q_\chi & Q_\chi \Phi_l^T \\ \Phi_l Q_\chi & Q_\chi \end{bmatrix} \ge 0, \; l \in \{1,\ldots,L\}. \quad (9.1.17)$$

Below we explain, when the state lies inside of $\varepsilon_{x,\chi}$, how to guarantee the satisfaction of hard constraints. When (9.1.17) is satisfied, $\varepsilon_{x,\chi}$ is an invariant ellipsoid. Let ξ_s be the s-th row of the q-ordered identity matrix, E_m be the first m rows of the mn_c-ordered identity matrix. We can make the following deductions:

$$\max_{i \ge 0} |\xi_s[Gx(k+i|k) + Du(k+i|k)]|$$

$$= \max_{i \ge 0} |\xi_s[(G+DK)x(k+i|k) + Dc(k+i|k)]|$$

$$= \max_{i \ge 0} |\xi_s[(G+DK)x(k+i|k) + DE_m f(k+i|k)]|$$

$$= \max_{i \ge 0} |\xi_s[G+DK \quad DE_m]\chi(k+i|k)|$$

$$= \max_{i \ge 0} \left|\xi_s[G+DK \quad DE_m]Q_\chi^{1/2} Q_\chi^{-1/2}\chi(k+i|k)\right|$$

$$\le \max_{i \ge 0} \left\|\xi_s[G+DK \quad DE_m]Q_\chi^{1/2}\right\| \left\|Q_\chi^{-1/2}\chi(k+i|k)\right\|$$

$$\le \left\|\xi_s[G+DK \quad DE_m]Q_\chi^{1/2}\right\|.$$

Hence, if the following is satisfied:

$$\left\|\xi_s[G+DK \quad DE_m]Q_\chi^{1/2}\right\| \le g_{s,\text{inf}} \Leftrightarrow g_{s,\text{inf}}^2 -$$

$$[(G+DK)_s \quad D_s E_m] Q_\chi [(G+DK)_s \quad D_s E_m]^T \ge 0, \; s \in \{1,\ldots,q\},$$
(9.1.18)

then $|\xi_s[Gx(k+i|k) + Du(k+i|k)]| \le g_{s,\text{inf}}, \; s \in \{1,\ldots,q\}$. In (9.1.18), $(G+DK)_s$ (D_s) is the s-th row of $G+DK$ (D).

Since the aim is to obtain a larger region of attraction, we can take the maximization of the volume of $\varepsilon_{x,\chi}$ as the criterion of MPC. Maximization of the volume of $\varepsilon_{x,\chi}$ is equivalent to the maximization of $\det(TQ_\chi T^T)$. Then, Q_χ can be computed by:

$$\min_{Q_\chi} \log \det(TQ_\chi T^T)^{-1}, \; \text{s.t. } (9.1.17) - (9.1.18). \quad (9.1.19)$$

Algorithm 9.2

Stage 1. Off-line, ignoring constraints, compute K so as to optimize certain robust performance (e.g., we can adopt (9.1.6) to compute a K). Obtain Q_χ by solving (9.1.19). Increase n_c, repeat (9.1.19), until $\varepsilon_{x,\chi}$ is satisfactory in size.

Stage 2. On-line, at each time k, perform the minimization:

$$\min_f f^T f, \text{ s.t. } \chi^T Q_\chi^{-1} \chi \leq 1. \tag{9.1.20}$$

and implement $u(k) = Kx(k) + c(k|k)$.

If $x(0) \in \varepsilon_{x,\chi}$, then $f(0) = -\hat{Q}_{22}^{-1}\hat{Q}_{21}x(0)$ is feasible for (9.1.20) since by this solution, $\chi(0)^T Q_\chi^{-1} \chi(0) \leq 1$ leads to

$$
\begin{aligned}
x(0)^T \hat{Q}_{11} x(0) &\leq 1 - 2f(0)^T \hat{Q}_{21} x(0) - f(0)^T \hat{Q}_{22} f(0) \\
&= 1 + x(0)^T \hat{Q}_{21}^T \hat{Q}_{22}^{-1} \hat{Q}_{21} x(0)
\end{aligned}
\tag{9.1.21}
$$

which is equivalent to $x(0)^T Q_{x,\chi}^{-1} x(0) \leq 1$. For $x(0) \notin \varepsilon_{x,\chi}$, there does not exist $f(0)$ such that $\chi(0)^T Q_\chi^{-1} \chi(0) \leq 1$. For $f(0) = 0$, $\chi(0)^T Q_\chi^{-1} \chi(0) \leq 1$ leads to $x(0)^T \hat{Q}_{11} x(0) \leq 1$.

Theorem 9.1.1. *(Stability) Suppose there exist K and n_c which make (9.1.19) feasible. When $x(0) \in \varepsilon_{x,\chi}$, by adopting Algorithm 9.2, the constraint (9.1.4) is always satisfied and the closed-loop system is asymptotically stable.*

Proof. Firstly, when $x(0) \in \varepsilon_{x,\chi}$, the feasible solution exists. Let $f(0)^*$ be the solution at the initial time. Since $\varepsilon_{x,\chi}$ is an invariant set, at time $k+1$, $f(1) = \Pi f(0)^*$ is a feasible solution which yields a smaller cost value (than that corresponding to $f(0)^*$). Certainly, $f(1) = \Pi f(0)^*$ is not necessarily the optimal solution. When the optimal solution $f(1)^*$ is obtained, a smaller cost value (than that corresponding to $f(1)$) is obtained. By analogy, we know that the feasibility of (9.1.20) at time k guarantees its feasibility at any $k > 0$, and the cost value is monotonically decreasing with the evolution of time. Therefore, by applying Algorithm 9.2, the perturbation item will gradually become zero, and the constraints will be always satisfied. When the perturbation item becomes zero, the control move becomes $u = Kx$ which can drive the state to the origin. $\square$

Algorithm 9.2 involves on-line optimization. Hence, its on-line computational burden is heavier than off-line approach. However, by properly tuning K and n_c, it is easy to make the region of attraction of Algorithm 9.2 larger than that of off-line approach (compared with respect to the volume). Moreover, (9.1.20) is easy to solve and its computational burden is greatly lower than that of on-line approach.

In general, on-line approach yields non-ellipsoidal region of attraction. By adopting Algorithm 9.2 or off-line approach, one can only obtain the ellipsoidal region of attraction. An ellipsoidal region of attraction is conservative in volume.

9.2 Triple-mode approach

Here, "mode" refers to the pattern of control move computation. Since the volume of $\varepsilon_{x,\chi}$ is conservative, when the state lies outside of $\varepsilon_{x,\chi}$, we can adopt the standard approach of predictive control.

Suppose $K = F_1$, where F_1 is the off-line state feedback gain in Algorithm 9.1. Then $\hat{\varepsilon}_1 = \{x | x^T \hat{Q}_{11} x \leq 1\}$ is a region of attraction for F_1, $\hat{\varepsilon}_1$ is invariant, and $Q(k) = \hat{Q}_{11}^{-1}$ is a feasible solution for (9.1.6) with $x(k) = x_1$. Note that the region of attraction of the previous off-line approach is $\varepsilon_{x,1}$ (rather than $\hat{\varepsilon}_1$), while the region of attraction of Algorithm 9.2 is $\varepsilon_{x,\chi}$.

Clearly, $\varepsilon_{x,\chi} \supseteq \hat{\varepsilon}_1$. However, we can neither ensure $\varepsilon_{x,1} \supseteq \hat{\varepsilon}_1$ nor guarantee $\varepsilon_{x,1} \subseteq \hat{\varepsilon}_1$. $\varepsilon_{x,\chi} \supseteq \varepsilon_{x,1}$ if and only if $\hat{Q}_{11} - \hat{Q}_{21}^T \hat{Q}_{22}^{-1} \hat{Q}_{21} \leq Q_1^{-1}$. If we choose $K = F_1$, then we can obtain $\varepsilon_{x,\chi} \supseteq \varepsilon_{x,1}$ by tuning n_c and Q_χ.

In triple-mode MPC, the free perturbation items will be utilized, which can enlarge the region of attraction of the closed-loop system. Hence, for simplicity, one does not have to select (9.1.19) to maximize $\varepsilon_{x,\chi}$; one can choose $K = F_1$ and maximize Q_χ in the following simple way:

$$\min_{\rho, Q_\chi} \rho, \text{ s.t. } (9.1.17) - (9.1.18) \text{ and,} \tag{9.2.1}$$

$$\rho T Q_\chi T^T - Q_1 \geq 0 \Leftrightarrow \begin{bmatrix} T Q_\chi T^T & * \\ Q_1^{1/2} & \rho I \end{bmatrix} \geq 0. \tag{9.2.2}$$

Note that (9.2.2) imposes $\varepsilon_{x,\chi} \supseteq \frac{1}{\rho} \cdot \varepsilon_{x,1}$. Thus, by minimizing ρ, $\varepsilon_{x,\chi}$ can be maximized in some sense. If $\rho < 1$, then $\varepsilon_{x,\chi} \supset \varepsilon_{x,1}$. Note that, by applying (9.1.19), it is not necessary that $\varepsilon_{x,\chi} \supset \varepsilon_{x,1}$, which is the main reason for us to adopt (9.2.1)-(9.2.2).

Define $\mathcal{A}(\cdot) = A(\cdot) + B(\cdot)K$. Then, the polytopic description of $[\mathcal{A}(\cdot) | B(\cdot)]$ can inherit that of $[A(\cdot) | B(\cdot)]$. The prediction of the state is given by:

$$\begin{bmatrix} x(k+1|k) \\ x(k+2|k) \\ \vdots \\ x(k+\bar{N}|k) \end{bmatrix} = \sum_{l_0 \cdots l_{\bar{N}-1}=1}^{L} \left\{ \left(\prod_{h=0}^{\bar{N}-1} \omega_{l_h}(k+h) \right) \begin{bmatrix} x^{l_0}(k+1|k) \\ x^{l_1 l_0}(k+2|k) \\ \vdots \\ x^{l_{\bar{N}-1} \cdots l_1 l_0}(k+\bar{N}|k) \end{bmatrix} \right\}, \tag{9.2.3}$$

where

$$
\begin{bmatrix}
x^{l_0}(k+1|k) \\
x^{l_1 l_0}(k+2|k) \\
\vdots \\
x^{l_{\bar{N}-1}\cdots l_1 l_0}(k+\bar{N}|k)
\end{bmatrix}
=
\begin{bmatrix}
\mathcal{A}_{l_0} \\
\mathcal{A}_{l_1}\mathcal{A}_{l_0} \\
\vdots \\
\prod_{i=0}^{\bar{N}-1}\mathcal{A}_{l_{\bar{N}-1-i}}
\end{bmatrix}
x(k)
$$

$$
+
\begin{bmatrix}
B_{l_0} & 0 & \cdots & 0 \\
\mathcal{A}_{l_1}B_{l_0} & B_{l_1} & \ddots & \vdots \\
\vdots & \vdots & \ddots & 0 \\
\prod_{i=0}^{\bar{N}-2}\mathcal{A}_{l_{\bar{N}-1-i}}B_{l_0} & \prod_{i=0}^{\bar{N}-3}\mathcal{A}_{l_{\bar{N}-1-i}}B_{l_1} & \cdots & B_{l_{\bar{N}-1}}
\end{bmatrix}
\begin{bmatrix}
c(k|k) \\
c(k+1|k) \\
\vdots \\
c(k+\bar{N}-1|k)
\end{bmatrix}.
$$

$$(9.2.4)$$

As in Chapter 8, $x^{l_{i-1}\cdots l_1 l_0}(k+i|k)$ $(i \in \{1,\cdots,\bar{N}\})$ are called "vertex state predictions."

Algorithm 9.3 (Triple-mode robust MPC)

Off-line, choose K, n_c and $\bar{N}$. Solve problem (9.2.1)-(9.2.2) to obtain matrices Q_χ, $Q_{x,\chi}$ and ellipsoidal set $\varepsilon_{x,\chi}$.

On-line, at each time k,

i) if $x(k) \in \varepsilon_{x,\chi}$, perform (9.1.20);

ii) if $x(k) \notin \varepsilon_{x,\chi}$, solve:

$$
\min_{c(k),c(k+1|k),\cdots,c(k+\bar{N}+n_c-1|k)} J(k) =
$$
$$
\left\| [c(k|k)^T \ c(k+1|k)^T \ \cdots \ c(k+\bar{N}+n_c-1|k)^T] \right\|_2^2, \qquad (9.2.5)
$$
$$
\text{s.t. } -\underline{g} \le (G+DK)x(k) + Dc(k) \le \bar{g},
$$
$$
-\underline{g} \le (G+DK)x^{l_{i-1}\cdots l_0}(k+i|k) + Dc(k+i|k) \le \bar{g},
$$
$$
\forall i \in \{1,\ldots,\bar{N}-1\}, \ l_{i-1} \in \{1,\ldots,L\}, \qquad (9.2.6)
$$
$$
\left\| [x(k+\bar{N}|k)^T \ c(k+\bar{N}|k)^T \ \cdots \ c(k+\bar{N}+n_c-1|k)^T] \right\|_{Q_\chi^{-1}}^2
$$
$$
\le 1, \ (9.2.4), \ \forall l_0,\cdots,l_{\bar{N}-1} \in \{1,\ldots,L\}. \qquad (9.2.7)
$$

Then, implement $u(k) = Kx(k) + c(k|k)$.

From Algorithm 9.3, we can see that triple-mode MPC belongs to MPC based on the partial closed-loop optimization.

Theorem 9.2.1. *(Stability) Suppose: (i) $x(0) \in \varepsilon_{x,\chi}$, or (ii) $x(0) \notin \varepsilon_{x,\chi}$ but (9.2.5)-(9.2.7) has a feasible solution. Then by applying Algorithm 9.3, the constraint (9.1.4) is always satisfied and the closed-loop system is asymptotically stable.*

Proof. The details for satisfaction of the constraint (9.1.4) are omitted here. Suppose (i) holds, then according to Theorem 9.1.1, the closed-loop system is stable. Suppose (ii) holds, and (9.2.5)-(9.2.7) has a feasible solution

$$\left\{ c(k|k)^*, \ c(k+1|k)^*, \ \cdots, \ c(k+\bar{N}+n_c-1|k)^* \right\} \qquad (9.2.8)$$

at time k. Denote the corresponding performance cost under (9.2.8) as $J^*(k)$. Due to (9.2.7), (9.2.8) guarantees $x(k+\bar{N}+j|k) \in \varepsilon_{x,\chi}, \ 0 \le j \le n_c$. Hence, at time $k+1$, the following solution is feasible for (9.2.5)-(9.2.7):

$$\left\{ c(k+1|k)^*, \ c(k+2|k)^*, \ \cdots, \ c(k+\bar{N}+n_c-1|k)^*, \ 0 \right\}. \qquad (9.2.9)$$

By applying (9.2.9) at time $k+1$ ((9.2.9) may not be actually adopted), the resultant performance cost is

$$J(k+1) = J^*(k) - c(k|k)^{*T} c(k|k)^*. \qquad (9.2.10)$$

After $J(k+1)$ is optimized, the optimum $J^*(k+1) \le J(k+1) \le J^*(k)$. Therefore, $J^*(k)$ will be monotonically decreasing such that the state will be finally driven into $\varepsilon_{x,\chi}$ inside of which (i) becomes true. □

The three modes of triple mode MPC in Algorithm 9.3 are: (i) $f = 0$, $u = Kx$ inside of $\hat{\varepsilon}_1$; (ii) $u = Kx + c$, $f \ne 0$ inside of $\varepsilon_{x,\chi} \backslash \hat{\varepsilon}_1$; (iii) $u = Kx + c$ outside of $\varepsilon_{x,\chi}$.

Proposition 9.2.1. *(Monotonicity) Consider Algorithm 9.3. By increasing either n_c or $\bar{N}$, the region of attraction of the overall closed-loop system will not shrink, i.e., the region of attraction by increasing n_c or $\bar{N}$ always includes the original.*

Proof. Suppose $\left\{ c(k|k)^*, \ c(k+1|k)^*, \ \cdots, \ c(k+\bar{N}+n_c-1|k)^* \right\}$ is a feasible solution for $\{\bar{N}, n_c\}$, then $\{c(k|k)^*, \ c(k+1|k)^*, \ \cdots, \ c(k+\bar{N}+n_c - 1|k)^*, 0\}$ is a feasible solution when $\{\bar{N}, n_c\}$ are replaced by $\{\bar{N}+1, n_c\}$ or $\{\bar{N}, n_c+1\}$. □

In general, it is not necessary to take a large n_c. For reasonable n_c, the on-line computation is efficient when $x(k) \in \varepsilon_{x,\chi}$. However, $\varepsilon_{x,\chi}$ may be unsatisfactory in volume since its shape is restricted as ellipsoid. In general, compared with increasing n_c, increasing $\bar{N}$ is more efficient for expanding the region of attraction.

However, the selection of $\bar{N}$ should be more careful in order to compromise between the desired region of attraction and the on-line computational burden. For any $\bar{N} \ge 1$, the overall region of attraction of the closed-loop system is not smaller than $\varepsilon_{x,\chi}$ in any direction. However, by increasing $\bar{N}$, the computational burden increases exponentially.

It is easy to transform (9.2.5)-(9.2.7) into an LMI optimization problem. By the fastest interior-point algorithms, the computational complexity involved in this LMI optimization is proportional to $\mathfrak{K}^3 \mathfrak{L}$ where $\mathfrak{K} =$

$(n_c + \bar{N})m + 1$, $\mathcal{L} = (n_c m + n + 1)L^{\bar{N}} + 2q \sum_{i=1}^{\bar{N}-1} L^i + (n_c + \bar{N})m + 2q + 1$. Hence, increasing n, q only linearly increases the computational complexity. For larger $\{n, q\}$ and smaller $\{L, \bar{N}, n_c\}$, the computation involved in this optimization can be less expensive than that involved in (9.1.6).

9.3   Mixed approach

After applying triple-mode control, there is still conservativeness since the single-valued perturbation items $c(k|k)$, $c(k+1|k)$, $\cdots$, $c(k+\bar{N}-1|k)$ have to deal with all possible state evolutions. In this section, in order to achieve larger region of attraction, when the state lies outside of $\varepsilon_{x,\chi}$ we adopt the "vertex perturbation items"; in order to lower down the on-line computational burden, inside of $\varepsilon_{x,1}$, the off-line approach is adopted. Through these two means, the achieved region of attraction can be mutual-complementary with that of on-line approach; at the same time, the on-line computational burden is much smaller than that of on-line approach.

Here, the so-called "mixed" means that there are both partial closed-loop optimization and closed-loop optimization, and there are both standard approach and off-line approach.

9.3.1   Algorithm

In Chapter 8, we have adopted the following "within-horizon feedback":

$$\widehat{u}\,(k) \;=\; \left\{ u(k|k), \; \sum_{l_0=1}^{L} \omega_{l_0}(k) u^{l_0}(k+1|k), \; \cdots, \right.$$

$$\left. \sum_{l_0 \cdots l_{\bar{N}-2}=1}^{L} \left(\prod_{h=0}^{\bar{N}-2} \omega_{l_h}(k+h) \right) u^{l_{\bar{N}-2} \cdots l_0}(k+\bar{N}-1|k) \right\}$$

and optimized

$$\tilde{u}(k) \;=\; \left\{ u(k|k), \; u^{l_0}(k+1|k), \; \cdots, \; u^{l_{\bar{N}-2} \cdots l_0}(k+\bar{N}-1|k)|l_0, \cdots, \right.$$
$$\left. l_{\bar{N}-2} = 1 \ldots L \right\}.$$

Similarly to Chapter 8, in the method of this chapter, outside of $\varepsilon_{x,\chi}$, the following "vertex perturbation item" and "parameter-dependent perturbation item" are utilized:

$$\tilde{c}(k) = \{c(k|k),\ c^{l_0}(k+1|k),\ \cdots,\ c^{l_{\bar{N}-2}\cdots l_0}(k+\bar{N}-1|k)|l_0,\cdots,$$
$$l_{\bar{N}-2} = 1\ldots L\}, \tag{9.3.1}$$

$$\widehat{c}(k) = \left\{ c(k|k),\ \sum_{l_0=1}^{L} \omega_{l_0}(k)c^{l_0}(k+1|k),\ \cdots, \right.$$

$$\left. \sum_{l_0\cdots l_{\bar{N}-2}=1}^{L} \left(\prod_{h=0}^{\bar{N}-2} \omega_{l_h}(k+h) \right) c^{l_{\bar{N}-2}\cdots l_0}(k+\bar{N}-1|k) \right\}. \tag{9.3.2}$$

Equation (9.3.2) takes perturbation items for all the vertices of the uncertain state predictions.

The prediction of the state is given by (9.2.3), where

$$\begin{bmatrix} x^{l_0}(k+1|k) \\ x^{l_1 l_0}(k+2|k) \\ \vdots \\ x^{l_{\bar{N}-1}\cdots l_1 l_0}(k+\bar{N}|k) \end{bmatrix} = \begin{bmatrix} A_{l_0} \\ A_{l_1}A_{l_0} \\ \vdots \\ \prod_{i=0}^{\bar{N}-1} A_{l_{\bar{N}-1-i}} \end{bmatrix} x(k)$$

$$+ \begin{bmatrix} B_{l_0} & 0 & \cdots & 0 \\ A_{l_1}B_{l_0} & B_{l_1} & \ddots & \vdots \\ \vdots & \vdots & \ddots & 0 \\ \prod_{i=0}^{\bar{N}-2} A_{l_{\bar{N}-1-i}}B_{l_0} & \prod_{i=0}^{\bar{N}-3} A_{l_{\bar{N}-1-i}}B_{l_1} & \cdots & B_{l_{\bar{N}-1}} \end{bmatrix}$$

$$\times \begin{bmatrix} c(k|k) \\ c^{l_0}(k+1|k) \\ \vdots \\ c^{l_{\bar{N}-2}\cdots l_1 l_0}(k+\bar{N}-1|k) \end{bmatrix}. \tag{9.3.3}$$

In view of (9.3.1), (9.3.2), (9.2.3) and (9.3.3), we revise problem (9.2.5)-(9.2.7) as:

$$\min_{\tilde{c}(k),c(k+\bar{N}|k),\cdots,c(k+\bar{N}+n_c-1|k)} \max_{[A(k+i)|B(k+i)]\in\Omega, i\in\{0,\ldots,N-1\}} J(k)$$

$$= \left\| [c(k|k)^T \cdots c(k+\bar{N}-1|k)^T \cdots c(k+\bar{N}+n_c-1|k)^T] \right\|_2^2, \tag{9.3.4}$$

s.t. $-\underline{g} \leq (G+DK)x(k) + Dc(k) \leq \bar{g},$

$\quad -\underline{g} \leq (G+DK)x^{l_{i-1}\cdots l_0}(k+i|k) + Dc^{l_{i-1}\cdots l_0}(k+i|k) \leq \bar{g},$

$\quad \forall i \in \{1,\ldots,\bar{N}-1\},\ l_{i-1} \in \{1,\ldots,L\}, \tag{9.3.5}$

$\left\| [x(k+\bar{N}|k)^T\ c(k+\bar{N}|k)^T\ \cdots\ c(k+\bar{N}+n_c-1|k)^T] \right\|_{Q_x^{-1}}^2$

$\quad \leq 1,\ \forall l_0,\cdots,l_{\bar{N}-1} \in \{1,\ldots,L\}. \tag{9.3.6}$

Note that in (9.3.4), $\{c(k+1|k),\cdots,c(k+\bar{N}-1|k)\}$ are parameter-dependent perturbation items, which are uncertain values. Hence, the optimization problem is of "min-max" form.

For solving (9.3.4)-(9.3.6), let us define

$$\left\|[\tilde{c}^{l_0\cdots l_{\bar{N}-2}}(k)^T \quad c(k+\bar{N}|k)^T \quad \cdots \quad c(k+\bar{N}+n_c-1|k)^T]\right\|_2^2 \le \eta,$$
$$\forall l_0,\cdots,l_{\bar{N}-2} \in \{1,\ldots,L\}, \tag{9.3.7}$$

where $\tilde{c}^{l_0\cdots l_{\bar{N}-2}}(k) = \left[c(k|k)^T, \ c^{l_0}(k+1|k)^T, \ \cdots, \ c^{l_{\bar{N}-2}\cdots l_0}(k+\bar{N}-1|k)^T\right]^T$, and η is a scalar. By applying (9.2.3), (9.3.3) and Schur complement, (9.3.6) and (9.3.7) can be transformed into the following LMIs:

$$\begin{bmatrix} 1 & * \\ [x^{l_{\bar{N}-1}\cdots l_1 l_0}(k+\bar{N}|k)^T \ c(k+\bar{N}|k)^T \ c(k+\bar{N}+n_c-1|k)^T]^T & Q_\chi \end{bmatrix} \ge 0,$$

$$x^{l_{\bar{N}-1}\cdots l_1 l_0}(k+\bar{N}|k) = \prod_{i=0}^{\bar{N}-1} A_{l_{\bar{N}-1-i}} x(k) + \prod_{i=0}^{\bar{N}-2} A_{l_{\bar{N}-1-i}} B_{l_0} c(k|k) + \cdots + B_{l_{\bar{N}-1}}$$

$$c^{l_{\bar{N}-2},\ \cdots\ l_0}(k+\bar{N}-1|k), \ \forall l_0,\cdots,l_{\bar{N}-1} = \{1,\ldots,L\}, \tag{9.3.8}$$

$$\begin{bmatrix} \eta & * \\ [\tilde{c}^{l_0\cdots l_{\bar{N}-2}}(k)^T \ c(k+\bar{N}|k)^T \ \cdots \ c(k+\bar{N}+n_c-1|k)^T]^T & I \end{bmatrix} \ge 0,$$
$$\forall l_0,\cdots,l_{\bar{N}-2} \in \{1,\ldots,L\}. \tag{9.3.9}$$

Constraint (9.3.5) can be transformed into the following LMI:

$$-\begin{bmatrix} \underline{g} \\ \underline{g} \\ \vdots \\ \underline{g} \end{bmatrix} \le \tilde{G} \begin{bmatrix} I \\ A_{l_0} \\ \vdots \\ \prod_{i=0}^{\bar{N}-2} A_{l_{\bar{N}-2-i}} \end{bmatrix} x(k)$$

$$+ \left\{ \tilde{G} \begin{bmatrix} 0 & 0 & \cdots & 0 \\ B_{l_0} & 0 & \ddots & \vdots \\ \vdots & \ddots & \ddots & 0 \\ \prod_{i=0}^{\bar{N}-3} A_{l_{\bar{N}-2-i}} B_{l_0} & \cdots & B_{l_{\bar{N}-2}} & 0 \end{bmatrix} + \tilde{D} \right\}$$

$$\times \begin{bmatrix} c(k|k) \\ c^{l_0}(k+1|k) \\ \vdots \\ c^{l_{\bar{N}-2}\cdots l_1 l_0}(k+\bar{N}-1|k) \end{bmatrix} \le \begin{bmatrix} \bar{g} \\ \bar{g} \\ \vdots \\ \bar{g} \end{bmatrix},$$

$$\tilde{G} = \begin{bmatrix} G+DK & 0 & \cdots & 0 \\ 0 & G+DK & \ddots & \vdots \\ \vdots & \ddots & \ddots & 0 \\ 0 & \cdots & 0 & G+DK \end{bmatrix}, \quad \tilde{D} = \begin{bmatrix} D & 0 & \cdots & 0 \\ 0 & D & \ddots & \vdots \\ \vdots & \ddots & \ddots & 0 \\ 0 & \cdots & 0 & D \end{bmatrix},$$

$$\forall l_0, \cdots, l_{\bar{N}-2} \in \{1, \ldots, L\}. \tag{9.3.10}$$

In this way, problem (9.3.4)-(9.3.6) is transformed into the following LMI optimization problem:

$$\min_{\eta, \tilde{c}(k), c(k+\bar{N}|k), \cdots, c(k+\bar{N}+n_c-1|k)} \eta, \text{ s.t. } (9.3.8) - (9.3.10). \tag{9.3.11}$$

Algorithm 9.4 (Mixed robust MPC)

Stage 1. See Stage 1 of Algorithm 9.1.

Stage 2. Off-line, choose $K = F_1$, n_c and $\bar{N}$. Solve problem (9.2.1)-(9.2.2) to obtain the matrix Q_χ, $Q_{x,\chi}$ and ellipsoidal set $\varepsilon_{x,\chi}$.

Stage 3. On-line, at each time k,

    (a) if $x(k) \in \varepsilon_{x,1}$, see Stage 2 of Algorithm 9.1;

    (b) if $x(k) \in \varepsilon_{x,\chi} \backslash \varepsilon_{x,1}$, then perform (9.1.20) and implement $u(k) = Kx(k) + c(k|k)$;

    (c) if $x(k) \notin \varepsilon_{x,\chi}$, then solve (9.3.11) and implement $u(k) = Kx(k) + c(k|k)$.

In Algorithm 9.4, $\varepsilon_{x,\chi} \supseteq \varepsilon_{x,1}$ can be guaranteed. However, by extensively choosing x_i, it may occur that $\varepsilon_{x,\chi}$ becomes very close to $\varepsilon_{x,1}$. In this case, we can remove Stage 3(b) from Algorithm 9.4 and revise Stage 3(c) accordingly.

Theorem 9.3.1. *(Stability) Suppose: (i) $x(0) \in \varepsilon_{x,1}$, or (ii) $x(0) \in \varepsilon_{x,\chi} \backslash \varepsilon_{x,1}$, or (iii) $x(0) \notin \varepsilon_{x,\chi}$ but (9.3.11) has a feasible solution. Then by applying Algorithm 9.4, the constraint (9.1.4) is always satisfied and the closed-loop system is asymptotically stable.*

Proof. The details for satisfaction of constraint are omitted here. Suppose (i) holds. Then according to Stage 3(a) and off-line approach in Chapter 7, the state will be driven to the origin. Suppose (ii) holds. Then the state will be driven into $\varepsilon_{x,1}$ according to Stage 3(b) and stability of Algorithm 9.2, and (i) becomes true.

Suppose (iii) holds and (9.3.11) has a feasible solution at time k:

$$\{\tilde{c}(k)^*, \ c(k+\bar{N}|k)^*, \ \cdots, \ c(k+\bar{N}+n_c-1|k)^*\}. \tag{9.3.12}$$

Denote the corresponding performance cost under (9.3.12) as $J^*(k)$. Due to (9.3.8), according to Algorithm 9.3 the solution (9.3.12) guarantees $x(k+\bar{N}+$

$j|k) \in \varepsilon_{x,\chi}$, $0 \leq j \leq n_c$. Hence, at time $k+1$, the following solution is feasible for (9.3.11):

$$\{\tilde{c}(k+1),\ c(k+\bar{N}+1|k)^*,\ \cdots,\ c(k+\bar{N}+n_c-1|k)^*,\ 0\} \qquad (9.3.13)$$

where, for constructing $\tilde{c}(k+1)$,

$$\tilde{c}^{l_0\cdots l_{\bar{N}-2}}(k+1) \quad = \quad \left[\sum_{l_0=1}^{L} \omega_{l_0}(k)c^{l_0}(k+1|k)^{*T},\ \cdots,\right.$$

$$\left. \sum_{l_0=1}^{L} \omega_{l_0}(k)c^{l_{\bar{N}-2}\cdots l_0}(k+\bar{N}-1|k)^{*T},\ c(k+\bar{N}|k)^{*T} \right]^T .$$

Note that $c(k+\bar{N}|k)^*$ can be expressed as $\sum_{l_0=1}^{L}\omega_{l_0}(k)c^{l_{\bar{N}-1}\cdots l_0}(k+\bar{N}|k)^*$ or $c(k+\bar{N}|k)^* = c^{l_{\bar{N}-1}\cdots l_0}(k+\bar{N}|k)^*$. By applying (9.3.13) at time $k+1$ (actually this is not applied), the resultant performance cost $J(k+1) = J^*(k) - c(k|k)^{*T}c(k|k)^*$.

After $J(k+1)$ is optimized at time $k+1$, the optimum $J^*(k+1) \leq J(k+1) \leq J^*(k)$. Therefore, $J^*(k)$ will be monotonically decreasing with evolution of k, such that the state will be finally driven into $\varepsilon_{x,\chi}$ inside of which (ii) becomes true.   □

Proposition 9.3.1. *(Monotonicity) Consider Algorithm 9.4. By increasing either n_c or $\bar{N}$, the region of attraction of the overall close-loop system will not shrink.*

Proof. The proof is the same as Proposition 9.2.1.   □

9.3.2   Joint superiorities

By the fastest interior-point algorithms, the computational complexity involved in (9.3.11) is proportional to $\mathfrak{K}^3\mathfrak{L}$, where $\mathfrak{K} = m\sum_{i=0}^{\bar{N}-1}L^i + n_cm+1$, $\mathfrak{L} = (n_cm+n+1)L^{\bar{N}} + (n_cm+\bar{N}m+2q+1)L^{\bar{N}-1} + 2q\sum_{i=0}^{\bar{N}-2}L^i$. Hence, increasing n or q only linearly increases the computational complexity. For larger $\{n,q\}$ and smaller $\{L,\bar{N},n_c\}$ the computation involved in this optimization can be less expensive than that involved in (9.1.6).

In Algorithm 9.4, the selection of $\bar{N}$ should compromise between the desired region of attraction and the on-line computational burden. By increasing $\bar{N}$, the computational burden will increase exponentially. For the same $\bar{N} > 1$, the computation involved in Algorithm 9.4 is more expensive than that involved in Algorithm 9.3. However, any region of attraction achievable via Algorithm 9.3 can also be achieved via Algorithm 9.4, while the region of attraction achievable via Algorithm 9.4 may not be achievable via Algorithm 9.3.

Denote the region of attraction by solving (9.2.5)-(9.2.7) alone as $\mathcal{P}$, and (9.3.11) alone as $\mathcal{P}_v$. Then for the same K, n_c and $\bar{N}$, $\mathcal{P}_v \supseteq \mathcal{P}$. Hence, for achieving a specified region of attraction, by utilizing (9.3.11) smaller $\bar{N}$ can be chosen. The reason is that the single-valued perturbation items $c(k|k)$, $c(k+1|k), \cdots, c(k+\bar{N}-1|k)$ in Algorithm 9.3 have to deal with all possible state evolutions; "the vertex perturbation items" in Algorithm 9.4, on the other hand, define different perturbation items for different vertices of the uncertainty evolution.

Consider on-line approach based on (9.1.6), off-line approach and Algorithm 9.2. None of them is superior to others in both computational efficiency and size of the region of attraction. Note that, in Algorithm 9.4, the computation of $\varepsilon_{x,\chi}$ can also adopt the procedure as in Algorithm 9.2. Algorithm 9.4 can inherit all the merits of Algorithms 9.1 and 9.2, and can achieve a region of attraction complementary with respect to on-line robust MPC (i.e., the region of attraction of Algorithm 9.4 does not necessarily include that of on-line approach, and vice versa). The average computational burden incurred by Algorithm 9.4 can be much smaller than on-line robust MPC.

Remark 9.3.1. In mixed MPC, it is not certain that $\varepsilon_{x,1}$ includes $\hat{\varepsilon}_1$, or $\hat{\varepsilon}_1$ includes $\varepsilon_{x,1}$. Hence, it is not guaranteed that the switching is continuous with respect to the system state. However, it is easy to modify the algorithm such that the switching is continuous with respect to the system state.

Remark 9.3.2. In triple-mode robust MPC and mixed robust MPC, the optimization of the volume of $\varepsilon_{x,\chi}$ has certain artificial features.

9.3.3 Numerical example

Consider

$$\left[\begin{array}{c} x^{(1)}(k+1) \\ x^{(2)}(k+1) \end{array} \right] = \left[\begin{array}{cc} 1-\beta & \beta \\ K(k) & 1-\beta \end{array} \right] \left[\begin{array}{c} x^{(1)}(k) \\ x^{(2)}(k) \end{array} \right] + \left[\begin{array}{c} 1 \\ 0 \end{array} \right] u(k),$$

where $K(k) \in [0.5,\ 2.5]$ is an uncertain parameter and β a constant. The constraint is $|u| \leq 2$. Take $W = I$ and $R = 1$. The true state is generated by $K(k) = 1.5 + \sin(k)$.

Case A: $\beta = 0$

Denote $x_1^{(1),\max}$ as the maximum value such that, when $x(k) = x_1 = [x_1^{(1)},\ 0]^T$, the corresponding optimization problem remains feasible. Then $x_1^{(1),\max} = 59.2$. In Algorithm 9.1, choose $x_i = [x_i^{(1)},\ 0]^T$, $x_i^{(1)} = 10, 18, 26, 34, 42, 50, 59.2$. The ellipsoidal regions of attraction by Algorithm 9.1 are shown in Figure 9.3.1 in dotted lines. By adopting $K = F_1$, choosing $n_c = 5$ and solving (9.2.1)-(9.2.2), we then find the ellipsoidal region of attraction $\varepsilon_{x,\chi}$, shown in Figure 9.3.1 in solid line. Further, choose $\bar{N} = 3$, then the non-ellipsoidal regions of attraction $\mathcal{P}$ (corresponding to (9.2.5)-(9.2.7)) and

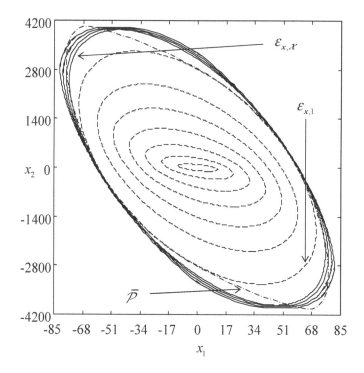

Figure 9.3.1: Regions of attraction when $\beta = 0$.

$\mathcal{P}_v$ (corresponding to (9.3.11)) are depicted in Figure 9.3.1 in dotted line and solid line, respectively.

In case A, $\mathcal{P}$ and $\mathcal{P}_v$ are nearly identical. In Figure 9.3.1, $\mathcal{P}$, $\mathcal{P}_v$ for $\bar{N} = 2$ and $\bar{N} = 1$ and $\bar{\mathcal{P}}$ are also given, $\bar{\mathcal{P}}$ denoting the region of attraction of on-line robust MPC based on (9.1.6) in dash-dotted line. The solid lines from outside to inside are: $\mathcal{P}_v$ ($\bar{N} = 3$), $\mathcal{P}_v$ ($\bar{N} = 2$), $\mathcal{P}_v = \mathcal{P}$ ($\bar{N} = 1$), $\varepsilon_{x,\chi}$.

We give the following conclusions about the regions of attraction which are general (not restricted to this example):

- $\bar{\mathcal{P}} \supseteq \varepsilon_{x,1}$.

- $\bar{\mathcal{P}}$ and $\varepsilon_{x,\chi}$ can be mutual-complementary ($\varepsilon_{x,\chi}$ is calculated by either (9.1.19) or (9.2.1)-(9.2.2)).

- $\bar{\mathcal{P}}$ and $\mathcal{P}$ can be mutual-complementary.

- $\bar{\mathcal{P}}$ and $\mathcal{P}_v$ can be mutual-complementary.

- For any $\bar{N} \geq 1$, $\mathcal{P}_v \supseteq \mathcal{P} \supseteq \varepsilon_{x,\chi}$.

- By increasing $\bar{N}$, it can result in $\mathcal{P}_v \supseteq \bar{\mathcal{P}}$ (however, the computational burden is prohibitive for larger $\bar{N}$).

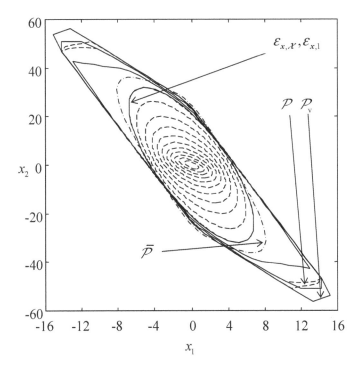

Figure 9.3.2: Regions of attraction when $\beta = 0.1$.

- $\mathcal{P}_v$ is not necessarily much larger than $\mathcal{P}$ (one can utilize single perturbation items instead of vertex ones in this case).

Choose $x(0) = [-60 \quad 3980]^T$ and compute $x(201)$. It takes 12 seconds by applying on-line robust MPC based on (9.1.6), and less than 1 second by applying Algorithm 9.4. In the simulation, we have utilized LMI Toolbox of Matlab 5.3 on our laptop (1.5G Hz Pentium IV CPU, 256 MB Memory).

Case B: $\beta = 0.1$

The details, if not specified, are the same as those in Case A. $x_1^{(1),\max} = 4.48$. Choose $x_i = [x_i^{(1)}, \ 0]^T$, $x_i^{(1)} = 1.0$, 1.4, 1.8, 2.2, 2.6, 3.0, 3.4, 3.8, 4.2, 4.48. The results are shown in Figure 9.3.2. In this case, $\varepsilon_{x,\chi}$ and $\varepsilon_{x,1}$ are nearly identical.

Choose $x(0) = [-7.5 \quad 35]^T$ and compute $x(201)$. It takes 13 seconds by applying on-line robust MPC based on (9.1.6), and less than 1 second by applying Algorithm 9.4.

Then, in Algorithm 9.1, let us choose $x_i = [x_i^{(1)}, \ 0]^T$, $x_i^{(1)} = 1.0$, 1.4, 1.8, 2.2, 2.6, 3.0, 3.4, 3.8, 4.2. For three initial states, the closed-loop state trajectories with Algorithm 9.4 are shown in Figure 9.3.3 in marked lines. The corresponding regions of attraction are also depicted in Figure 9.3.3.

Case C: $\beta = -0.1$

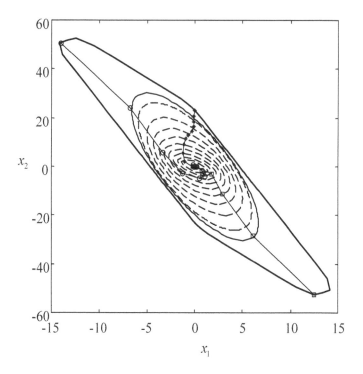

Figure 9.3.3: Closed-loop state trajectories by applying Algorithm 9.4 when $\beta = 0.1$.

The details, if not specified, are the same as those in Case A and Case B. $x_1^{(1),\max} = 2.12$. Choose $x_i = [x_i^{(1)},\ 0]^T$, $x_i^{(1)} = 1.0$, 1.15, 1.3, 1.45, 1.6, 1.75, 1.9, 2.12. The results are shown in Figure 9.3.4. In this case, $\varepsilon_{x,\chi}$ and $\varepsilon_{x,1}$ are nearly identical.

Choose $x(0) = [-2.96 \quad 10.7]^T$ and compute $x(201)$. It takes 12 seconds by applying on-line robust MPC based on (9.1.6), and less than 1 second by applying Algorithm 9.4.

9.4 Approach based on single-valued open-loop optimization and its deficiencies

In section 8.5 we have given an on-line MPC for systems with polytopic description, where the following is defined:

$$u(k+i|k) = F(k+i|k)x(k+i|k) + c(k+i|k),\ F(\cdot|0) = 0, \qquad (9.4.1)$$

where $F(k+i|k)$, $k > 0$, $i \in \{0, 1, \dots, N-1\}$ are always carried over from the previous sampling time, i.e., for $k > 0$, take $F(k+i|k) = F(k+i|k-1)$, $i \in \{0, 1, \dots, N-2\}$, $F(k+N-1|k) = F(k-1)$.

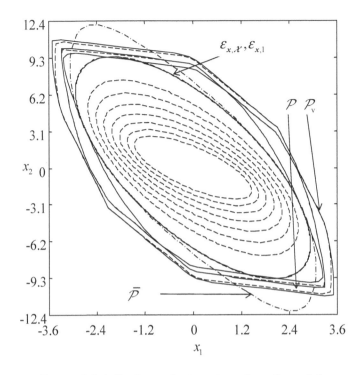

Figure 9.3.4: Regions of attraction when $\beta = -0.1$.

For $N > 1$, it is apparent that, at $k = 0$, this approach is open-loop MPC; for $k \geq N$, this approach becomes partial feedback MPC. For $0 < k < N$, this approaches changes gradually from open-loop MPC to partial feedback MPC.

In open-loop MPC, if $u(k|k), u(k + 1|k), \cdots, u(k + N - 1|k)$ are single-valued, then the corresponding is the single-valued open-loop MPC. The so-called single-valued means that, each control move $u(k + i|k)$ within the switching horizon is a fixed value. In section 8.6, another open-loop MPC is addressed, but where $u(k + 1|k), \cdots, u(k + N - 1|k)$ are parameter dependent. Due to the parameter-dependent nature, there are an infinite number of possible values of $u(k + 1|k), \cdots, u(k + N - 1|k)$, which are convex combined values of the "vertex values." For $N > 1$, we call the strategy in section 8.6 parameter-dependent MPC.

Let us still consider the system, constraints and optimization problem as in section 8.5. We can give single-valued MPC ($N > 1$), the only difference with the method in section 8.5 being the substitution of $c(k|k), c(k+1|k), \cdots, c(k+N-1|k)$ by $u(k|k), u(k+1|k), \cdots, u(k+N-1|k)$. Simply, the optimization problem can be approximated by the following optimization problem:

$$\min_{u(k|k),\cdots,u(k+N-1|k),\gamma_i,\gamma,Q,Y,Z,\Gamma} \sum_{i=0}^{N-1} \gamma_i + \gamma, \text{ s.t. } (9.4.3) - (9.4.9), \qquad (9.4.2)$$

$$\begin{bmatrix} Z & Y \\ Y^T & Q \end{bmatrix} \geq 0, \; Z_{jj} \leq \bar{u}^2_{j,\mathrm{inf}}, \; j \in \{1, \ldots, m\}, \tag{9.4.3}$$

$$\begin{bmatrix} Q & * & * & * \\ A_l Q + B_l Y & Q & * & * \\ W^{1/2} Q & 0 & \gamma I & * \\ R^{1/2} Y & 0 & 0 & \gamma I \end{bmatrix} \geq 0, \; \forall l \in \{1, \ldots, L\}, \tag{9.4.4}$$

$$\begin{bmatrix} Q & * \\ \Psi \left(A_l Q + B_l Y \right) & \Gamma \end{bmatrix} \geq 0, \; \Gamma_{ss} \leq \bar{\psi}^2_{s,\mathrm{inf}}, \; \forall l \in \{1, \ldots, L\}, \; s \in \{1, \ldots, q\}, \tag{9.4.5}$$

$$\begin{bmatrix} \gamma_0 & * \\ u(k|k) & R^{-1} \end{bmatrix} \geq 0, \; \begin{bmatrix} \gamma_i & * & * \\ v_{l_{i-1}\cdots l_1 l_0}(k+i|k) & W^{-1} & * \\ u(k+i|k) & 0 & R^{-1} \end{bmatrix} \geq 0,$$

$$l_0, l_1, \cdots l_{i-1} \in \{1, 2, \ldots, L\}, \; i \in \{1, \ldots, N-1\}, \tag{9.4.6}$$

$$\begin{bmatrix} 1 & * \\ v_{l_{N-1}\cdots l_1 l_0}(k+N|k) & Q \end{bmatrix} \geq 0, \; l_0, l_1, \cdots l_{N-1} \in \{1, 2, \ldots, L\}, \tag{9.4.7}$$

$$-\underline{u} \leq u(k+i|k) \leq \bar{u}, \; i \in \{0, 1, \ldots, N-1\}, \tag{9.4.8}$$

$$-\underline{\psi} \leq \Psi v_{l_{i-1}\cdots l_1 l_0}(k+i|k) \leq \bar{\psi}, \; l_0, l_1, \cdots l_{i-1} \in \{1, 2, \ldots, L\},$$

$$i \in \{1, 2, \ldots, N\}. \tag{9.4.9}$$

Notice that, in (9.4.2), $v_{l_{i-1}\cdots l_1 l_0}(k+i|k)$ should be expressed as function of $u(k|k), u(k+1|k), \cdots, u(k+i-1|k)$ (see Lemma 8.5.1).

The only deficiency by applying (9.4.2) is that closed-loop stability is not guaranteed, and stability cannot be proved (this issue has been discussed in [55], and is also considered in [9]). Suppose (9.4.2) is feasible at time k (denoted by $*$). Then whether or not

$$\begin{aligned} u(k+i|k+1) &= u^*(k+i|k), i \in \{1, 2, \ldots, N-1\}; \\ u(k+N|k+1) &= F^*(k)x^*(k+N|k) \end{aligned} \tag{9.4.10}$$

is a feasible solution at time $k+1$? The answer is negative. When $N > 1$, $x^*(k+N|k)$ is an uncertain value, and the value given by $F^*(k)x^*(k+N|k)$ is uncertain. According to the definition of $u(k+N|k+1)$, $u(k+N|k+1)$ is the control move and is a deterministic value. This rationale has been discussed in "feedback MPC" of Chapter 6. Although systems with disturbance (rather than polytopic uncertain systems) are discussed in Chapter 6, the results are suitable for all uncertain system descriptions.

Then, does this mean that the proving method has not been found? The answer is, in some situations stability cannot be guaranteed. Notice that, here, the so-called stability guarantee is based on the fact that "the optimization problem is feasible for all future time if it is feasible at the initial time" (so-called "recursive feasibility"). The example for not satisfying this "recursive feasibility" can be found.

Remark 9.4.1. For open-loop stable systems, we can take

$$F(k|k) = F(k+1|k) = \cdots = F(k+N-1|k) = F(k) = 0$$

(hence, $Y = 0$). Thus, partial feedback MPC is equivalent to open-loop MPC, i.e., by adopting open-loop MPC, closed-loop stability can be guaranteed.

9.5   Approach based on parameter-dependent open-loop optimization and its properties

The approach in section 8.6 is better than feedback MPC with respect to feasibility and optimality. From the side of feasibility, both feedback MPC and single-valued open-loop MPC are special cases of parameter-dependent open-loop MPC.

Suppose, within the switching horizon, we define

$$u(k+i|k) = K(k+i|k)x(k+i|k), \ i \in \{1, \dots, N-1\} \tag{9.5.1}$$

and after the switching horizon, we define

$$u(k+i|k) = F(k)x(k+i|k), \ \forall i \geq N. \tag{9.5.2}$$

Then, on-line feedback MPC solves the following optimization problem at each time k:

$$\min_{\{u(k|k),K(k+1|k),K(k+2|k),\cdots,K(k+N-1|k),F(k)\}} \max_{[A(k+i)|B(k+i)]\in\Omega,i\geq 0} J_\infty(k),$$

s.t. $(9.5.1) - (9.5.2), (9.5.4),$ \hfill $(9.5.3)$

$$-\underline{u} \leq u(k+i|k) \leq \bar{u}, \ -\underline{\psi} \leq \Psi x(k+i+1|k) \leq \bar{\psi}, \ i \geq 0, \tag{9.5.4}$$

where

$$J_\infty(k) = \sum_{i=0}^{\infty} \left[\|x(k+i|k)\|_W^2 + \|u(k+i|k)\|_R^2 \right].$$

On-line parameter-dependent open-loop MPC solves the following problem at each time k:

$$\min_{\{\tilde{u}(k|k),F(k)\}} \max_{[A(k+i)|B(k+i)]\in\Omega,i\geq 0} J_\infty(k), \ \text{s.t. } (8.6.1), (8.6.5), (9.5.2), (9.5.4),$$

$$\tag{9.5.5}$$

where

$$\tilde{u}(k) \triangleq \{u(k|k), u^{l_0}(k+1|k), \cdots, u^{l_{N-2}\cdots l_0}(k+N-1|k)|l_0, \cdots, l_{N-2} = 1 \dots L\}.$$

Proposition 9.5.1. *Consider $N \geq 2$. For the same state $x(k)$, feasibility of* (9.5.3) *implies feasibility of* (9.5.5).

Proof. For feedback MPC, applying (9.5.1) and the definition of polytopic description yields

$$u(k+i|k) = K(k+i|k) \sum_{l_0 \cdots l_{i-1}=1}^{L} \{\omega_{l_{i-1}}(k+i-1)[A_{l_{i-1}} + B_{l_{i-1}}K(k+i-1|k)]$$

$$\times \cdots \times \omega_{l_1}(k+1)[A_{l_1} + B_{l_1}K(k+1|k)] \times \omega_{l_0}(k)[A_{l_0}x(k)$$
$$+ B_{l_0}u(k|k)]\}, i \in \{1,\ldots,N-1\}.$$

Apparently, $u(k+i|k), \forall i \in \{1,\ldots,N-1\}$ is the convex combination of the following L^i control moves:

$$\bar{u}^{l_{i-1}\cdots l_0}(k+i|k) = K(k+i|k) \times [A_{l_{i-1}} + B_{l_{i-1}}K(k+i-1|k)]$$
$$\times \cdots \times [A_{l_1} + B_{l_1}K(k+1|k)] \times [A_{l_0}x(k) + B_{l_0}u(k|k)],$$
$$l_0, \cdots, l_{i-1} \in \{1,\ldots,L\}, \tag{9.5.6}$$

i.e.,

$$u(k+i|k) = \sum_{l_0 \cdots l_{i-1}=1}^{L} \left(\left(\prod_{h=0}^{i-1} \omega_{l_h}(k+h) \right) \bar{u}^{l_{i-1}\cdots l_0}(k+i|k) \right),$$

$$\sum_{l_0 \cdots l_{i-1}=1}^{L} \left(\prod_{h=0}^{i-1} \omega_{l_h}(k+h) \right) = 1, \ i \in \{1,\ldots,N-1\}.$$

Define

$$\tilde{\tilde{u}}(k) \triangleq \{u(k|k), \bar{u}^{l_0}(k+1|k), \cdots, \bar{u}^{l_{N-2}\cdots l_0}(k+N-1|k)|l_0, \cdots, l_{N-2} = 1 \ldots L\}.$$

Then (9.5.3) can be equivalently written as the following optimization problem:

$$\min_{\tilde{\tilde{u}}(k),F(k),K(k+1|k)\cdots K(k+N-1|k)} \max_{[A(k+i)|B(k+i)]\in\Omega,i\geq0} J_\infty(k),$$
s.t. $(8.6.1),(8.6.5),(9.5.2),(9.5.4),(9.5.6)$, with $\tilde{u}(k)$ replaced with $\tilde{\tilde{u}}(k)$. (9.5.7)

Notice that (9.5.1) and (9.5.6) are equivalent and, hence, (9.5.1) is omitted in (9.5.7).

In (9.5.7), $\{K(k+1|k), \cdots, K(k+N-1|k)\}$ and $\tilde{\tilde{u}}(k)$ are connected via (9.5.6). Removing (9.5.6) from (9.5.7) yields

$$\min_{\tilde{\tilde{u}}(k),F(k)} \max_{[A(k+i)|B(k+i)]\in\Omega,i\geq0} J_\infty(k),$$
s.t. $(8.6.1),(8.6.5),(9.5.2),(9.5.4)$, with $\tilde{u}(k)$ replaced by $\tilde{\tilde{u}}(k)$. (9.5.8)

Notice that, in (9.5.2), (9.5.4), (8.6.1) and (8.6.5), $\{K(k+1|k), \cdots, K(k+N-1|k)\}$ is not involved. Hence, in the decision variables of (9.5.8), there is no $\{K(k+1|k), \cdots, K(k+N-1|k)\}$.

Now, observe (9.5.8) and (9.5.5); we know that they are only different in notation, i.e., (9.5.8) uses $\tilde{\tilde{u}}(k)$ while (9.5.5) uses $\tilde{u}(k)$. Hence, (9.5.8) and (9.5.5) are equivalent. Moreover, compared with (9.5.8), there is one more constraint (9.5.6) in (9.5.7). Therefore, (9.5.8) is easier to be feasible than (9.5.7). □

In parameter-dependent open-loop MPC, if we take all the vertex control moves for the same $k+i|k$ equal, then we obtain single-valued open-loop MPC; if we add the constraint (9.5.6), then we obtain feedback MPC. That is to say, parameter-dependent open-loop MPC is easier to be feasible than both single-valued open-loop MPC and feedback MPC.

Notice that, if we add the constraint (9.5.6), then the obtained feedback MPC cannot be solved via LMI toolbox as that in section 8.6. Hence, parameter-dependent open-loop MPC is a computational outlet for feedback MPC.

By applying the interior point algorithm, the computational complexities of partial feedback MPC in section 8.5, open-loop MPC in section 8.6 and the optimization problem (9.4.2) are all proportional to $\mathfrak{K}^3 \mathfrak{L}$ (refer to [31]). Denote

$$a = 2 + mn + \frac{1}{2}m(m+1) + \frac{1}{2}q(q+1) + \frac{1}{2}n(n+1),$$

$$b = (1+n)L^N + 2q \sum_{j=1}^{N} L^j + [1 + Nm + (N-1)n]L^{N-1}$$
$$+ (4n + m + q)L + n + 2m + q,$$

$$M = \sum_{j=1}^{N} L^{j-1}.$$

Then, for (9.4.2), $\mathfrak{K} = a + mN$, $\mathfrak{L} = b + 2mN$; for partial feedback MPC in section 8.5 (when $k \geq N$), $\mathfrak{K} = a + mN$, $\mathfrak{L} = b + 2mM$; for open-loop MPC in section 8.6, $\mathfrak{K} = a + mM$, $\mathfrak{L} = b + 2mM$. (Note that, in order for the comparison to be made on the same basis, for (9.4.2) and partial feedback MPC in section 8.5, a single γ_1 is utilized rather than a set of $\{\gamma_0, \gamma_1, \cdots, \gamma_{N-1}\}$. Of course, this small revision is not intrinsic.)

In general, on-line parameter-dependent open-loop MPC involves with very heavy computational burden.

Remark 9.5.1. By listing the various methods, with respect to performance, from the worst to best, we obtain: single-valued open-loop MPC, single-valued partial feedback MPC, feedback MPC, parameter-dependent open-loop MPC. By listing these methods, with respect to computational burden, from the smallest to the largest, we obtain: single-valued open-loop MPC, single-valued partial feedback MPC, parameter-dependent open-loop MPC, feedback MPC. However, the list should be re-considered in the concrete situation. In general, parameter-dependent open-loop MPC and parameter-dependent partial feedback MPC are equivalent, since the vertex state predictions and vertex control moves are all single-valued.

Remark 9.5.2. Consider Algorithm 9.2. If, in (9.1.20), we substitute the original single-valued perturbation items with the parameter-dependent perturbation items, then we obtain the same result as Algorithm 9.2. That is to say, in Algorithm 9.2, it is not necessary to adopt the parameter-dependent perturbation items. Thus, although Algorithm 9.2 has the appearance of single-valued partial feedback MPC, it is actually parameter-dependent partial feedback MPC. If, Remark 9.5.1 is also considered, then Algorithm 9.2 is actually feedback MPC; an extra constraint is added in this feedback MPC problem, i.e., the region of attraction is an invariant ellipsoid.

Remark 9.5.3. The varying horizon off-line approach in Chapter 7 can be regarded as feedback MPC. For nominal system, open-loop MPC and feedback MPC are equivalent.

Remark 9.5.4. The various classifications of MPC can sometimes be blurry, i.e., there is no clear boundary.

9.6   Approach with unit switching horizon

By taking $N = 1$ (see [47]), we always obtain feedback MPC. Then, it should optimize $u(k|k)$, and there is no necessity to define $u(k|k) = F(k|k) + c(k|k)$; and parameter-dependent open-loop MPC cannot be adopted.

Although it is simple, MPC with $N = 1$ has a number of advantages. The region of attraction with $N = 1$ includes that with $N = 0$ (for $N = 0$ refer to Chapter 7) (certainly, the basis for comparison is that the ways for computing the three ingredients should be the same).

Consider $A(k) = \begin{bmatrix} 1 & 0.1 \\ \mu(k) & 1 \end{bmatrix}$, $B(k) = \begin{bmatrix} 1 \\ 0 \end{bmatrix}$, $\mu(k) \in [0.5, 2.5]$. The input constraint is $|u(k)| \leq 1$. The weighting matrices are $W = I$ and $R = 1$. For $N > 1$, by applying single-valued open-loop MPC, the regions of attraction are shown in Figure 9.6.1 with dotted lines. For $N > 1$, by applying parameter-dependent open-loop MPC, the regions of attraction are shown in Figure 9.6.1 with solid lines. For $N = 1, 0$, by applying feedback MPC, the regions of attraction are shown in Figure 9.6.1 with solid and dash-dotted lines. The region of attraction with $N = 1$ includes that with $N = 0$; for single-valued open-loop MPC, the region of attraction with $N = 3$ ($N = 2$) does not include that with $N = 2$ ($N = 1$). For parameter-dependent open-loop MPC, the region of attraction with $N = 3$ ($N = 2$) includes that with $N = 2$ ($N = 1$).

Feedback MPC is also referred to in [59]. For partial feedback MPC, one can also refer to [44].

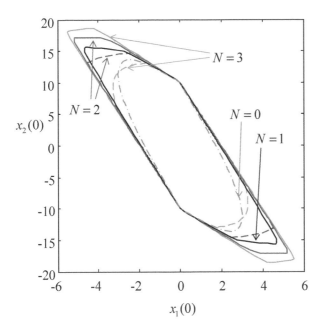

Figure 9.6.1: Regions of attraction for single-valued open-loop min-max MPC, parameter-dependent open-loop min-max MPC and robust MPC for $N = 1, 0$.

Remark 9.6.1. Usually, in the industrial applications, MPC is based on the open-loop optimization, i.e., at each sampling time a sequence of single-value control moves are calculated. When MPC is applied to the industrial processes, the "transparent control" is usually adopted; since the plant has been pre-stabilized by PID, it is easy to increase the region of attraction by increasing N.

Chapter 10

Output feedback synthesis approaches

In DMC, MAC and GPC, direct output feedback is adopted, which can also be regarded as state feedback, where the state is composed of the past input, past and current outputs. In real applications, if one wants to adopt state space model and synthesis approaches of MPC, it would be better to select the measurable states when he/she sets up the state space model.

If one cannot make all the state measurable, but wishes to adopt state space model in synthesis approaches of MPC, he/she can utilize state observer (when noise is considered, in general called state estimator). Considering the input/output nonlinear systems and the general polytopic description, this chapter gives output feedback MPC techniques based on the state estimator. For the input/output nonlinear model, the linear difference inclusion technique is firstly adopted to obtain the polytopic description.

Sections 10.1-10.3 are referred to in [11]. Sections 10.3-10.5 are referred to in [22].

10.1 Optimization problem: case systems with input-output (I/O) nonlinearities

Consider the following system represented by the Hammerstein-Wiener model

$$x(k+1) = Ax(k) + Bv(k) + Dw(k), \ v(k) = f(u(k)),$$

$$y(k) = h(z(k)) + Ew(k), \ z(k) = C\,\widehat{z}\,(k), \ \widehat{z}\,(k) = \phi(x(k)), \ w(k) \in \mathcal{W}$$
$$(10.1.1)$$

where $u \in \mathfrak{R}^{n_u}$, $x \in \mathfrak{R}^{n_x}$, $y \in \mathfrak{R}^{n_y}$ and $w \in \mathfrak{R}^{n_w}$ are input, unmeasurable state, output and stochastic disturbance/noise, respectively; $v \in \mathfrak{R}^{n_u}$, $z \in \mathfrak{R}^{n_y}$ and $\widehat{z} \in \mathfrak{R}^{n_{\widehat{z}}}$ are unmeasurable intermediate variables; f and h are invertible nonlinearities; $\mathcal{W} \in Co\{\mathcal{W}_1, \mathcal{W}_2, \cdots, \mathcal{W}_{m_w}\} \supseteq \{0\}$, i.e., $\mathcal{W}$ is a convex polyhedral set that includes the origin as an interior point. ϕ allows $\hat{z}$ and x to have different dimensions.

The input and output constraints are

$$\underline{u} \leq u(k+i) \leq \bar{u}, \ \forall i \geq 0, \tag{10.1.2}$$

$$\underline{y} \leq y(k+i+1) \leq \bar{y}, \ \forall i \geq 0. \tag{10.1.3}$$

Construct $g(\cdot)$ as the inverse (or the approximate inverse) of $f(\cdot)$. For the Hammerstein nonlinearity, we apply the technique of "nonlinear removal" (see Chapter 5). First, we consider

$$\begin{aligned} x(k+1) &= Ax(k) + Bf \circ g(v^L(k)) + Dw(k), \\ y(k) &= h(C\phi(x(k))) + Ew(k), \ w(k) \in \mathcal{W}, \end{aligned} \tag{10.1.4}$$

with constraints (10.1.2) and (10.1.3), where $v^L(k)$ can be interpreted as the "desired intermediate variable." We will design the following output feedback controller:

$$\hat{x}(k+1) = \hat{A}(k)\hat{x}(k) + \hat{L}(k)y(k), \ \forall k \geq 0, \ v^L(k) = F(k)\hat{x}(k) \tag{10.1.5}$$

where $\hat{x}$ is the estimator state. Then, the actual control input is given by

$$u(k) = g(v^L(k)). \tag{10.1.6}$$

By applying (10.1.6), $v(k) = f(u(k)) = f \circ g(v^L(k))$, and (10.1.4) is justified.

Clearly, the input nonlinearity is completely removed if $f \circ g = 1$. When $f \circ g \neq 1$, $f \circ g$ should be a weaker nonlinearity than f. By combining (10.1.4) and (10.1.5), the augmented closed-loop system is given by

$$\begin{cases} x(k+1) = Ax(k) + Bf \circ g(F(k)\hat{x}(k)) + Dw(k) \\ \hat{x}(k+1) = \hat{A}(k)\hat{x}(k) + \hat{L}(k)h(C\phi(x(k))) + \hat{L}(k)Ew(k) \end{cases}. \tag{10.1.7}$$

Assumption 10.1.1. *There exists $\bar{v} > 0$, such that constraint (10.1.2) is satisfied whenever*

$$-\bar{v} \leq v^L(k+i) \leq \bar{v}, \ \forall i \geq 0. \tag{10.1.8}$$

Assumption 10.1.2. *There exists $\bar{z} > 0$, such that constraint (10.1.3) is satisfied for all $w(k+i+1) \in \mathcal{W}$ whenever*

$$-\bar{z} \leq z(k+i+1) \leq \bar{z}, \forall i \geq 0. \tag{10.1.9}$$

By substituting (10.1.2) and (10.1.3) with (10.1.8) and (10.1.9), we can consider weaker nonlinearities in dealing with input/output constraints.

Assumption 10.1.3. *For all $v^L(k)$ satisfying $-\bar{v} \leq v^L(k) \leq \bar{v}$, $f \circ g(\cdot) \in \Omega^{fg} = Co\{\Pi_1, \Pi_2, \cdots, \Pi_{m_{fg}}\}$, that is, $v(k)$ can be incorporated by $v(k) = \Pi(k)v^L(k)$, $\Pi(k) \in \Omega^{fg}$.*

Assumption 10.1.4. *For all $x(k) \in S = \{x \in \Re^{n_x}| - \bar{\theta} \le \Theta x \le \bar{\theta}\}$ (where $\Theta \in \Re^{q \times n_x}$, $\bar{\theta} > 0$) $h(C\phi(\cdot)) \in \Omega^{h\phi} = Co\{\Psi_1, \Psi_2, \cdots, \Psi_{m_{h\phi}}\}$, that is, $h(C\phi(x(k)))$ can be incorporated by $h(C\phi(x(k))) = \Psi(k)x(k)$, $\Psi(k) \in \Omega^{h\phi}$.*

Assumption 10.1.5. *For all $x(k) \in S$, $\phi(\cdot) \in \Omega^\phi = Co\{\Xi_1, \Xi_2, \cdots, \Xi_{m_\phi}\}$, that is, $\widehat{z}(k)$ can be incorporated by $\widehat{z}(k) = \Xi(k)x(k)$, $\Xi(k) \in \Omega^\phi$.*

Assumptions 10.1.3-10.1.5 have utilized the technique of linear difference inclusion, which incorporates the nonlinearity by polytope. For polytopic description, robust control techniques can be adopted.

If $x(k) \in S$ and $-\bar{v} \le v^L(k) \le \bar{v}$, then (10.1.7) is linearly included by

$$\left[\begin{array}{c} \hat{x}(k+1) \\ e(k+1) \end{array} \right] = \mathcal{A}(k) \left[\begin{array}{c} \hat{x}(k) \\ e(k) \end{array} \right] + \mathcal{D}(k)w(k), \qquad (10.1.10)$$

where $\mathcal{A}(k) = \left[\begin{array}{cc} \hat{A}(k) + \hat{L}(k)\Psi(k) & \hat{L}(k)\Psi(k) \\ A - \hat{A}(k) + B\Pi(k)F(k) - \hat{L}(k)\Psi(k) & A - \hat{L}(k)\Psi(k) \end{array} \right]$,

$\mathcal{D}(k) = \left[\begin{array}{c} \hat{L}(k)E \\ D - \hat{L}(k)E \end{array} \right]$; and $e(k) = x(k) - \hat{x}(k)$ is the estimation error.

In output feedback MPC based on the state estimator, a key issue is how to handle the estimation error. Here, we bound the estimation error consistently for the entire time horizon, that is impose

$$-\bar{e} \le e(k+i) \le \bar{e}, \ \forall i \ge 0, \qquad (10.1.11)$$

where $0 < \bar{e}_s < \infty$, $s \in \{1, 2, \ldots, n_x\}$. Technically, we can express (10.1.11) by

$$e(k+i) \in Co\{\epsilon_1, \epsilon_2, \cdots, \epsilon_{2^{n_x}}\}, \ \forall i \ge 0 \qquad (10.1.12)$$

where ϵ_j $(j \in \{1, \ldots, 2^{n_x}\})$ has its s-th $(s \in \{1, \ldots, n_x\})$ element being $-\bar{e}_s$ or $\bar{e}_s$, i.e., each ϵ_j is a vertex of the region $\{e \in \Re^{n_x}| - \bar{e} \le e \le \bar{e}\}$.

At each time k, the controller/estimator parameters are obtained by solving the following optimization problem:

$$\min_{\hat{A}(k), \hat{L}(k), F(k)} \max_{\Psi(k+i) \in \Omega^{h\phi}, \Pi(k+i) \in \Omega^{fg}, \Xi(k+i) \in \Omega^\phi, i \ge 0} J_\infty(k)$$

$$= \sum_{i=0}^{\infty} \left[\|y_u(k+i|k)\|_W^2 + \|F(k)\hat{x}_u(k+i|k)\|_R^2 \right], \qquad (10.1.13)$$

s.t. $-\bar{e} \le e(k+i+1|k) \le \bar{e}$, $-\bar{v} \le F(k)\hat{x}(k+i|k) \le \bar{v}$, $-\bar{z}$

$\le z(k+i+1|k) \le \bar{z}, \hat{x}(k+i+1|k) \in S, x(k+i+1|k) \in S, \forall i \ge 0$,

$$(10.1.14)$$

where $W > 0$ and $R > 0$ are weighting matrices; y_u is the prediction of the uncorrupted output,

(i) $y_u(k+i|k) = \Psi(k+i)x_u(k+i|k)$, $x_u(\cdot) = \hat{x}_u(\cdot) + e_u(\cdot)$,

$[\hat{x}_u(\cdot)^T, e_u(\cdot)^T]^T = \tilde{x}_u(\cdot)$, $\tilde{x}_u(k+i+1|k) = \mathcal{A}(k, i)\tilde{x}_u(k+i|k)$, $\tilde{x}_u(k|k) = \tilde{x}(k)$,

$$\mathcal{A}(k,i) = \begin{bmatrix} \hat{A}(k) + \hat{L}(k)\Psi(k+i) & \hat{L}(k)\Psi(k+i) \\ A - \hat{A}(k) + B\Pi(k+i)F(k) - \hat{L}(k)\Psi(k+i) & A - \hat{L}(k)\Psi(k+i) \end{bmatrix},$$

(ii) $z(k+i|k) = C\Xi(k+i)x(k+i|k)$, $x(\cdot) = \hat{x}(\cdot) + e(\cdot)$, $[\hat{x}(\cdot)^T, e(\cdot)^T]^T = \tilde{x}(\cdot)$,
$\tilde{x}(k+i+1|k) = \mathcal{A}(k,i)\tilde{x}(k+i|k) + \mathcal{D}(k)w(k+i)$, $\tilde{x}(k|k) = \tilde{x}(k)$.

10.2   Conditions for stability and feasibility: case systems with I/O nonlinearities

For problem (10.1.13)-(10.1.14), stability means the convergence of the augmented state $\tilde{x}$ towards a neighborhood of the origin $\tilde{x} = 0$ with the evolution of time, and feasibility means satisfaction of constraints in (10.1.14) for any $k \geq 0$. For deriving conditions of stability and feasibility, we utilize ellipsoidal confinement, that is, restrict the augmented state within an ellipsoid.

Lemma 10.2.1. *(Invariance) Suppose at time k, there exist properly dimensional matrices $\left\{ M, \hat{L}, Y, \tilde{G} = \begin{bmatrix} G & 0 \\ G_{12} & G_2 \end{bmatrix} \right\}$, symmetric positive-definite matrix $\tilde{Q} = \begin{bmatrix} Q_{11} & Q_{12}^T \\ Q_{12} & Q_{22} \end{bmatrix}$ and scalar η, $0 < \eta < 1$, such that the following inequalities are satisfied:*

$$\begin{bmatrix} 1 & * & * \\ \hat{x}(k) & Q_{11} & * \\ e(k) & Q_{12} & Q_{22} \end{bmatrix} \geq 0, \tag{10.2.1}$$

$$\begin{bmatrix} (1-\eta)^2 & * & * \\ \hat{L}EW_l & Q_{11} & * \\ (D - \hat{L}E)W_l & Q_{12} & Q_{22} \end{bmatrix} \geq 0, \ l \in \{1,2,\ldots,m_w\}, \tag{10.2.2}$$

$$\begin{bmatrix} \eta^2(G + G^T - Q_{11}) & * & * & * \\ \eta^2(G_{12} - Q_{12}) & \eta^2(G_2 + G_2^T - Q_{22}) & * & * \\ \hat{L}\Psi_s(G + G_{12}) + M & \hat{L}\Psi_s G_2 & Q_{11} & * \\ (A - \hat{L}\Psi_s)(G + G_{12}) - M + B\Pi_l Y & (A - \hat{L}\Psi_s)G_2 & Q_{12} & Q_{22} \end{bmatrix} > 0,$$
$$s \in \{1,2,\ldots,m_{h\phi}\}, \ l \in \{1,2,\ldots,m_{fg}\}. \tag{10.2.3}$$

Then, with $u(k+i|k) = g(v^L(k+i|k))$, $\forall i \geq 0$ and $\{\hat{A}(k) = MG^{-1}, \hat{L}(k) = \hat{L}\}$ (where $v^L(\cdot) = F(k)\hat{x}(\cdot)$, $F(k) = YG^{-1}$) being applied, the following inequality holds:

$$\tilde{x}(k+i|k)^T \tilde{Q}^{-1} \tilde{x}(k+i|k) \leq 1, \ \forall i \geq 0. \tag{10.2.4}$$

Proof. We will use a property: suppose $X > 0$ is a symmetric matrix, a and b are vectors with appropriate dimensions; then, the following inequality holds for any scalar $\delta > 0$:

$$(a+b)^T X(a+b) \leq (1+\delta)a^T Xa + (1+\frac{1}{\delta})b^T Xb. \tag{10.2.5}$$

Condition (10.2.2) guarantees

$$w(k+i)^T \mathcal{D}(k)^T \tilde{Q}^{-1} \mathcal{D}(k) w(k+i) \leq (1-\eta)^2, \ \forall i \geq 0. \tag{10.2.6}$$

By applying (10.2.5) and (10.2.6), it follows that

$$\tilde{x}(k+i+1|k)^T \tilde{Q}^{-1} \tilde{x}(k+i+1|k)$$
$$\leq (1+\delta)\tilde{x}(k+i|k)^T \mathcal{A}(k,i)^T \tilde{Q}^{-1} \mathcal{A}(k,i)\tilde{x}(k+i|k) + \left(1+\frac{1}{\delta}\right)(1-\eta)^2. \tag{10.2.7}$$

Impose

$$\mathcal{A}(k,i)^T \tilde{Q}^{-1} \mathcal{A}(k,i) < \eta^2 \tilde{Q}^{-1}. \tag{10.2.8}$$

By substituting (10.2.8) into (10.2.7) and choosing $\delta = \frac{1}{\eta} - 1$, it follows that:

$$\tilde{x}(k+i+1|k)^T \tilde{Q}^{-1} \tilde{x}(k+i+1|k) \leq \eta\tilde{x}(k+i|k)^T \tilde{Q}^{-1}\tilde{x}(k+i|k) + (1-\eta). \tag{10.2.9}$$

With (10.2.1) satisfied, by applying (10.2.9) recursively $(i = 0, 1, \ldots)$, (10.2.4) can be verified.

Due to $(\tilde{G} - \tilde{Q})^T \tilde{Q}^{-1}(\tilde{G} - \tilde{Q}) \geq 0$, the following inequality holds:

$$\tilde{G} + \tilde{G}^T - \tilde{Q} \leq \tilde{G}^T \tilde{Q}^{-1} \tilde{G}. \tag{10.2.10}$$

By multiplying left and right sides of (10.2.8) by $\tilde{G}^T$ and $\tilde{G}$, respectively, and applying Schur complement, (10.2.10), $F(k) = YG^{-1}$, $\{\hat{A}(k) = MG^{-1}, \hat{L}(k) = \hat{L}\}$ and convexity of the polytopic descriptions, it can be shown that (10.2.3) guarantees satisfaction of (10.2.8).   □

Equation (10.2.3) guarantees $\mathcal{A}(k,i)^T \tilde{Q}^{-1} \mathcal{A}(k,i) < \eta^2 \tilde{Q}^{-1}$, which means that (10.2.3) guarantees exponential stability of $\tilde{x}_u(k+i+1|k) = \mathcal{A}(k,i)\tilde{x}_u(k+i|k)$ and $\lim_{i\to\infty} \tilde{x}_u(k+i|k) = 0$. Hence, with (10.2.3) satisfied, $\tilde{x}(k+i|k)$ will lie in a neighborhood of the origin $\tilde{x} = 0$ for properly large i. Based on the invariance property stated above, the following conclusion can be obtained.

Lemma 10.2.2. *(Satisfaction of constraints) Suppose at time k, there exist properly dimensional matrices* $\left\{ M, \hat{L}, Y, \tilde{G} = \begin{bmatrix} G & 0 \\ G_{12} & G_2 \end{bmatrix} \right\}$, *symmetric positive-definite matrices* $\left\{ \tilde{Q} = \begin{bmatrix} Q_{11} & Q_{12}^T \\ Q_{12} & Q_{22} \end{bmatrix}, \Lambda, U, Z, \Upsilon, \Gamma \right\}$, *scalar η,*

$0 < \eta < 1$ *and*

$$\zeta_{1j}^2 = \frac{1}{\bar{e}_j^2} \max_{l \in \{1,2,\dots,m_w\}} \left| \xi_{1j}(D - \hat{L}E)\mathcal{W}_l \right|^2 < 1, \ j \in \{1,\dots,n_x\}, \quad (10.2.11)$$

$$\zeta_{2j}^2 = \frac{1}{\bar{z}_j^2} \max_{s \in \{1,\dots,m_\phi\}, l \in \{1,\dots,m_w\}} |\xi_{2j} C \Xi_s D \mathcal{W}_l|^2 < 1, \ j \in \{1,\dots,n_y\},$$
$$(10.2.12)$$

$$\zeta_{3j}^2 = \frac{1}{\bar{\theta}_j^2} \max_{l \in \{1,2,\dots,m_w\}} \left| \xi_{3j} \Theta \hat{L} E \mathcal{W}_l \right|^2 < 1, \ j \in \{1,\dots,q\}, \quad (10.2.13)$$

$$\zeta_{4j}^2 = \frac{1}{\bar{\theta}_j^2} \max_{l \in \{1,2,\dots,m_w\}} |\xi_{3j} \Theta D \mathcal{W}_l|^2 < 1, \ j \in \{1,\dots,q\}, \quad (10.2.14)$$

such that (10.2.1)-(10.2.3) and the following inequalities are satisfied:

$$\begin{bmatrix} G + G^T - Q_{11} & * & * \\ G_{12} - Q_{12} & G_2 + G_2^T - Q_{22} & * \\ (A - \hat{L}\Psi_s)(G + G_{12}) - M + B\Pi_l Y & (A - \hat{L}\Psi_s)G_2 & \Lambda \end{bmatrix} \geq 0,$$
$$\Lambda_{jj} \leq (1 - \zeta_{1j})^2 \bar{e}_j^2, \ s \in \{1, \cdots, m_{h\phi}\}, \ l \in \{1,\dots,m_{fg}\}, \ j \in \{1,\dots,n_x\},$$
$$(10.2.15)$$

$$\begin{bmatrix} G + G^T - Q_{11} & * & * \\ G_{12} - Q_{12} & G_2 + G_2^T - Q_{22} & * \\ Y & 0 & U \end{bmatrix} \geq 0, \ U_{jj} \leq \bar{v}_j^2, \ j \in \{1,\dots,n_u\},$$
$$(10.2.16)$$

$$\begin{bmatrix} G + G^T - Q_{11} & * & * \\ G_{12} - Q_{12} & G_2 + G_2^T - Q_{22} & * \\ C\Xi_s A(G + G_{12}) + C\Xi_s B\Pi_l Y & C\Xi_s AG_2 & Z \end{bmatrix} \geq 0,$$
$$Z_{jj} \leq (1 - \zeta_{2j})^2 \bar{z}_j^2, \ s \in \{1,\dots,m_\phi\}, \ l \in \{1,\dots,m_{fg}\}, \ j \in \{1,\dots,n_y\},$$
$$(10.2.17)$$

$$\begin{bmatrix} G + G^T - Q_{11} & * & * \\ G_{12} - Q_{12} & G_2 + G_2^T - Q_{22} & * \\ \Theta M + \Theta\hat{L}\Psi_s(G + G_{12}) & \Theta\hat{L}\Psi_s G_2 & \Upsilon \end{bmatrix} \geq 0,$$
$$\Upsilon_{jj} \leq (1 - \zeta_{3j})^2 \bar{\theta}_j^2, \ s \in \{1,\dots,m_{h\phi}\}, \ j \in \{1,\dots,q\}, \quad (10.2.18)$$

$$\begin{bmatrix} G + G^T - Q_{11} & * & * \\ G_{12} - Q_{12} & G_2 + G_2^T - Q_{22} & * \\ \Theta A(G + G_{12}) + \Theta B\Pi_l Y & \Theta AG_2 & \Gamma \end{bmatrix} \geq 0,$$
$$\Gamma_{jj} \leq (1 - \zeta_{4j})^2 \bar{\theta}_j^2, \ l \in \{1,\dots,m_{fg}\}, \ j \in \{1,\dots,q\}, \quad (10.2.19)$$

where ξ_{1j} (ξ_{2j}, ξ_{3j}) is the j-th row of the n_x (n_y, q) ordered identity matrix; Λ_{jj} (U_{jj}, Z_{jj}, Υ_{jj}, Γ_{jj}) is the j-th diagonal element of Λ (U, Z, Υ, Γ). Then, (10.1.14) holds by applying $u(k + i|k) = g(v^L(k + i|k))$, $\forall i \geq 0$ and $\{\hat{A}(k) = MG^{-1}, \ \hat{L}(k) = \hat{L}\}$ (where $v^L(\cdot) = F(k)\hat{x}(\cdot)$, $F(k) = YG^{-1}$).

Proof. Equation (10.2.15) guarantees

$$\left\|\xi_{1j}[A - \hat{A} + B\Pi(k+i)F - \hat{L}\Psi(k+i) \quad A - \hat{L}\Psi(k+i)]\tilde{Q}^{1/2}\right\|^2 \leq (1-\zeta_{1j})^2\bar{e}_j^2.$$
(10.2.20)

Applying (10.2.5), (10.2.11), (10.2.20) and (10.2.4) we have

$$|\xi_{1j}e(k+i+1|k)|^2 = |\xi_{1j}[0 \quad I]\tilde{x}(k+i+1|k)|^2$$
$$= |\xi_{1j}[A - \hat{A} + B\Pi(k+i)F - \hat{L}\Psi(k+i) \quad A$$
$$- \hat{L}\Psi(k+i)]\tilde{x}(k+i|k) + \xi_{1j}(D - \hat{L}E)w(k+i)|^2$$

$$\leq (1+\delta_{1j})\left\|\xi_{1j}[A - \hat{A} + B\Pi(k+i)F - \hat{L}\Psi(k+i) \quad A - \hat{L}\Psi(k+i)]\tilde{x}(k+i|k)\right\|^2$$
$$+ \left(1 + \frac{1}{\delta_{1j}}\right)\zeta_{1j}^2\bar{e}_j^2$$
$$\leq (1+\delta_{1j})(1 - \zeta_{1j})^2\bar{e}_j^2 + \left(1 + \frac{1}{\delta_{1j}}\right)\zeta_{1j}^2\bar{e}_j^2.$$

By choosing $\delta_{1j} = \zeta_{1j}/(1 - \zeta_{1j})$, it follows that $|\xi_{1j}e(k+i+1|k)|^2 \leq \bar{e}_j^2$, $\forall i \geq 0$.

Define ξ_j as the j-th row of the n_u-ordered identity matrix. Equation (10.2.16) guarantees

$$\left\|\xi_j[F \quad 0]\tilde{Q}^{1/2}\right\|^2 \leq \bar{v}_j^2. \qquad (10.2.21)$$

Applying (10.2.5),(10.2.21) and (10.2.4) we have

$$\max_{i\geq 0}|\xi_j F\hat{x}(k+i|k)|^2 = \max_{i\geq 0}|\xi_j[F \quad 0]\tilde{x}(k+i|k)|^2 \leq \left\|\xi_j[F \quad 0]\tilde{Q}^{1/2}\right\|^2 \leq \bar{v}_j^2.$$

Equation (10.2.17) guarantees

$$\left\|\xi_{2j}C\Xi(k+i)[A + B\Pi(k+i)F \quad A]\tilde{Q}^{1/2}\right\|^2 \leq (1 - \zeta_{2j})^2\bar{z}_j^2. \qquad (10.2.22)$$

Applying (10.2.5), (10.2.12), (10.2.22) and (10.2.4) we have

$$|\xi_{2j}z(k+i+1|k)|^2 = |\xi_{2j}C\Xi(k+i)[I \quad I]\tilde{x}(k+i+1|k)|^2$$
$$= |\xi_{2j}C\Xi(k+i)[A + B\Pi(k+i)F \quad A]\tilde{x}(k+i|k) + \xi_{2j}C\Xi(k+i)Dw(k+i)|^2$$
$$\leq (1+\delta_{2j})\left\|\xi_{2j}C\Xi(k+i)[A + B\Pi(k+i)F \quad A)]\tilde{x}(k+i|k)\right\|^2 + \left(1 + \frac{1}{\delta_{2j}}\right)\zeta_{2j}^2\bar{z}_j^2$$
$$\leq (1+\delta_{2j})(1 - \zeta_{2j})^2\bar{z}_j^2 + \left(1 + \frac{1}{\delta_{2j}}\right)\zeta_{2j}^2\bar{z}_j^2.$$

By choosing $\delta_{2j} = \zeta_{2j}/(1 - \zeta_{2j})$, it follows that $|\xi_{2j}z(k+i+1|k)|^2 \leq \bar{z}_j^2$, $\forall i \geq 0$.

Equation (10.2.18) guarantees

$$\left\| \xi_{3j}\Theta[\ \hat{A}+\hat{L}\Psi(k)\quad \hat{L}\Psi(k)\ ]\tilde{Q}^{1/2}\right\|^2 \le (1-\zeta_{3j})^2\bar{\theta}_j^2. \tag{10.2.23}$$

Applying (10.2.5), (10.2.13), (10.2.23) and (10.2.4) we have

$$|\xi_{3j}\Theta\hat{x}(k+i+1|k)|^2 = |\xi_{3j}\Theta[I\ \ 0]\tilde{x}(k+i+1|k)|^2$$

$$= \left|\xi_{3j}\Theta[\ \hat{A}+\hat{L}\Psi(k)\quad \hat{L}\Psi(k)\ ]\tilde{x}(k+i|k)+\xi_{3j}\Theta\hat{L}Ew(k+i)\right|^2$$

$$\le (1+\delta_{3j})\left\|\xi_{3j}\Theta[\ \hat{A}+\hat{L}\Psi(k)\quad \hat{L}\Psi(k)\ ]\tilde{x}(k+i|k)\right\|^2 + \left(1+\frac{1}{\delta_{3j}}\right)\zeta_{3j}^2\bar{\theta}_j^2$$

$$\le (1+\delta_{3j})(1-\zeta_{3j})^2\bar{\theta}_j^2 + \left(1+\frac{1}{\delta_{3j}}\right)\zeta_{3j}^2\bar{\theta}_j^2.$$

By choosing $\delta_{3j}=\zeta_{3j}/(1-\zeta_{3j})$, it follows that $|\xi_{3j}\Theta\hat{x}(k+i+1|k)|^2 \le \bar{\theta}_j^2$, $\forall i \ge 0$.

Equation (10.2.19) guarantees

$$\left\| \xi_{3j}\Theta[\ A+B\Pi(k+i)F\quad A\ ]\tilde{Q}^{1/2}\right\|^2 \le (1-\zeta_{4j})^2\bar{\theta}_j^2. \tag{10.2.24}$$

Applying (10.2.5), (10.2.14), (10.2.24) and (10.2.4) we have

$$|\xi_{3j}\Theta x(k+i+1|k)|^2 = |\xi_{3j}\Theta[I\ \ I]\tilde{x}(k+i+1|k)|^2$$

$$= \left|\xi_{3j}\Theta[\ A+B\Pi(k+i)F\quad A\ ]\tilde{x}(k+i|k)+\xi_{3j}\Theta Dw(k+i)\right|^2$$

$$\le (1+\delta_{4j})\left\|\xi_{3j}\Theta[\ A+B\Pi(k+i)F\quad A\ )]\tilde{x}(k+i|k)\right\|^2 + \left(1+\frac{1}{\delta_{4j}}\right)\zeta_{4j}^2\bar{\theta}_j^2$$

$$\le (1+\delta_{4j})(1-\zeta_{4j})^2\bar{\theta}_j^2 + \left(1+\frac{1}{\delta_{4j}}\right)\zeta_{4j}^2\bar{\theta}_j^2.$$

By choosing $\delta_{4j}=\zeta_{4j}/(1-\zeta_{4j})$, it follows that $|\xi_{3j}\Theta x(k+i+1|k)|^2 \le \bar{\theta}_j^2$, $\forall i \ge 0$. □

10.3   Realization algorithm: case systems with I/O nonlinearities

10.3.1   General optimization problem

Define a positive-definite quadratic function $V(i,k) = \tilde{x}_u(k+i|k)^T X(k)\tilde{x}_u(k+i|k)$, and impose the following optimality requirement:

$$V(i+1,k) - V(i,k) \le -\|y_u(k+i|k)\|_W^2 - \|F(k)\hat{x}_u(k+i|k)\|_R^2, \ \forall k, i \ge 0 \tag{10.3.1}$$

which is guaranteed by

$$\|\mathcal{A}(k,i)\tilde{x}_u(k+i|k)\|^2_{X(k)} - \|\tilde{x}_u(k+i|k)\|^2_{X(k)}$$
$$\leq - \|\Psi(k+i)[I\ \ I]\tilde{x}_u(k+i|k)\|^2_W - \left\|[\ F(k)\ \ 0\ ]\tilde{x}_u(k+i|k)\right\|^2_R.$$
$$(10.3.2)$$

Equation (10.3.2) is satisfied if and only if:

$$\mathcal{A}(k,i)^T X(k)\mathcal{A}(k,i) - X(k) \leq - [I\ \ I]^T \Psi(k+i)^T W\Psi(k+i)[I\ \ I]$$
$$- [\ F(k)\ \ 0\ ]^T R [\ F(k)\ \ 0\ ].\quad (10.3.3)$$

Define $F(k) = YG^{-1}$ and $\tilde{Q} = \beta X(k)^{-1}$. Similarly using the deductions as in Lemma 10.2.1, (10.3.3) can be guaranteed by the following inequality:

$$\begin{bmatrix} G + G^T - Q_{11} & * & * & * & * & * \\ G_{12} - Q_{12} & G_2 + G_2^T - Q_{22} & * & * & * & * \\ \hat{L}\Psi_s(G + G_{12}) + M & \hat{L}\Psi_s G_2 & Q_{11} & * & * & * \\ (A - \hat{L}\Psi_s)(G + G_{12}) - M + B\Pi_l Y & (A - \hat{L}\Psi_s)G_2 & Q_{12} & Q_{22} & * & * \\ W^{1/2}\Psi_s\left(A(G + G_{12}) + B\Pi_l Y\right) & W^{1/2}\Psi_s A G_2 & 0 & 0 & \beta I & * \\ R^{1/2}Y & 0 & 0 & 0 & 0 & \beta I \end{bmatrix} \geq 0,$$

$$s \in \{1, \ldots, m_{h\phi}\},\ l \in \{1, \ldots, m_{fg}\}.\quad (10.3.4)$$

If $w(k + i) = 0,\ \forall i \geq 0$, then $\hat{x}(\infty|k) = 0$, $e(\infty|k) = 0$ and $V(\infty, k) = 0$. Summing (10.3.1) from $i = 0$ to $i = \infty$ leads to

$$J_\infty(k) \leq \tilde{x}(k)^T X(k)\tilde{x}(k) \leq \beta.\quad (10.3.5)$$

The right side inequality in (10.3.5) can be transformed into (10.2.1) and, according to (10.1.12), (10.2.1) is guaranteed by the following LMI:

$$\begin{bmatrix} 1 & * & * \\ \hat{x}(k) & Q_{11} & * \\ \epsilon_j & Q_{12} & Q_{22} \end{bmatrix} \geq 0,\ j \in \{1, \ldots, 2^{n_x}\}.\quad (10.3.6)$$

Let us also consider the following condition, the satisfaction of which renders satisfaction of $x(k) \in \mathcal{S}$ and $\hat{x}(k) \in \mathcal{S}$:

$$-\bar{\theta} \leq \Theta(\hat{x}(k) + \epsilon_j) \leq \bar{\theta},\ j \in \{1, 2, \ldots, 2^{n_x}\}.\quad (10.3.7)$$

Thus, if (10.2.12) and (10.2.14) are satisfied, then problem (10.1.13)-(10.1.14) can be approximated by:

$$\min_{\eta,\beta,M,\hat{L},Y,G,G_{12},G_2,Q_{11},Q_{12},Q_{22},\Lambda,U,Z,\Upsilon,\Gamma} \beta,\ \text{s.t. (10.2.2)} - (10.2.3),$$

$$(10.2.11), (10.2.13), (10.2.15) - (10.2.19), (10.3.4), (10.3.6) - (10.3.7).$$
$$(10.3.8)$$

Equations (10.2.3), (10.2.15), (10.2.18) and (10.3.4) are not LMIs, which means that in most cases there may not exist a polynomial time algorithm for solving the optimization problem.

10.3.2 Linear matrix inequality optimization problem

In order to simplify the computation, we can fix $\hat{L}$, η a priori. If $\hat{L}$, η are fixed a priori and (10.2.11)-(10.2.14) are satisfied,then problem (10.1.13)-(10.1.14) can be approximated by:

$$\min_{\beta,M,Y,G,G_{12},G_2,Q_{11},Q_{12},Q_{22},\Lambda,U,Z,\Upsilon,\Gamma} \beta, \text{ s.t. } (10.2.2) - (10.2.3),$$

$$(10.2.15) - (10.2.19), (10.3.4), (10.3.6) - (10.3.7). \qquad (10.3.9)$$

Equation (10.3.9) is an LMI optimization problem. The complexity of solving (10.3.9) is polynomial-time, which is proportional to $\mathfrak{K}^3\mathfrak{L}$, where
$\mathfrak{K} = 1 + \frac{13}{2}n_x^2 + \frac{3}{2}n_x + n_x n_u + \frac{1}{2}n_u^2 + \frac{1}{2}n_u + \frac{1}{2}n_y^2 + \frac{1}{2}n_y + q^2 + q,$
$\mathfrak{L} = (2n_x + 2q + 1)2^{n_x} + (2n_x + n_y)m_\phi m_{fg} + (11n_x + n_u + n_y)m_{h\phi}m_{fg}$
$\quad + (2n_x + q)(m_{h\phi} + m_{fg}) + (2n_x + 1)m_w + 3n_x + 2n_u + n_y + 2q.$
Hence, by increasing m_ϕ ($m_{h\phi}$, m_{fg}, m_w), the computational burden is increased linearly; by increasing n_u (n_y, q), the computational burden is increased with a power law; by increasing n_x, the computational burden is increased exponentially. For $n_x = n_u = n_y = q = m_{h\phi} = m_\phi = m_{fg}$, preserving the most influential parameters in $\mathfrak{K}$ and $\mathfrak{L}$ we can say that the computational complexity for solving (10.3.9) is proportional to $(4n_x + 1)n_x^6 2^{n_x-3} + 2n_x^9$.
Further, let us consider $w(k) = 0$, $\forall k \geq 0$; then (10.3.9) can be approximated by

$$\min_{\beta,M,Y,G,G_{12},G_2,Q_{11},Q_{12},Q_{22},\Lambda,U,Z,\Upsilon,\Gamma} \beta,$$

s.t. $(10.2.15) - (10.2.19)$, $(10.3.4), (10.3.6) - (10.3.7)$,

where $\Lambda_{jj} \leq (1 - \zeta_{1j})^2 \bar{e}_j^2$, $Z_{jj} \leq (1 - \zeta_{2j})^2 \bar{z}_j^2$, $\Upsilon_{jj} \leq (1 - \zeta_{3j})^2 \bar{\theta}_j^2$

and $\Gamma_{jj} \leq (1 - \zeta_{4j})^2 \bar{\theta}_j^2$ are replaced by:

$$\Lambda_{jj} \leq \bar{e}_j^2, Z_{jj} \leq \bar{z}_j^2, \Upsilon_{jj} \leq \bar{\theta}_j^2 \text{ and } \Gamma_{jj} \leq \bar{\theta}_j^2. \qquad (10.3.10)$$

Equation (10.3.10) is the simplification of (10.3.9); $\mathfrak{K}$ is the same as that of (10.3.9), and
$\mathfrak{L} = (2n_x + 2q + 1)2^{n_x} + (2n_x + n_y)m_\phi m_{fg} + (7n_x + n_u + n_y)m_{h\phi}m_{fg}$
$\quad + (2n_x + q)(m_{h\phi} + m_{fg}) + 3n_x + 2n_u + n_y + 2q.$
For $n_x = n_u = n_y = q = m_{h\phi} = m_\phi = m_{fg}$, preserving the most influential parameters in $\mathfrak{K}$ and $\mathfrak{L}$ we can say that the computational complexity for solving (10.3.10) is proportional to $(4n_x + 1)n_x^6 2^{n_x-3} + \frac{3}{2}n_x^9$.

Remark 10.3.1. One can calculate $\Psi_0 = \sum_{l=1}^{m_{h\phi}} \Psi_l / m_{h\phi}$ and design $\hat{L}$ by the usual pole placement method with respect to $A - \hat{L}\Psi_0$. In case the performance (region of attraction, optimality) is not satisfactory, one can change the poles and re-design $\hat{L}$. If $\hat{L} = \hat{L}_0$ is feasible but not satisfactory, one can choose $\hat{L} = \rho\hat{L}_0$ and search over the scalar ρ. One can also choose η as an optimization variable in solving (10.3.9). By line searching η over the interval $(0, 1)$, (10.3.9) can still be solved via LMI techniques. By line searching η over the interval

$(0, 1)$, we can find η such that β is minimized (the computational burden is increased in this situation).

Algorithm 10.1 Fix $\hat{L}$, η a priori. At the initial time $k = 0$, choose an appropriate $\hat{x}(0)$ such that $-\bar{e} \leq e(0) \leq \bar{e}$. At any time $k \geq 0$, solve (10.3.10) to obtain $\{F(k), \hat{A}(k)\}$. For $k > 0$, if (10.3.10) is infeasible, then choose $\{F(k), \hat{A}(k)\} = \{F(k-1), \hat{A}(k-1)\}$.

Theorem 10.3.1. *(Stability) Consider system (10.1.1) with $w(k) = 0$, $\forall k \geq 0$. Suppose $-\bar{e} \leq e(0) \leq \bar{e}$ and (10.3.10) is feasible at $k = 0$. Then with Algorithm 10.1 applied, under the control law (10.1.5)-(10.1.6), the constraints (10.1.2)-(10.1.3) are always satisfied and $\lim_{k \to \infty} \tilde{x}(k) = 0$.*

Proof. The important feature of our OFMPC is that the estimation error is consistently bounded, and $x \in \mathcal{S}$ satisfied, for the infinite-time horizon. At the initial time $k = 0$, if there is a feasible solution to problem (10.3.10), then at any time $k > 0$ it is reasonable to adopt (10.3.6) and (10.3.7) in the optimization. However, since (10.3.6) is a stronger condition than (10.2.1), and (10.3.7) is stronger than $-\bar{\theta} \leq \Theta x(k) \leq \bar{\theta}$, it may occur that at some sampling instants, problem (10.3.10) becomes infeasible. According to Algorithm 10.1, in this situation $\{F(k-1), \hat{A}(k-1)\}$ will be applied. According to Lemma 10.2.2, constraints (10.1.2) and (10.1.3) will be satisfied for all $k \geq 0$. The proof of stability of $\{F(k), \hat{A}(k)\}$ for any k is an extension of the state feedback case (and hence is omitted here). Due to the utilization of (10.3.1), both the estimator state and estimation error will converge towards the origin with the evolution of time. □

Remark 10.3.2. Let us define the region of attraction for Algorithm 10.1 as $\mathfrak{D}$ (the closed-loop system is asymptotically stable whenever $\hat{x}(0) \in \mathfrak{D}$). Suppose at time $k > 0$, (10.3.10) is infeasible, then $\hat{x}(k) \notin \mathfrak{D}$. However, in this situation, $\tilde{x}(k)^T \tilde{Q}(k-1)^{-1}\tilde{x}(k) \leq 1$ is satisfied according to Lemma 10.2.1. Hence, with $\{F(k-1), \hat{A}(k-1)\}$ applied at time k, the input/state constraints are satisfied (Lemma 10.2.2) and $\tilde{x}(k+1)^T \tilde{Q}(k-1)^{-1}\tilde{x}(k+1) \leq 1$ (Lemma 10.2.1). The convergence of the augmented state towards the origin is due to (10.3.1), which forces $V(k+1)$ to decrease at least by the stage cost $\|y_u(k)\|_W^2 + \|F(k)\hat{x}_u(k)\|_R^2$ than $V(k)$.

Assumption 10.3.1. *The choice of $\hat{L}$, together with the selections of error bounds in (10.1.11) and $\mathcal{S}$ in Assumption 10.1.4, renders satisfaction of (10.2.11) and (10.2.13).*

Assumption 10.3.2. *The selection Ω^ϕ in Assumption 10.1.5 renders satisfaction of (10.2.12).*

Assumption 10.3.3. *The selection of $\mathcal{S}$ in Assumption 10.1.4 renders satisfaction of (10.2.14).*

Assumptions 10.3.1-10.3.3 reflect the ability of OFMPC to handle disturbance/noise. If $w(k) = 0$, $\forall k \geq 0$, then Assumptions 10.3.1-10.3.3 are trivially satisfied.

Algorithm 10.2 Fix $\hat{L}$, η a priori. At the initial time $k = 0$, choose an appropriate $\hat{x}(0)$ such that $-\bar{e} \leq e(0) \leq \bar{e}$. At any time $k \geq 0$, solve (10.3.9) to obtain $\{F(k), \hat{A}(k)\}$. For $k > 0$, if (10.3.9) is infeasible, then choose $\{F(k), \hat{A}(k)\} = \{F(k-1), \hat{A}(k-1)\}$.

Theorem 10.3.2. *(Stability) Consider system* (10.1.1). *Suppose Assumptions 10.3.1-10.3.3 hold,* $-\bar{e} \leq e(0) \leq \bar{e}$ *and* (10.3.9) *is feasible at $k = 0$. Then, with Algorithm 10.2 applied, under the control law* (10.1.5)-(10.1.6), *the constraints* (10.1.2)-(10.1.3) *are always satisfied and there exists a region $\mathfrak{D}^0$ about $\tilde{x} = 0$ such that* $\lim_{k \to \infty} \tilde{x}(k) \in \mathfrak{D}^0$.

Proof. This is an extension of Theorem 10.3.1. In the presence of nonzero disturbance/noise $w(k)$, the estimator state and estimation error will not converge to the origin.   □

10.3.3   Summary of the idea

In the above, OFMPC for input/output nonlinear models is considered, where the technique of linear difference inclusion is applied to obtain the polytopic description. The whole procedure is not direct. To make clear, it is necessary to set forth the whole procedure. Given (10.1.1)-(10.1.3), the following steps can be followed to implement OFMPC by (10.3.9):

Off-line stage:

Step 1. Define $\eta = 0$, $\bar{\eta} = 1$. Substitute (10.1.3) with (10.1.9).

Step 2. Construct $g(\cdot)$ as the inverse (or the approximate inverse) of $f(\cdot)$.

Step 3. Substitute (10.1.2) with (10.1.8).

Step 4. Transform $f \circ g(\cdot)$ into a polytopic description.

Step 5. Select $\bar{e}$ to obtain (10.1.12).

Step 6. Select $\mathcal{S}$. Transform $h(C\phi(\cdot))$ and $\phi(\cdot)$ into polytopic descriptions.

Step 7. Check if (10.2.12) is satisfied. If not, go back to Step 6.

Step 8. Check if (10.2.14) is satisfied. If not, go back to Step 6.

Step 9. Select $\hat{L}$ (considering Remark 10.3.1) to satisfy (10.2.11). If (10.2.11) cannot be satisfied, then re-select $\bar{e}$.

Step 10. Check if (10.2.13) is satisfied. If (10.2.13) cannot be satisfied, then go back to Step 6 (or go back to Step 9 to re-select $\hat{L}$).

Step 11. Select $\hat{x}(0) = 0$.

Step 12. Search η over the interval $(\underline{\eta}, \bar{\eta})$ to refresh $\underline{\eta}$ and $\bar{\eta}$, such that whenever $\underline{\eta} < \eta < \bar{\eta}$, (10.3.9) is feasible.

Step 13. If (10.3.9) is not feasible for any $\eta \in (\underline{\eta}, \bar{\eta})$, then go back to Step 2.

Step 14. Select several $\hat{x}(0) \neq 0$ over a user-specified region. For each $\hat{x}(0)$, go back to Step 12.

Step 15. For $\hat{x}(0) = 0$, search η over the interval $(\underline{\eta}, \bar{\eta})$, such that a minimum β is obtained.

On-line Stage:
At each time k, proceed with Algorithm 10.2.

In the above Off-line Stage, we have selected η and meant to fix it in the On-line Stage. If η is chosen as an optimization variable in solving (10.3.9) (see Remark 10.3.1), then Step 15 can be ignored and Step 12 can be revised as:

Step 12. Search η over the interval $(0, 1)$ such that (10.3.9) is feasible.

10.4 Optimization problem: case systems with polytopic description

Consider the following uncertain time-varying system

$$
\begin{aligned}
x(k+1) &= A(k)x(k) + B(k)u(k) + D(k)\bar{w}(k), \quad y(k) \\
&= C(k)x(k) + E(k)w(k), \quad k \geq 0,
\end{aligned} \tag{10.4.1}
$$

where $u \in \Re^{n_u}$, $x \in \Re^{n_x}$, $\bar{w} \in \Re^{n_{\bar{w}}}$, $y \in \Re^{n_y}$ and $w \in \Re^{n_w}$ are input, unmeasurable state, state disturbance, output and measurement noise, respectively. The state disturbance and measurement noise are persistent, bounded and satisfy

$$
\bar{w}(k) \in Co\{\bar{\mathcal{W}}_1, \bar{\mathcal{W}}_2, \cdots, \bar{\mathcal{W}}_{m_{\bar{w}}}\} \supseteq \{0\}, \; w(k) \in Co\{\mathcal{W}_1, \mathcal{W}_2, \cdots, \mathcal{W}_{m_w}\} \supseteq \{0\}. \tag{10.4.2}
$$

The input and state constraints are

$$
-\bar{u} \leq u(k) \leq \bar{u}, \; -\bar{\psi} \leq \Psi x(k+1) \leq \bar{\psi}, \tag{10.4.3}
$$

which has the same meaning as in the former chapters. Suppose

$$
[A(k)|B(k)|C(k)] \in \Omega = Co\{[A_1|B_1|C_1], [A_2|B_2|C_2], \cdots, [A_L|B_L|C_L]\}, \tag{10.4.4}
$$

$$
[D(k)|E(k)] \in Co\{[D_1|E_1], [D_2|E_2], \cdots, [D_p|E_p]\}. \tag{10.4.5}
$$

For the above system (10.4.1)-(10.4.5), our output feedback controller is of the following form

$$\hat{x}(k+1) = A_o(k)\hat{x}(k) + B_o u(k) + L_o y(k), \ u(k) = F(k)\hat{x}(k), \quad (10.4.6)$$

where $\hat{x}$ is the estimator state; $A_o(k)$, $F(k)$ are matrices to be designed; B_o, L_o are pre-specified matrices. Here, the controller that will be given is a more general case of that in section 10.3.2, i.e., rather than considering the polytope obtained from nonlinearity, here we directly consider polytopic description.

Remark 10.4.1. In Remark 10.3.1, the pole-placement scheme for designing $\hat{L}$ has been given. Now, we give a more conservative (but safer) scheme for designing L_o. Select L_o such that there should exist a positive-definite matrix P such that $P - (A_l - L_o C_l)^T P (A_l - L_o C_l) > 0$, $\forall l \in \{1, \dots, L\}$. By defining $\hat{B} = \frac{1}{L} \sum_{l=1}^{L} B_l$, we can simply choose $B_o = \hat{B}$.

In order to conveniently implement OFRMPC in a receding horizon manner, let the estimation error $e = x - \hat{x}$ satisfy $-\bar{e} \le e(k) \le \bar{e}$, where $\bar{e}_j > 0$, $j \in \{1, \dots, n_x\}$. Technically, we can express this estimation error constraint by

$$e(k) \in \Omega^e = Co\{\epsilon_1, \epsilon_2, \cdots, \epsilon_{2^{n_x}}\}, \ \forall k. \quad (10.4.7)$$

For simplicity, denote $\tilde{x} = [\hat{x}^T \ e^T]^T$, $\tilde{w} = [\bar{w}^T \ w^T]^T$.

We aimed at synthesizing an OFRMPC that brings system (10.4.1)-(10.4.6) to a bounded target set about $\{\hat{x}, e, u\} = 0$, by solving the following optimization problem at each time k:

$$\min_{A_o, F, Q, \gamma} \max_{[A(k+i)|B(k+i)|C(k+i)] \in \Omega, \ i \ge 0} J_\infty(k)$$

$$= \sum_{i=0}^{\infty} \left[\|\tilde{x}_u(k+i|k)\|_W^2 + \|F(k)\hat{x}_u(k+i|k)\|_R^2 \right], \quad (10.4.8)$$

s.t. $\tilde{x}_u(k+i+1|k) = T_o(k+i)\tilde{x}_u(k+i|k), \ \tilde{x}_u(k|k) = \tilde{x}(k), \quad (10.4.9)$

$\tilde{x}(k+i+1|k) = T_o(k+i)\tilde{x}(k+i|k) + H_o(k+i)\tilde{w}(k+i), \ \tilde{x}(k|k) = \tilde{x}(k),$

$$\quad (10.4.10)$$

$-\bar{e} \le e(k+i+1|k) \le \bar{e}, \ -\bar{u} \le u(k+i|k) = F\hat{x}(k+i|k) \le \bar{u},$

$-\bar{\psi} \le \Psi x(k+i+1|k) \le \bar{\psi}, \quad (10.4.11)$

$\|\tilde{x}_u(k+i+1|k)\|_{Q^{-1}}^2 - \|\tilde{x}_u(k+i|k)\|_{Q^{-1}}^2 \le -1/\gamma \|\tilde{x}_u(k+i|k)\|_W^2$

$-1/\gamma \|F\hat{x}_u(k+i|k)\|_R^2, \quad (10.4.12)$

$\|\tilde{x}(k+i|k)\|_{Q^{-1}}^2 \le 1, \ Q = Q^T > 0, \quad (10.4.13)$

where $\tilde{x}_u = [\hat{x}_u^T \ e_u^T]^T$ and $\hat{x}_u$ (e_u) is the prediction of the estimator state (estimation error) not corrupted by the disturbance/noise;

$$T_o(k+i) =$$
$$\begin{bmatrix} A_o(k) + B_o F(k) + L_o C(k+i) & L_o C(k+i) \\ A(k+i) - A_o(k) + (B(k+i) - B_o)F(k) - L_o C(k+i) & A(k+i) - L_o C(k+i) \end{bmatrix},$$
$$H_o(k+i) = \begin{bmatrix} 0 & L_o E(k+i) \\ D(k+i) & -L_o E(k+i) \end{bmatrix}, \quad W = \begin{bmatrix} W_1 & 0 \\ 0 & W_2 \end{bmatrix};$$

$W_1 > 0$, $W_2 > 0$ and $R > 0$ are symmetric weighting matrices; (10.4.12) is to ensure cost monotonicity; (10.4.13) is to ensure invariance of the augmented state $\tilde{x}$.

10.5 Optimality, invariance and constraint handling: case systems with polytopic description

Considering (10.4.13) ,with $i = 0$, yields

$$[\hat{x}(k)^T \quad e(k)^T] Q^{-1} [\hat{x}(k)^T \quad e(k)^T]^T \le 1. \tag{10.5.1}$$

To achieve an asymptotically stable closed-loop system, $\lim_{i \to \infty} \tilde{x}_u(k+i|k) = 0$. By summing (10.4.12) from $i = 0$ to $i = \infty$ and applying (10.5.1), it follows that $J_\infty(k) \le \gamma$. Thus, by minimizing γ subject to (10.4.12) and (10.5.1), the performance cost is optimized with respect to the worst-case of the polytopic description Ω. By applying (10.4.9), it follows that (10.4.12) is satisfied if and only if

$$T_o(k+i)^T Q^{-1} T_o(k+i) - Q^{-1} \le -1/\gamma W - 1/\gamma [F(k) \quad 0]^T R[F(k) \quad 0], \quad i \ge 0. \tag{10.5.2}$$

Proposition 10.5.1. *(Optimality) Suppose there exist a scalar γ, matrices Q_{12}, G, G_{12}, G_2, Y, M and symmetric matrices Q_{11}, Q_{22} such that the following LMIs are satisfied:*

$$\begin{bmatrix} G + G^T - Q_{11} & * & * & * & * & * & * \\ G_{12} - Q_{12} & G_2 + G_2^T - Q_{22} & * & * & * & * & * \\ L_o C_l(G + G_{12}) + M + B_o Y & L_o C_l G_2 & Q_{11} & * & * & * & * \\ (A - L_o C_l)(G + G_{12}) - M + (B_l - B_o)Y & (A_l - L_o C_l)G_2 & Q_{12} & Q_{22} & * & * & * \\ W_1^{1/2} G & 0 & 0 & 0 & \gamma I & * & * \\ W_2^{1/2} G_{12} & W_2^{1/2} G_2 & 0 & 0 & 0 & \gamma I & * \\ R^{1/2} Y & 0 & 0 & 0 & 0 & 0 & \gamma I \end{bmatrix} \ge 0,$$
$$l \in \{1, \ldots, L\}, \tag{10.5.3}$$

$$\begin{bmatrix} 1 & * & * \\ \hat{x}(k) & Q_{11} & * \\ \epsilon_j & Q_{12} & Q_{22} \end{bmatrix} \ge 0, \quad j \in \{1, \ldots, 2^{n_x}\}. \tag{10.5.4}$$

Then, (10.5.1)-(10.5.2) hold by parameterizing

$$Q = \begin{bmatrix} Q_{11} & Q_{12}^T \\ Q_{12} & Q_{22} \end{bmatrix} \geq 0, \ F(k) = YG^{-1}, \ A_o(k) = MG^{-1}. \quad (10.5.5)$$

Proof. Denote $\tilde{G} = \begin{bmatrix} G & 0 \\ G_{12} & G_2 \end{bmatrix}$. By multiplying the left and right sides of (10.5.2) by $\tilde{G}^T$ and $\tilde{G}$, respectively, and applying Schur complement, (10.5.5), utilizing the fact that $\tilde{G} + \tilde{G}^T - Q \leq \tilde{G}^T Q^{-1} \tilde{G}$ and the convexity of the polytopic description, it can be shown that (10.5.3) guarantees (10.5.2). Moreover, by applying (10.4.7) and Schur complement, it is shown that (10.5.4) guarantees (10.5.1). $\qquad \square$

Proposition 10.5.2. *(Invariance) Suppose there exist a scalar γ, matrices Q_{12}, G, G_{12}, G_2, Y, M and symmetric matrices Q_{11}, Q_{22} such that (10.5.4) and the following LMIs are satisfied:*

$$\begin{bmatrix} \theta(G + G^T - Q_{11}) & * & * & * \\ \theta(G_{12} - Q_{12}) & \theta(G_2 + G_2^T - Q_{22}) & * & * \\ L_oC_l(G + G_{12}) + M + B_oY & L_oC_lG_2 & Q_{11} & * \\ (A - L_oC_l)(G + G_{12}) - M + (B_l - B_o)Y & (A_l - L_oC_l)G_2 & Q_{12} & Q_{22} \end{bmatrix} \geq 0,$$

$$l \in \{1, \ldots, L\}, \quad (10.5.6)$$

$$\begin{bmatrix} (1 - \theta^{1/2})^2 & * & * \\ L_oE_t\mathcal{W}_h & Q_{11} & * \\ D_t\bar{\mathcal{W}}_s - L_oE_t\mathcal{W}_h & Q_{12} & Q_{22} \end{bmatrix} \geq 0,$$

$$t \in \{1, \ldots, p\}, \ s \in \{1, \ldots, m_{\bar{w}}\}, \ h \in \{1, \ldots, m_w\}, \quad (10.5.7)$$

where θ is a pre-specified scalar, $0 < \theta < 1$. Then, (10.4.13) holds by the parameterization (10.5.5).

Proof. Eq. (10.5.6) guarantees (analogously to Proposition 10.5.1):

$$T_o(k + i)^T Q^{-1} T_o(k + i) \leq \theta Q^{-1}. \quad (10.5.8)$$

Let $\zeta > 0$ be a scalar such that

$$\tilde{w}(k + i)^T H_o(k + i)^T Q^{-1} H_o(k + i)\tilde{w}(k + i) \leq \zeta. \quad (10.5.9)$$

Applying (10.4.10), (10.2.5), (10.5.8) and (10.5.9) yields

$$\|\tilde{x}(k + i + 1|k)\|_{Q^{-1}}^2 \leq (1 + \delta)\theta\|\tilde{x}(k + i|k)\|_{Q^{-1}}^2 + (1 + 1/\delta)\zeta. \quad (10.5.10)$$

Suppose

$$(1 + \delta)\theta + (1 + 1/\delta)\zeta \leq 1. \quad (10.5.11)$$

With (10.5.1) and (10.5.11) satisfied, by applying (10.5.10) recursively (for $i = 0, 1, 2, \ldots$), (10.4.13) can be verified. The maximum allowable ζ satisfying (10.5.11) is $\zeta = (1 - \theta^{1/2})^2$. Hence, by applying (10.4.2) and (10.4.10), it is shown that (10.5.7) guarantees (10.5.9). $\qquad \square$

Proposition 10.5.3. *(Constraints handling) For each $j \in \{1, \ldots, n_x\}$, obtain ζ_{1j} by solving*

$$\min_{\zeta_{1j}} \zeta_{1j}, \text{ s.t. } \begin{bmatrix} \zeta_{1j} & * \\ D_{tj}\bar{W}_s - L_{oj}E_t\mathcal{W}_h & 1 \end{bmatrix} \geq 0,$$

$$t \in \{1, \ldots, p\}, \ s \in \{1, \ldots, m_{\bar{w}}\}, \ h \in \{1, \ldots, m_w\}, \quad (10.5.12)$$

where D_{tj} (L_{oj}) is the j-th row of D_t (L_o); for each $j \in \{1, \ldots, q\}$, obtain ζ_{2j} by solving

$$\min_{\zeta_{2j}} \zeta_{2j}, \text{ s.t. } \begin{bmatrix} \zeta_{2j} & * \\ \Psi_j D_t \bar{W}_s & 1 \end{bmatrix} \geq 0, \ t \in \{1, \ldots, p\}, \ s \in \{1, \ldots, m_{\bar{w}}\},$$

$$(10.5.13)$$

where Ψ_j is the j-th row of Ψ. Suppose there exist a scalar γ, matrices Q_{12}, G, G_{12}, G_2, Y, M and symmetric matrices Q_{11}, Q_{22}, Ξ, Z, Γ such that (10.5.4), (10.5.6), (10.5.7) and the following LMIs are satisfied:

$$\begin{bmatrix} G + G^T - Q_{11} & * & * \\ G_{12} - Q_{12} & G_2 + G_2^T - Q_{22} & * \\ (A_l - L_oC_l)(G + G_{12}) - M + (B_l - B_o)Y & (A_l - L_oC_l)G_2 & \Xi \end{bmatrix} \geq 0,$$

$$l \in \{1, \ldots, L\}, \ \Xi_{jj} \leq \tilde{e}_j^2, \ j \in \{1, \ldots, n_x\}, \quad (10.5.14)$$

$$\begin{bmatrix} G + G^T - Q_{11} & * & * \\ G_{12} - Q_{12} & G_2 + G_2^T - Q_{22} & * \\ Y & 0 & Z \end{bmatrix} \geq 0, \ Z_{jj} \leq \bar{u}_j^2, \ j \in \{1, \ldots, n_u\},$$

$$(10.5.15)$$

$$\begin{bmatrix} G + G^T - Q_{11} & * & * \\ G_{12} - Q_{12} & G_2 + G_2^T - Q_{22} & * \\ \Psi(A_lG + A_lG_{12} + B_lY) & \Psi A_lG_2 & \Gamma \end{bmatrix} \geq 0,$$

$$l \in \{1, \ldots, L\}, \ \Gamma_{jj} \leq \tilde{\psi}_j^2, \ j \in \{1, \ldots, q\}, \quad (10.5.16)$$

where $\tilde{e}_j = \bar{e}_j - \sqrt{\zeta_{1j}} > 0$, $\tilde{\psi}_j = \bar{\psi}_j - \sqrt{\zeta_{2j}} > 0$; Ξ_{jj} (Z_{jj}, Γ_{jj}) is the j-th diagonal element of Ξ (Z, Γ). Then, (10.4.11) is guaranteed through the parameterization (10.5.5).

Proof. Define ξ_j as the j-th row of the n_x-ordered identity matrix. LMIs in (10.5.12) guarantee $\max_{i \geq 0} |\xi_j[0 \ I]H_o(k+i)\tilde{w}(k+i)|^2 \leq \zeta_{1j}$. Applying (10.4.10), (10.2.5) and Proposition 10.5.2 yields, for any $\delta_{1j} > 0$,

$$\max_{i \geq 0} |\xi_j e(k+i+1|k)|^2 \leq (1 + \delta_{1j}) \max_{i \geq 0} \|\xi_j[0 \ I]T_o(k+i)Q^{1/2}\|^2 + (1 + 1/\delta_{1j})\zeta_{1j}.$$

$$(10.5.17)$$

Suppose

$$\max_{i \geq 0} \|\xi_j[0 \ I]T_o(k+i)Q^{1/2}\|^2 \leq \tilde{e}_j^2, \ j \in \{1, \ldots, n_x\}. \quad (10.5.18)$$

By considering (10.5.17) and (10.5.18), the estimation error constraint in (10.4.11) is satisfied if $(1 + \delta_{1j})\tilde{e}_j^2 + (1 + 1/\delta_{1j})\zeta_{1j} \leq \bar{e}_j^2$. By solving $\tilde{e}_j^2 =$

$\max_{\delta_{1j}} \{1/(1 + \delta_{1j})[\bar{e}_j^2 - (1 + 1/\delta_{1j})\zeta_{1j}]\}$, the best (maximum) choice of $\tilde{e}_j$ is obtained as $\tilde{e}_j = \bar{e}_j - \sqrt{\zeta_{1j}}$. Similarly to the state feedback case, (10.5.14) can guarantee (10.5.18).

Define ξ_j as the j-th row of the n_u-ordered identity matrix. Applying (10.5.5) and Proposition 10.5.2 yields $\max_{i \geq 0} |\xi_j F(k)\hat{x}(k + i|k)|^2 \leq \|\xi_j [Y \ \ 0]\tilde{G}^{-1}Q^{1/2}\|^2$. Hence, the input constraint in (10.4.11) is satisfied if $\|\xi_j [Y \ \ 0]\tilde{G}^{-1}Q^{1/2}\|^2 \leq \bar{u}_j^2$, which is in turn guaranteed by (10.5.15).

Define ξ_j as the j-th row of the q-ordered identity matrix. LMIs in (10.5.13) guarantee $\max_{i \geq 0} |\xi_j [I \ \ I]H_o(k + i)\tilde{w}(k + i)|^2 \leq \zeta_{2j}$. Applying (10.4.10), (10.2.5) and Proposition 10.5.2 yields, for any $\delta_{2j} > 0$,

$$\max_{i \geq 0} |\xi_j \Psi x(k + i + 1|k)|^2 \leq (1 + \delta_{2j}) \max_{i \geq 0} \|\xi_j \Psi [I \ \ I]T_o(k + i)Q^{1/2}\|^2$$
$$+ (1 + 1/\delta_{2j})\zeta_{2j}. \tag{10.5.19}$$

Suppose

$$\max_{i \geq 0} \|\xi_j \Psi [I \ \ I]T_o(k + i)Q^{1/2}\|^2 \leq \tilde{\psi}_j^2, \ j \in \{1, \ldots, q\}. \tag{10.5.20}$$

By considering (10.5.19) and (10.5.20), the state constraint in (10.4.11) is satisfied if $(1 + \delta_{2j})\tilde{\psi}_j^2 + (1 + 1/\delta_{2j})\zeta_{2j} \leq \bar{\psi}_j^2$. By solving $\tilde{\psi}_j^2 = \max_{\delta_{2j}} \{1/(1 + \delta_{2j})[\bar{\psi}_j^2 - (1 + 1/\delta_{2j})\zeta_{2j}]\}$, the best (maximum) choice of $\tilde{\psi}_j$ is obtained as $\tilde{\psi}_j = \bar{\psi}_j - \sqrt{\zeta_{2j}}$. Similarly to the state feedback case, (10.5.16) can guarantee (10.5.20). $\square$

10.6   Realization algorithm: case systems with polytopic description

By considering Propositions 10.5.1, 10.5.2, 10.5.3, problem (10.4.8)-(10.4.13) can be solved by LMI optimization problem:

$$\min_{\gamma, M, Y, G, G_{12}, G_2, Q_{11}, Q_{12}, Q_{22}, \Xi, Z, \Gamma} \gamma, \text{ s.t. } (10.5.3) - (10.5.4),$$
$$(10.5.6) - (10.5.7), (10.5.14) - (10.5.16). \tag{10.6.1}$$

The complexity of solving (10.6.1) is polynomial-time, which is proportional to $\mathfrak{K}^3 \mathfrak{L}$, where
$\mathfrak{K} = \frac{1}{2}(13n_x^2 + n_u^2 + q^2) + n_x n_u + \frac{1}{2}(3n_x + n_u + q) + 1$,
$\mathfrak{L} = (15n_x + n_u + q)L + (1 + 2n_x)(2^{n_x} + p m_{\tilde{w}} m_w) + 3n_x + 2n_u + q$.
One can also take θ as a degree-of-freedom in the optimization. By line searching θ over the interval $(0, 1)$, (10.6.1) can be iteratively solved by LMI technique.

In the following, we give the off-line approach based on (10.6.1).

Algorithm 10.3 Off-line, choose $\hat{x}_h$, $h \in \{1, \ldots, N\}$. For each h, substitute $\hat{x}(k)$ in (10.5.4) by $\hat{x}_h$, and solve the optimization problem (10.6.1) to obtain the corresponding matrix $A_o^h = M_h G_h^{-1}$, feedback gain $F_h = Y_h G_h^{-1}$ and region $\varepsilon_h = \{\hat{x} \in \Re^{n_x} | [\hat{x}^T \;\; \epsilon_j^T] Q_h^{-1} [\hat{x}^T \;\; \epsilon_j^T]^T \leq 1, j = 1, 2, \ldots, 2^{n_x}\}$ (an intersection of 2^{n_x} ellipsoidal regions). Note that $\hat{x}_h$ should be chosen such that $\varepsilon_{h+1} \subset \varepsilon_h$, $\forall h \neq N$.

On-line, at each time k, perform the following steps:

i) At $k = 0$, choose $\hat{x}(0)$; otherwise ($k > 0$), evaluate $\hat{x}(k) = A_o(k-1)\hat{x}(k-1) + B_o u(k-1) + L_o y(k-1)$.

ii) Choose $\{F(k), A_o(k)\} = \{F_1, A_o^1\}$. If $\hat{x}(k) \in \varepsilon_N$, then adopt $\{F(k), A_o(k)\} = \{F_N, A_o^N\}$; for $k > 1$, if $\hat{x}(k) \notin \varepsilon_1$, then adopt $\{F(k), A_o(k)\} = \{F(k-1), A_o(k-1)\}$; otherwise, if $\hat{x}(k) \in \varepsilon_h \backslash \varepsilon_{h+1}$, then adopt $\{F(k), A_o(k)\} = \{F_h, A_o^h\}$.

iii) Evaluate $u(k) = F(k)\hat{x}(k)$ and implement $u(k)$.

Theorem 10.6.1. *For system* (10.4.1)-(10.4.5), *Algorithm 10.3 is adopted. If* $\hat{x}(0) \in \varepsilon_1$ *and* $-\bar{e} \leq e(0) \leq \bar{e}$, *then there exists a region* $\mathfrak{D}$ *about* $\{\hat{x}, e, u\} = 0$ *such that* $\lim_{k \to \infty}\{\hat{x}(k), e(k), u(k)\} \in \mathfrak{D}$, *and the input/state constraints are satisfied for all* $k \geq 0$.

Proof. Consider $N = 1$. Then, $\{F(k), A_o(k)\} = \{F_1, A_o^1\}$, $\forall k \geq 0$. For some $k > 0$, it may happen that $\hat{x}(k) \notin \varepsilon_1$. However, $\{F(k), A_o(k)\} = \{F_1, A_o^1\}$ is still utilized. According to Proposition 10.5.3, the input/state constraints will be satisfied. According to Proposition 10.5.2, $T_o(k)$ is exponentially stable and $\tilde{x}$ will converge. In the presence of nonzero disturbance/noise, $\tilde{x}$ will not settle at 0. Instead, $\tilde{x}$ will converge to a region $\tilde{\mathfrak{D}}$ about $\tilde{x} = 0$ and stay within $\tilde{\mathfrak{D}}$ thereafter. Since $\tilde{x}$ converges to $\tilde{\mathfrak{D}}$, $u = F_1\hat{x}$ will converge to a region about $u = 0$, i.e., $\{\hat{x}, e, u\}$ will converge to $\mathfrak{D}$.

Consider $N > 1$. According to Proposition 10.5.3, the estimation error will satisfy $-\bar{e} \leq e(k) \leq \bar{e}$. Therefore, $\hat{x}(k) \in \varepsilon_h$ implies $\{\hat{x}(k), e(k)\} \in \varepsilon_h \times \Omega^e$ and the control law can be switched according to the location of $\hat{x}(k)$. At time k, if $\{F(k), A_o(k)\} = \{F_h, A_o^h\}$ is applied, then $T_o(k)$ is exponentially stable, (10.4.3) is satisfied and $\tilde{x}(k+1)$ will converge. At last, $\{\hat{x}, e, u\}$ will converge to a region $\mathfrak{D}$ and stay within $\mathfrak{D}$ thereafter. $\qquad\square$

For each $h \in \{1, \ldots, N\}$ in Algorithm 10.3, let us give

$$\hat{x}(k+1) = [A_o^h + B_o F_h + L_o C(k)]\hat{x}(k) + L_o C(k)e(k) + L_o E(k)w(k), \quad (10.6.2)$$

where $e(k) \in \Omega^e$, $w(k) \in Co\{W_1, W_2, \cdots, W_{m_w}\}$, $C(k) \in Co\{C_1, C_2, \cdots, C_L\}$, $E(k) \in Co\{E_1, E_2, \cdots, E_p\}$. Considering (10.6.2), there exists a region $\hat{\mathfrak{D}}_h$ about $\hat{x} = 0$ such that $\lim_{k \to \infty} \hat{x}(k) \in \hat{\mathfrak{D}}_h$. $\hat{\mathfrak{D}}_h$ is bounded since $e(k)$, $w(k)$ are bounded and $A_o^h + B_o F_h + L_o C(k)$ is asymptotically stable. The following conclusion can be easily obtained by considering Theorem 10.6.1:

Corollary 10.6.1. *(Stability) For system* (10.4.1)-(10.4.5), *Algorithm 10.3 is adopted. Suppose* $\varepsilon_N \supseteq \hat{\mathfrak{D}}_h \supseteq \{0\}$, $\forall h \in \{1, \dots, N\}$, $\hat{x}(0) \in \varepsilon_1$ *and* $-\bar{e} \le e(0) \le \bar{e}$. *Then,* $\lim_{k \to \infty}\{\hat{x}(k), e(k), u(k)\} \in \hat{\mathfrak{D}}_N \times \Omega^e \times F_N \hat{\mathfrak{D}}_N \supseteq \{0\}$, $\lim_{k \to \infty}\{F(k), A_o(k)\} = \{F_N, A_o^N\}$, *and the input/state constraints are satisfied for all* $k \ge 0$.

Remark 10.6.1. Both (10.1.11) and (10.4.7) are artificial. Treating the estimation error similarly to the input/state constraints brings conservativeness. Moreover, using the polytope to restrict and define the estimation error also increases the computational burden.

For the case when there is no disturbance/noise, [28] gives a simpler method.

Bibliography

[1] H.H.J. Bloemen, T.J.J. van de Boom, and H.B. Verbruggen. Model-based predictive control for Hammerstein-Wiener systems. *International Journal of Control*, 74:482–495, 2001.

[2] H.H.J. Bloemen, T.J.J. van de Boom, and H.B. Verbruggen. Optimizing the end-point state-weighting matrix in model-based predictive control. *Automatica*, 38:1061–1068, 2002.

[3] S. Boyd, L. El Ghaoui, E. Feron, and V. Balakrishnan. *Linear matrix inequalities in system and control theory*. SIAM Studies in Applied Mathematics. SIAM, Philadelphia, PA, 1994.

[4] M. Cao, Z. Wu, B. Ding, and C. Wang. On the stability of two-step predictive controller based-on state observer. *Journal of Systems Engineering and Electronics*, 17:132–137, 2006.

[5] H. Chen and F. Allgower. A quasi-infinite horizon nonlinear model predictive control scheme with guaranteed stability. *Automatica*, 34:1205–1217, 1998.

[6] D.W. Clarke, C. Mohtadi, and P.S. Tuffs. Generalized predictive control, Part I: Basic algorithm and Part II: Extensions and interpretations. *Automatica*, 23:137–160, 1987.

[7] D.W. Clarke and R. Scattolini. Constrained receding-horizon predictive control. *IEE Control Theory and Applications*, 138:347–354, 1991.

[8] B. Ding. *Methods for stability analysis and synthesis of predictive control*. PhD thesis, Shanghai Jiaotong University, China, 2003 (in Chinese).

[9] B. Ding and B. Huang. Constrained robust model predictive control for time-delay systems with polytopic description. *International Journal of Control*, 80:509–522, 2007.

[10] B. Ding and B. Huang. New formulation of robust MPC by incorporating off-line approach with on-line optimization. *International Journal of Systems Science*, 38:519–529, 2007.

[11] B. Ding and B. Huang. Output feedback model predictive control for nonlinear systems represented by Hammerstein-Wiener model. *IET Control Theory and Applications*, 1(5):1302–1310, 2007.

[12] B. Ding and S. Li. Design and analysis of constrained nonlinear quadratic regulator. *ISA Transactions*, 42(3):251–258, April 2003.

[13] B. Ding, S. Li, and Y. Xi. Robust stability analysis for predictive control with input nonlinearity. In *Proceedings of the American Control Conference*, pages 3626–3631. Denver, CO, 2003.

[14] B. Ding, S. Li, and Y. Xi. Stability analysis of generalized predictive control with input nonlinearity based-on Popov's theorem. *ACTA AUTOMATICA SINICA*, 29(4):582–588, 2003.

[15] B. Ding, S. Li, P. Yang, and H. Wang. Multivariable GPC and Kleinman's controller: stability and equivalence. In *Proceedings of the 3rd International Conference on Machine Learning and Cybernetics*, volume 1, pages 329–333. Shanghai, 2004.

[16] B. Ding, H. Sun, P. Yang, H. Tang, and B. Wang. A design approach of constrained linear time-varying quadratic regulation. In *Proceedings of the 43rd IEEE Conference on Decision and Control*, volume 3, pages 2954–2959. Atlantis, Paradise Island, Bahamas, 2004.

[17] B. Ding and J.H. Tang. Constrained linear time-varying quadratic regulation with guaranteed optimality. *International Journal of Systems Science*, 38:115–124, 2007.

[18] B. Ding and Y. Xi. Stability analysis of generalized predictive control based on Kleinman's controllers. *Science in China Series F-Information Science*, 47(4):458–474, 2004.

[19] B. Ding and Y. Xi. Design and analysis of the domain of attraction for generalized predictive control with input nonlinearity. *ACTA AUTOMATICA SINICA*, 30(6):954–960, 2004 (in Chinese).

[20] B. Ding and Y. Xi. A two-step predictive control design for input saturated Hammerstein systems. *International Journal of Robust and Nonlinear Control*, 16:353–367, 2006.

[21] B. Ding, Y. Xi, M.T. Cychowski, and T. O'Mahony. Improving off-line approach to robust MPC based-on nominal performance cost. *Automatica*, 43:158–163, 2007.

[22] B. Ding, Y. Xi, M.T. Cychowski, and T. O'Mahony. A synthesis approach of output feedback robust constrained model predictive control. *Automatica*, 44(1):258–264, 2008.

[23] B. Ding, Y. Xi, and S. Li. Stability analysis on predictive control of discrete-time systems with input nonlinearity. *ACTA AUTOMATICA SINICA*, 29(6):827–834, 2003.

[24] B. Ding, Y. Xi, and S. Li. On the stability of output feedback predictive control for systems with input nonlinearity. *Asian Journal of Control*, 6(3):388–397, 2004.

[25] B. Ding and P. Yang. Synthesizing off-line robust model predictive controller based-on nominal performance cost. *ACTA AUTOMATICA SINICA*, 32(2):304–310, 2006 (in Chinese).

[26] B. Ding, P. Yang, X. Li, H. Sun, and J. Yuan. Stability analysis of input nonlinear predictive control systems based-on state observer. In *Proceedings of the 23rd Chinese Control Conference*, pages 659–663. Wuxi, 2004 (in Chinese).

[27] B. Ding and J. Yuan. Steady properties of nonlinear removal generalized predictive control. *Control Engineering*, 11(4):364–367, 2004 (in Chinese).

[28] B. Ding and T. Zou. Synthesizing output feedback predictive control for constrained uncertain time-varying discrete systems. *ACTA AUTOMATICA SINICA*, 33(1):78–83, 2007 (in Chinese).

[29] B. Ding, T. Zou, and S. Li. Varying-horizon off-line robust predictive control for time-varying uncertain systems. *Control Theory and Applications*, 23(2):240–244, 2006 (in Chinese).

[30] K.P. Fruzzetti, A. Palazoglu, and K.A. Mcdonald. Nonlinear model predictive control using Hammerstein models. *Journal of Process Control*, 7(1):31–41, 1997.

[31] P. Gahinet, A. Nemirovski, A.J. Laub, and M. Chilali. *LMI control toolbox for use with Matlab, User's guide*. The MathWorks Inc., Natick, MA, 1995.

[32] E.G. Gilbert and K.T. Tan. Linear systems with state and control constraints: the theory and application of maximal output admissible sets. *IEEE Transactions on Automatic Control*, 36:1008–1020, 1991.

[33] T. Hu and Z. Lin. Semi-global stabilization with guaranteed regional performance of linear systems subject to actuator saturation. In *Proceedings of the American Control Conference*, pages 4388–4392. Chicago, IL, 2000.

[34] T. Hu, D.E. Miller, and L. Qiu. Controllable regions of LTI discrete-time systems with input nonlinearity. In *Proceedings of the 37th IEEE Conference on Decision and Control*, pages 371–376. Tampa, FL, 1998.

[35] T.A. Johansen. Approximate explicit receding horizon control of constrained nonlinear systems. *Automatica*, 40:293–300, 2004.

[36] D.L. Kleinman. Stabilizing a discrete, constant, linear system with application to iterative methods for solving the Riccati equation. *IEEE Transactions on Automatic Control*, 19:252–254, 1974.

[37] M.V. Kothare, V. Balakrishnan, and M. Morari. Robust constrained model predictive control using linear matrix inequalities. *Automatica*, 32:1361–1379, 1996.

[38] B. Kouvaritakis, J.A. Rossiter, and J. Schuurmans. Efficient robust predictive control. *IEEE Transactions on Automatic Control*, 45:1545–1549, 2000.

[39] W.H. Kwon. *Receding horizon control: model predictive control for state space model*. Springer-Verlag, 2004.

[40] W.H. Kwon and D.G. Byun. Receding horizon tracking control as a predictive control and its stability properties. *International Journal of Control*, 50:1807–1824, 1989.

[41] W.H. Kwon, H. Choi, D.G. Byun, and S. Noh. Recursive solution of generalized predictive control and its equivalence to receding horizon tracking control. *Automatica*, 28:1235–1238, 1992.

[42] W.H. Kwon and A.E. Pearson. On the stabilization of a discrete constant linear system. *IEEE Transactions on Automatic Control*, 20:800–801, 1975.

[43] J.W. Lee. Exponential stability of constrained receding horizon control with terminal ellipsoidal constraints. *IEEE Transactions on Automatic Control*, 45:83–88, 2000.

[44] X. Li, B. Ding, and Y. Niu. A synthesis approach of constrained robust regulation based-on partial closed-loop optimization. In *Proceedings of the 18th Chinese Control and Decision Conference*, pages 133–136. China, 2006 (in Chinese).

[45] Z. Lin and A. Saberi. Semi-global exponential stabilization of linear systems subject to input saturation via linear feedback. *Systems and Control Letters*, 21:225–239, 1993.

[46] Z. Lin, A. Saberi, and A.A. Stoorvogel. Semi-global stabilization of linear discrete-time systems subject to input saturation via linear feedback — an ARE-based approach. *IEEE Transactions on Automatic Control*, 41:1203–1207, 1996.

[47] Y. Lu and Y. Arkun. Quasi-Min-Max MPC algorithms for LPV systems. *Automatica*, 36:527–540, 2000.

[48] L. Magni and R. Sepulchre. Stability margins of nonlinear receding-horizon control via inverse optimality. *Systems and Control Letters*, 32:241–245, 1997.

[49] D.Q. Mayne, J.B. Rawlings, C.V. Rao, and P.O.M. Scokaert. Constrained model predictive control: stability and optimality. *Automatica*, 36:789–814, 2000.

[50] M. Morari and N.L. Ricker. *Model predictive control toolbox for use with Matlab: User's guide, version 1*. The MathWorks Inc., Natick, MA, 1995.

[51] E. Mosca and J. Zhang. Stable redesign of predictive control. *Automatica*, 28:1229–1233, 1992.

[52] Y. Niu, B. Ding, and H. Sun. Robust stability of two-step predictive control for systems with input nonlinearities. *Control and Decision*, 21(4):457–461, 2006 (in Chinese).

[53] J.M. Ortega and W.C. Rheinboldt. *Iterative solutions of nonlinear equations in several variables*. Academic Press, New York, 1970.

[54] R.K. Pearson and M. Pottmann. Gray-box identification of block-oriented nonlinear models. *Journal of Process Control*, 10:301–315, 2000.

[55] B. Pluymers, J.A.K. Suykens, and B. de Moor. Min-max feedback MPC using a time-varying terminal constraint set and comments on "Efficient robust constrained model predictive control with a time-varying terminal constraint set." *Systems and Control Letters*, 54:1143–1148, 2005.

[56] M.A. Poubelle, R.R. Bitmead, and M.R. Gevers. Fake algebraic Riccati techniques and stability. *IEEE Transactions on Automatic Control*, 33:379–381, 1988.

[57] J. Richalet, A. Rault, J.L. Testud, and J. Papon. Model predictive heuristic control: application to industrial processes. *Automatica*, 14:413–428, 1978.

[58] J. Schuurmans and J.A. Rossiter. Robust predictive control using tight sets of predicted states. *IEE Control Theory and Applications*, 147(1):13–18, 2000.

[59] P.O.M. Scokaert and D.Q. Mayne. Min-max feedback model predictive control for constrained linear systems. *IEEE Transactions on Automatic Control*, 43:1136–1142, 1998.

[60] P.O.M. Scokaert and J.B. Rawlings. Constrained linear quadratic regulation. *IEEE Transactions on Automatic Control*, 43:1163–1169, 1998.

[61] Z. Wan and M.V. Kothare. An efficient off-line formulation of robust model predictive control using linear matrix inequalities. *Automatica*, 39:837–846, 2003.

[62] Y.J. Wang and J.B. Rawlings. A new robust model predictive control method I: theory and computation. *Journal of Process Control*, 14:231–247, 2004.

[63] Y. Xi. *Predictive control*. National Defence Industry Press, Beijing, China, 1991 (in Chinese).

[64] E. Yaz and H. Selbuz. A note on the receding horizon control method. *International Journal of Control*, 39:853–855, 1984.

[65] Q. Zhu, K. Warwick, and J.L. Douce. Adaptive general predictive controller for nonlinear systems. *IEE Control Theory and Applications*, 138:33–40, 1991.

Index